Lecture Notes in Management and Industrial Engineering

Series Editor

Adolfo López-Paredes, INSISOC, University of Valladolid, Valladolid, Spain

This book series provides a means for the dissemination of current theoretical and applied research in the areas of Industrial Engineering and Engineering Management. The latest methodological and computational advances that can be widely applied by both researchers and practitioners to solve new and classical problems in industries and organizations contribute to a growing source of publications written for and by our readership.

The aim of this book series is to facilitate the dissemination of current research in the following topics:

- Strategy and Entrepreneurship
- Operations Research, Modelling and Simulation
- Logistics, Production and Information Systems
- Quality Management
- Product Management
- Sustainability and Ecoefficiency
- Industrial Marketing and Consumer Behavior
- Knowledge and Project Management
- Risk Management
- Service Systems
- Healthcare Management
- Human Factors and Ergonomics
- Emergencies and Disaster Management
- Education

More information about this series at https://link.springer.com/bookseries/11786

Ronggui Ding · Reinhard Wagner ·
Constanta-Nicoleta Bodea
Editors

Research on Project, Programme and Portfolio Management

Projects as an Arena for Self-Organizing

Springer

Editors
Ronggui Ding
School of Management
Shandong University
Jinan, Shandong, China

Reinhard Wagner
Tiba Managementberatung GmbH
Munich, Germany

Constanta-Nicoleta Bodea
Economic Informatics and Cybernetics
Bucharest University of Economic Studies
Bucharest, Romania

ISSN 2198-0772 ISSN 2198-0780 (electronic)
Lecture Notes in Management and Industrial Engineering
ISBN 978-3-030-86247-3 ISBN 978-3-030-86248-0 (eBook)
https://doi.org/10.1007/978-3-030-86248-0

This Springer imprint is published by the registered company Springer Nature Switzerland AG
The registered company address is: Gewerbestrasse 11, 6330 Cham, Switzerland

Foreword

In social science, "self-organization" is also referred to as "spontaneous order" and means a process in which a form of general "order" emerges from local interactions between parts of an originally "disordered" system in the sense of systems theory. This process can be spontaneous if there is sufficient "energy" (i.e., an individual and/or group critical mass) and no control by an external actor is needed. It is often triggered by seemingly random fluctuations that are amplified by positive feedback, consequently the resulting organization is completely decentralized and distributed across all components of the system; as such, the organization is typically robust and able to survive significant disturbances or repair itself. Chaos theory discusses self-organization in terms of islands of "predictability" in a sea of chaotic unpredictability.

Self-organization is an approach that can address development potentials and areas of application in project management because self-organization is to be anticipated in every human group or society. It presupposes (right) behavior as an elementary competence and connects the individual and the group. This implies self-referentiality because both the individual and the group are connected through self-producing communication: Communication produces further and further communication. Thus, as a social system, the group can reproduce itself as long as there is a vibrant mode of communication. Self-organization is also related to learning: enabling others to learn about learning. The "S.O.L Self Organized Learning" concept takes place throughout life as a collaborative process.

IPMA® contributes through the IPMA Global Standards competence-based to the development of self-organization potentials and application areas in project management: for individuals, for projects and programs and for organizations.

The IPMA Individual Competence Baseline in its version 4.0 (ICB4®) offers various approaches to the development of individual competences such as: self-development, peer development, education and training, coaching and mentoring, simulation and gaming to help stakeholders in their competence development, especially in the area of Perspective Competence with the competence elements "governance, structures and processes" and "culture and values", in the area of People Competence with the competence elements "self-reflection and self-management", "personal integrity and reliability", "personal communication", "relationship and engagement", "leadership" and "teamwork", and in the area of Practice Competence with the competence elements: "organisation and information", "stakeholder" and "change and transformation".

The IPMA Organizational Competence Baseline (OCB®) describes the role of key stakeholders. The main purpose is to clearly show what role an organization has in managing its project, program or portfolio related work. It describes the concept of organizational competence in managing projects and how it should be used to achieve the organization's vision, mission, and strategic goals in a sustainable manner.

Finally, the IPMA Project Excellence Baseline (PEB®) was developed to promote excellence in managing projects and programs. It complements our previous standards for individuals (ICB) and for organizations (OCB) and serves as a guide for organizations in assessing the ability of their projects and programs to achieve project excellence. The Baseline is designed to be used in any context, i.e., it is an adaptable and open assessment method designed for multiple purposes in the three areas of people and purpose, processes and resources, and project results, ensuring close interaction between key areas in any organization such as performance, effectiveness and efficiency, reliability, flexibility, continuous improvement, scalability, and sustainability.

This book covers the best papers from the 8th IPMA Research Conference on "Projects as arena for self-organizing", held in September 2020, which identifies multiple dimensions of self-organization in projects on the individual, team, organizational, and societal level.

At each level, the degree and pattern of SO self-organization depends on the complexity of the SO's system subject, e.g., the project organization itself; on the characteristics of the components and working relationships of the SO's system subject, e.g. the competences required; on the use of innovative methods and tools (e.g. Artificial Intelligence). These linkages across different SO levels are important because they can act as drivers or barriers for extending the SO scale. They can lead to specific SO patterns that influence performance (benefits delivery) and learning of self-organized systems concerned with developing competences and performance across all levels. These relationships across different SO levels should be explored and assessed to understand and change them to achieve a desired impact on self-organized systems that manage and control projects, programs, and portfolios.

The book includes 17 chapters in three sections written by authors from 16 countries. The coverage of this geographic and cultural diversity contributes to a better understanding of research context, approaches and results. And, of course, the book invites further research on project, program, and portfolio management as an arena for self-organizing.

Jesús Martínez-Almela
IPMA President (2018–2020)
International Project Management
Association
Amsterdam, The Netherlands

Preface

Self-organizing in and through projects represents an increasing interest topic for project, programme and portfolio management. Project management professionals became increasingly interested in work autonomy and self-organizing through projects to fulfil their expectations in the workspace. Projects can be perceived as arena for self-organizing. Agile approaches build on or value the willingness and the capabilities of individuals and teams to self-organize in projects.

Considering these trends in project management, International Project Management Association—IPMA (www.ipma.world) organized between 9th–11th September 2020 its 8th Research Conference with the theme Projects as an arena for self-organizing. During this event that was held online, more than 150 researchers and practionnairs debate on the main topics related to self-organizing in projects. The book includes the extended version of the papers presented at the Conference.

The book is structured into four parts. Part I intends to set the stage of self-organizing, by introducing the main principles, concepts and approaches in self-organizing. The research literature and the global standards in project, programme and portfolio management, such as IPMA Individual Competence Baseline consider self-determination and self-management and self-control as important competences for people engaged in projects. However, there is still a need for better understanding the conditions, prerequisites, potentials and threats of self-organizing in the field of projects. Part II of the book offers a valuable resource for researchers interested in the relevance of competences in working with people for assuring a successful self-organizing in projects. Part III of the book is grouping several chapters revealing the impact of technologies in project organization, especially for self-organizing. Part IV of the book includes research on self-organizing in projects for different type of projects and industries, especially in construction and telecommunication.

Part I of the book includes five chapters:

Chapter 1, by *Reinhard Wagner* is entitled *Unleashing Hidden Potential by Enabling Self-Organization in Projects*. The author points out essential backgrounds, drivers and approaches of self-organization in projects with their impact on individuals, project teams and the ambient organization. The author argues that in order organizations to become more adaptable and resilient to the ever-changing context they are operating in, this requires to move from the principle of hierarchy and empowering employees to work voluntarily and self-determined.

In Chap. 2, *Alfred Oswald* regards self-organization as a universal principle that creates something new in evolutionary terms and whose basic concepts are applied in nature, in the psychological and social realms, and in technology. In this regards, self-organization can be considered as a common basic concept that helps us to understand, shape and lead the development and transformation of individuals, teams, organizations, and societies. In his chapter, the author outlines the characteristics of a project as a teal temporary organization.

Chapter 3 by *Hubertus C. Tuczek, Agnetha Flore, Helge F. R. Nuhn and Norbert Schaffitzel* present a holistic approach for the entire company, starting with the management mindset and the governance structure of self-organization, is the way forward for a consistently agile approach.

In Chap. 4, *Karyne C. S. Ang, Lars Kristian Hansen and Per Svejvig* present an in-depth single case study of an organisation that adopted agile practices in Project Portfolio Management (PPM). The study traces how an agile portfolio of projects underpinned by value is managed through short iterative cycles. The authors analyse how the sequences of the agile cycles and decision-making events assist the organisations to understand how thevalue can be embedded in project and portfolio management practices.

Chapter 5 addresses the topic of the new work. *Eckhard Heidling and Nick Kratzer* discuss the connections between new work and self-organization in work. A sociology-of-work perspective is applied for the new work discourse, meaning flexible work in terms of space and time, new spaces of work, (agile) project work and self-organization. The focus is set on the respective potentials, limits, and approaches for the design of our present world of work.

Part II of the book includes the following two chapters:

Chapter 6 by *Raji Sivaraman and Michal Raczka* is presenting the Personal Agility Lighthouse Model (PALHTM) model developed by the authors, in order to explore the path to self-organization in projects. The model consists of seven agilities and the authors investigatewhether these personal agilities are necessary for assuring the organizational agility.

In Chap. 7, *Rodney Turner* considers client experience on projects. The author reviewed that literature and also interviewed several clients, having experience of interacting with contractors through the project life-cycle. The conclusion was that clients on projects have similar experiences to customers in retail, even though on projects the client controls the interactions whereas in retail it is the vendor that controls the interactions. Based on these findings, Rodney Turner recommends to consider not only the client experience but also the contractor experience, in order to achieve a better performance in projects.

Part III of the book includes four chapters:

Chapter 8 is concerned with the *Projects Organization and Intelligent Technologies. Ronggui Ding, Constanta-Nicoleta Bodea* portray projects as social-technical systems including groups of persons and technologies which are executing together processes for achieving specific objectives. The authors argue that self-organizing in projects is directly related to the technologies applied in projects and discuss the potential of some intelligent technologies in transforming the projects towards a self-organizing arena.

In Chap. 9 *Hiroshi Ohtaka, Motomu Koumura and Masahiro Isokawa* presents how the organizational issues related to the digital transformation can be identified by analyzing IT dispute cases. The authors propose a new method based on *Kaizen* for the analysis of IT dispute cases of actual IT projects, where recent technologies of package software and agile are introduced for quick response to individual new challenge. The chapter also discuss the development of improved management to cope with the threat of the visualized business risk, from the aspect of organization.

In Chap. 10 *Zhixue Liu, Xinyi Song, Lei Wang, Rui Song and Itai Lishner* proposed the usage of Social Network Analysis for identification of Governance Structures for Private Public Partnership (PPP) Projects. The project stakeholders were characterised into four basics roles and established four basic project governance structures through different role combination. The project governance risks were identified by analyzing the project governance structure through Social Network Analysis (SNA). This approach was validated through a case study of a PPP project in China.

In Chap. 11 *Anne-Sofie Hansen, Per Svejvig and Lars K. Hansen* seek to evaluate the Diamond Model in different settings. A mixed-methods approach is applied and data from 62 projects in 16 project-based organizations are evaluated. The study points to several ways to upgrade the model, such as splitting the pace dimension into two dimensions: pace (time) and impact.

The last section of the book addresses the perspectives of self-organizing in different type of projects and industries:

In Chap. 12 *Lukas Beladi Sihombing and Jiwat Ram* present a new disaster recovery model that can be used in situations such as COVID-19 pandemic in the construction industry. For developing the model, the authors collected data through two surveys and they analyzed the collected data by using frequency analysis. The disaster recovery model that was developed will allow projects to be run smoothly during a crisis by taking into consideration several aspects such as: prioritizing business functions; having a communication plan (e.g. for employees, customers, vendors, the public, and the media); having prevention and mitigation strategies; having an emergency response checklist; determining the potential threats, vulnerabilities, and risks; backing up data periodically, both online and off-site; conducting routine tests of disaster possibilities; arranging for working off-site; and having insurance

Chapter 13 discusses the significance of integration of self-organisation and stakeholders into the development of infrastructure projects, for achieving more creativity and satisfying solutions in projects. Considering the factors that increase the likelihood of creative and good solutions as well as on definitions of self-organisation,

Pia Herrmann, Reiner Singer, Philipp Kaufmann and Konrad Spang consider that self-organisation can support the required creativity. Based on literature concerning self-organisation and governance and coordination of self-organisation, the authors discussed how to integrate self-organisation into the development of public infrastructure projects, outlined two possible applications and shared suggestions for further research.

In Chap. 14 *Yongkui Li and Yilong Han* consider the importance of self-organization, dynamic meta-governance and value creation in megaprojects. Considering the strategically importance of megaprojects, the authors consider that it is a need to re-examine these complex systems and to propose new governance strategies. Based on this perspective, the authors analyzed the organizational behavior and self-organization phenomenon in megaprojects under the background of complexity. They also developed a megaproject governance portfolio strategy for self-organization and heter-organizations, and proposed a new path and direction towards more sustainable value creation in megaprojects for addressing the current and future challenges of megaprojects.

Chapter 15 by *Önder Ökmen, Marian Bosch-Rekveldt and Hans Bakker* discusses whether and how "traditional" Critical Path Method (CPM) based schedules allow for flexibility in project planning and management. The managerial flexibilities provided by CPM were evaluated at different levels, and two CPM schedules from different projects were examined. By investigating the flexible features of CPM in its traditional form, the authors propose a more flexible schedule management approach based on CPM and its extensions, which future self-organizing teams can adjust or apply.

Chapter 16 by *Irina Nechaeva* addresses how the construction projects can be agile. The chapter presents the comparison of typical construction project design phase and new product development and a system hybrid project management approach with mix of agile and waterfall project management is proposed.

In Chap. 17 *Jovana Mihailovic, Marija Todorovic and Vladimir Obradovic* present the main challenges of open innovation in the telecommunication sector. The chapter provides a general overview of management practises on open innovation. It identifies challenges that occur in practice during open business model implementation and highlights the relationship between strategy and open innovation in telecommunication sector. The authors conducted a case study in a telecommunication company in order to analyze the bond between open innovation activities, self-organization, strategic orientation and innovation performances in practice.

The book is addressed to researchers and practitioners in project, programme and portfolio management, to educators and postgraduate students, in general for all those interested to better understand how self-organizing impact project organization. The readers of the book will find answers to several questions, such as: What is the impact of self-organizing on the organizational structures, processes, cultures and leadership? How to define the transformative power of self-organization? What is the motivation of individuals to perform activities, to engage with others and organizations in order to get things done? Which kind of leadership supports self-organizing in projects? Howwe can achieve the changes necessary from the traditional set-up towards new ways of leading?

Thank you for reading!

Jinan, China — Ronggui Ding
Munich, Germany — Reinhard Wagner
Bucharest, Romania — Constanta-Nicoleta Bodea

Contents

Contributors

Karyne C. S. Ang School of Project Management, Faculty of Engineering, University of Sydney, Forest Lodge, New South Wales, Australia

Hans Bakker Delft University of Technology, Delft, The Netherlands

Constanta-Nicoleta Bodea Bucharest University of Economic Studies, Bucharest, Romania

Marian Bosch-Rekveldt Delft University of Technology, Delft, The Netherlands

Ronggui Ding School of Management, Shandong University, Jinan, China

Agnetha Flore OFFIS – Institute for Information Technology, Oldenburg, Germany

Y. Han Tongji University, Shanghai, China

Anne-Sofie Hansen Aarhus Business and Social Sciences, Aarhus V, Denmark

Lars K. Hansen Aarhus Business and Social Sciences, Aarhus V, Denmark

Lars Kristian Hansen Business and Social Sciences, Department of Management, Aarhus University, Aarhus, Denmark

Eckhard Heidling Institut für Sozialwissenschaftliche Forschung – ISF München, München, Germany

Pia Herrmann University of Kassel, Kassel, Germany

Masahiro Isokawa IT Mieruka Institute, Tokyo, Japan

Philipp Kaufmann University of Kassel, Kassel, Germany

Motomu Koumura IT Mieruka Institute, Tokyo, Japan

Nick Kratzer Institut für Sozialwissenschaftliche Forschung – ISF München, München, Germany

Y. Li Tongji University, Shanghai, China

Itai Lishner Technion—Israel Institute of Technology, Haifa, Israel

Zhixue Liu Shandong University, Jinan, China

Jovana Mihailovic A1 Srbija, Bulevar Milutina Milankovica 1Z, Belgrade, Serbia

Irina Nechaeva National Research University Higher School of Economics, Moscow, Russian Federation

Helge F. R. Nuhn Wilhelm Büchner Hochschule Darmstadt, Darmstadt, Germany

Vladimir Obradovic Faculty of Organizational Sciences, University of Belgrade, Belgrade, Serbia

Hiroshi Ohtaka IT Mieruka Research, Yokohama, Kanagawa-ken, Japan

Önder Ökmen Delft University of Technology, Delft, The Netherlands

Alfred Oswald IFST-Institute for Social Technologies GmbH, Stolberg, Germany

Michal Raczka mBank S.A., Warsaw, Poland

Jiwat Ram Excelia Group, La Rochelle, France

Norbert Schaffitzel DB Systel GmbH, Frankfurt, Germany

Lukas Beladi Sihombing University of Pelita Harapan, Tangerang, Banten, Indonesia

Reiner Singer University of Kassel, Kassel, Germany

Raji Sivaraman ASBA LLC, Singapore/USA, Singapore;
Feliciano School of Business, Montclair State University, Montclair, USA

Rui Song College of Design, Georgia Institute of Technology, Atlanta, GA, USA

Xinyi Song College of Design, Georgia Institute of Technology, Atlanta, GA, USA

Konrad Spang University of Kassel, Kassel, Germany

Per Svejvig Business and Social Sciences, Department of Management, Aarhus University, Aarhus, Denmark

Marija Todorovic Faculty of Organizational Sciences, University of Belgrade, Belgrade, Serbia

Hubertus C. Tuczek Hochschule Landshut, Landshut, Germany

J. R. Turner Europrojex, East Horsley, Surrey, UK;
Department of Civil Engineering, University of Leeds, Leeds, UK

Reinhard Wagner Tiba Managementberatung GmbH, Munich, Germany

Lei Wang Shandong University, Jinan, China

Part I
Setting the Stage of Self-Organizing

Chapter 1
Unleashing Hidden Potential by Enabling Self-Organization in Projects

Reinhard Wagner

Abstract Organizations today are facing a complex environment that challenges them in many ways. This causes organizations to consider new approaches of organizing and of the way people collaborate, in general and particularly in projects. Concepts of self-organization are often used unconsciously, e.g., in the context of agile project management approaches. This chapter therefore prepares the topic fundamentally, points out essential backgrounds, drivers, and approaches of self-organization in projects with their impact on individuals, project teams, and the ambient organization. The effects of more self-organization in projects are promising according to the pertinent literature. For example, individuals can develop passion in what they do through more freedom for self-actualization. Diversified project teams can achieve peak performance through emergence as well, helping organizations become more adaptable and resilient to the ever-changing context they are operating in. However, this requires a departure from the traditional principles of an organization, including moving away from the principle of hierarchy and empowering employees to work voluntarily and self-determined.

Keywords Self-organization · Projects · Hidden potential · Leadership

1.1 Introduction

Since the beginning of time, people and what they do have been organized in a certain way. Harari [1] impressively describes the history of mankind and the successes that have been achieved over the millennia through skillful organizing. Certainly archaic forms of agriculture, urban and cultural development, craftsmanship, and early forms of industrialization are of great interest. However, it is primarily the optimization efforts of the early twentieth century in industry, also called "Taylorism" after the main protagonist Taylor [2], as well as the more theoretical reflections on social order

---

R. Wagner (✉)
Tiba Managementberatung GmbH, Perchtinger Straße 10, 81379 Munich, Germany
e-mail: reinhard.wagner@tiba.de

R. Ding et al. (eds.), *Research on Project, Programme and Portfolio Management*, Lecture Notes in Management and Industrial Engineering,
https://doi.org/10.1007/978-3-030-86248-0_1

in the form of bureaucracy and subordination by Weber [3], that still influence our considerations on meaningful forms of organization or organizing.

Taylor was concerned with continuously increasing the productivity of the manufacturing processes. In doing so, he took little account of the interests of the workers. They were simply factors of production and had to be completely subordinated to the "scientific" considerations of the process planners and the production sequence they planned for mass production. Weber built his theory on the Protestant ethics of asceticism and subordination to a given leadership, rules, and functionally specialized roles.

Although the term "project" has been used to denote the completion of demanding tasks since the late seventeenth century [4], the project-based form of work or organization has experienced a renaissance, starting from the middle of the twentieth century [5]. In the beginning, it was mainly heroic acts of individuals for the realization of projects, whereas later the focus shifted to the management of projects, inspired by the doctrines of Taylor and Weber. Projects served the realization of extraordinary assignments, i.e., those ventures that could not be accomplished well enough by the line organization, which was designed for efficiency.

Due to numerous developments in the marketplace, changing customer and employee expectations as well as technological innovations, to name just a few underlying trends, the focus of activities in many organizations has shifted. Number and importance of projects have increased significantly. For many organizations, projects are the norm rather than the exception. This trend is called "projectification" and is changing not only management, but also the way we organize work. Projects play an essential role in the transformation of major industries, such as Automotive [6].

Organizations' objectives today are less aligned with the paradigm of continuously improving efficiency and more focused on effectiveness, adaptability, and resilience as well as sustainability. Diverse external influences demand agile behavior, which is less likely to be pre-planned and not at all anticipated by one or a few executives. What is needed today, above all, are motivated and competent teams of people who see meaning and fulfillment in their task and do it in the context of an organization that does not demand subordination, but helps to unleash people's potential.

In the following, we will describe how the above-mentioned goals can be achieved with the help of self-organization, what this means for individuals, leaders, and organizing, and what the long-term results might be.

1.2 The Main Drivers of Change Toward Self-Organization

A multitude of challenges exert pressure on how organizations are structured and what organizational form is chosen for certain activities, such as projects. On the one hand, external forces have an impact on the organization, including those from society, both sales and supply markets, and also from technology. On the other hand, internal forces also add pressure on managers to organize differently. This can be, among other factors, a strategic realignment in response to external challenges, or the expectation of employees to reflect on new values, structures and teamwork; or the executive leadership sees the need to renew organizationally. The following will elaborate on some of the aspects without going into too much details.

1.2.1 Drivers Related to Society and the Individual

From a societal perspective, the role of work has certainly changed a lot for people. The economic success of the past and the resulting prosperity in many nations has ensured that society's values are changing. People want to reconcile work better with family and personal life, to achieve greater self-fulfillment, and generally to find a purpose in what they do at work. This includes not only their own well-being, but also opportunities for personal development and co-creation [7].

Generation Z in particular, i.e., those born after 1997, is looking for a working environment in which they can realize the above-mentioned aspects. Organizations must therefore make an effort in the "War for Talents" to be attractive to young talent. In addition to meaningful work content, young employees today want an employer to offer them space to maneuver, flexibility in terms of working hours, work location, and work equipment, as well as opportunities for personal development, support through continuing education, and coaching. Hamel and Zanini therefore call for a complete restructuring of organizations to make them more human and thus fit for the future [8].

1.2.2 Drivers Related to Business and the Organization

Customers today have become more demanding and are no longer satisfied with mass products. Products are becoming more individualized, with services tailored to specific target groups. As a result, the organization designed for efficiency and focused on repetition is also losing its advantage. Nowadays, the emphasis is on the organization that is completely focused on the customer, is able to realize a 1-piece flow, and offers services that are as scalable as possible. Global competition puts additional pressure on organizations. Services have to be delivered to markets quickly and cost-effectively, because increasingly "the winner takes it all" applies. In "Accelerate", John Kotter challenges organizations to become much more agile and adaptive in order to cope with major changes in their environment: "The world is now changing at a rate at which the basic systems, structures, and cultures built over the past century cannot keep up with the demands being placed on them" [9].

Increasing volatility, uncertainty, complexity, and ambiguity (VUCA) is a recurring theme in the literature when discussing the requirement for organizations to become more adaptive and resilient. In doing so, it is argued that organizations need to become both internally networked and better connected to the environment. Niels

Pfläging therefore argues for a consistent de-centralization of organizations and the empowerment of teams that work directly with customers and external partners [10]. This is a clear departure from Max Weber's principles, which were briefly addressed above. Yet, very few organizations have consistently moved in this direction. A lot is currently being tried out, some of it becomes rather fad and fashion.

1.2.3 Drivers Related to Technology and the Capabilities

Many of the developments outlined above would be inconceivable without technological progress. Klaus Schwab describes the developments as revolutionary, different in scale, scope and complexity, and foresees serious consequences for all disciplines, economies, industries, and governments [11]. For example, the digitization of many areas of life offers the opportunity to access a large amount of data, analyze it and use it as the basis for new business models. The "Internet of Things (IoT)", "Artificial Intelligence (AI)" as well as "Augmented Reality" are drastically changing the world of work. The International Project Management Association (IPMA) found in a recent study that AI technologies ranked with high or highest potential for improving management and delivery of projects are machine learning (78%), diagnosis (76%), and deep learning (74%) [12]. Increased flexibility and responsiveness are among the benefits modern technologies like AI bring to an organization.

The pandemic in particular has shown how technology can be used to continue working from home. Virtual conferences, online training, and many other forms of collaboration using digital tools have changed the world of work dramatically in recent months and years. Digital business models are booming at a time when the pandemic is forcing people to stay at home or severely restricting the freedom to travel. On the other hand, this also means new capability needs for people as well as organizations. Above all, it is the handling of the new technologies that employees as well as managers have to learn. Organizations are obliged to observe information security and data protection while at the same time allowing mobile working via the cloud and intensive networking across all barriers [13]. The technologically driven asynchronous way of working requires an increasingly decentralized organization with more liberties for individuals to organize themselves, respectively, in projects with others. Last but not least, the question arises as to whether an organization still needs permanent employees at all for the completion of certain tasks, or whether it should work with temporary employment contracts or project-related freelancers.

1.3 Self-Organization in Projects

Concepts of self-organization have been discussed since ancient times. The main focus is on the processual phenomenon that contributes to the emergence of new forms of biological, ecological, societal, and cultural structures. With the beginning of systems theory studies in the USA during the mid-1950s, the concept of self-organization was also developed in the context of self-regulation and evolution of biological systems, especially by Von Bertalanffy [14]. In this phase, the search for an equilibrium-oriented, cybernetic understanding was in the foreground, whereupon later, the disequilibrium, transformation, instability, and structural metamorphosis were primarily elaborated.

Throughout this discussion, it was realized that easily changeable systems in a dynamic and complex environment cannot be planned and controlled, but by a series of interventions be influenced in a certain direction. Thus, Haken and Schiepek [15] use the example of the laser to describe how desired results are achieved with the help of order, setting, and control parameters. Oswald et al. [16] apply this to social systems and propose the following definition: "Self-organization creates, under certain conditions (setting, control and order parameters) and through the interaction of system elements on the micro-level, emergent properties on a macro-tier. The consequence of a self-organizing process is an ordered pattern, collective behavior, or a new structure".

1.3.1 Self-Determination as a Prerequisite for Self-Organization in Projects

If we look at self-organization in a social system, then in contrast to the biological, physical, or chemical system, there are people at work who bring their own will to create such systems. Ryan and Deci describe in their "self-determination theory" three basic needs of people, which are expressed by corresponding actions, these are "competence", "relatedness", and "autonomy" [17]. This means that they are intrinsically motivated to keep improving their self-efficacy and to constantly move forward. Ideally, this happens together with other people, on the one hand in order to learn from the ideas of others and to incorporate them into their own behavior, and on the other hand to have their own achievements recognized by other people. If we think about the principles of organizations described by Taylor and Weber, then the need for "autonomy" of people is obviously completely contrary to the traditional way of thinking of organizations. For self-organization in projects, this need is certainly a decisive one, since it helps to unleash the potential of individuals for the benefit of the organization.

In this respect, organizations and their leadership are currently at a crossroads. The question is how much planning and control should still take place centrally, or to what extent this should be discontinued and people granted more autonomy or considerable space to maneuver. Revolutionary management systems such as Holacracy grant employees substantial autonomy when it comes to decision-making [18] and are highly decentralized. Nevertheless, there are explicit rules and principles on how collaboration works within the organization, e.g., through self-organized teams ("Circles").

1.3.2 Collective Mind, Emergence, and Radical Collaboration

If one sees organizations as social systems, then of course it is not only about the self-determination of the individual, but the self-organization of several people, for example, in a project team, in a department or a business unit. Collaboration will only occur if everyone works toward agreed upon objectives and shares the same values, beliefs, and basic assumptions. This may sound trivial at first, but it is the essential task of a manager to achieve this. The difference, however, is whether employees are allowed to make a self-confident decision to commit to the mutually agreed purpose of the project or they have been assigned to the project by a superior and handle it without intrinsic motivation. Balve and Schaffitzel advocate for fostering a "Collective Mind" that covers principal criteria for successful collaboration, including but not limited to the values and beliefs of the actors, their perceived identity during collaboration, the knowledge of their common bond, the purpose and raison d'être of their mission [19].

Despite uniformity in terms of the shared purpose, it is important for a team to be as diverse as possible in order to cope with the ambient complexity of a project and to bring enough creativity to meet the challenges. The ability of a team to make more of itself than the sum of its individual abilities is also referred to as "emergence", is only achieved with the help of a systematic team development process and enables high-performance teams [16].

Sennet [20] argues that cooperation is embedded in human nature, yet we find it typically difficult to cooperate with people who are different from us. He portrays cooperation as hard work that builds on dialogic skills, including listening well, behaving tactfully, finding points of agreement, and managing disagreement. Tamm and Luyet [21] connote cooperation with the adjective "radical," meaning fundamental, favoring basic change, as in the social structure. From their point of view, five skills are essential, namely, collaborative intention, openness, self-accountability, self-awareness, and awareness of others and negotiating as well as problem solving. Although most, if not all, of these skills lie within the realm of the individual or team, the surrounding organization and its leadership can certainly make an important contribution to successful cooperation.

1.3.3 Organizational Structures and Leadership That Fosters Self-Organization in Projects

Self-organization aims to create its own order for a suitable cooperation, one that meets the requirements of the task in its setting. For this, however, it is necessary that the ambient organization leaves enough freedom for self-organization and does not suppress self-organization from the beginning with an overwhelming order of its own. Heinz von Foerster considers it one of the most essential tasks of the leadership of an organization "to organize self-organization", for example, by replacing the hierarchical system with a heterarchical network [22]. Heterarchy means the rule of many and thus a balance of power in contrast to the rule of few in hierarchy. This new order is created primarily through interpersonal interaction or communication. Whereas in traditional organizations this has been regarded as an informal, parallel, or even secondary organization, this view is now coming to the fore for self-organization.

Since projects are a temporary overlay to the regular form of organization, the sponsors of the project can grant a higher degree of autonomy to the project team than might be the case with other activities in the organization. In this context, a multitude of events take place within the project team itself and involve stakeholders in the closer as well as the broader context of the project. Multiple overlapping networks, cultural influences, and the respective language and metaphors come into play [23]. Niels Pfläging sums up the relationship between self-organizing teams and the ambient organization as follows: "Self-organization is very different from command-and-control. It is, one might say, about disciplined togetherness, if anything, self-organization is about acknowledging outside-in market-pull, which relentlessly applies its forces on any organization today. Then, within an organization, these pull-forces from outside markets must be responded to, constructively, through inside-out value creation" [24].

The question remains as to what role leadership plays in the context of self-organization. At first glance, this is certainly much less than in hierarchical systems, where leadership plays a key role in prescribing goals and plans, setting rules for activities, providing instructions, making decisions, and so on and so forth. Leadership is enabling, at the individual, team, and organizational levels. Everybody has to lead themselves, certainly not an easy task given the complexity involved. Leadership will also be important at the team level, but this can fluctuate, i.e., change from one project phase to the next, and always be assigned to the person with the right competencies. Leadership will therefore come less from a formal and hierarchical role within the organization. Leaders at the organizational level are only responsible at this level, ensuring strategic goals based on organizational learning, prioritizing the portfolio of the various activities, providing necessary resources, and supporting the project teams in their daily activities. Thus, they are facilitators, mediators, and coaches rather than leading through command and control [25].

Summarizing the above, we conclude that self-organization in projects is self-determined collaboration of people in pursuit of mutually agreed upon objectives in concert with the immediate and wider project environment.

1.4 The Hidden Potential and How It Can Be Unleashed

What does it lead to when self-organization is enabled in projects, or what potential can be unleashed in that way? There are certainly many possible answers to this question. The four most important ones together with the path to achieving them will be presented below.

1.4.1 Making Sense of Organizations

In the context of the question of self-organization and the emancipation of the individual from the hierarchical order, the question certainly arises as to why people join an organization and comply with its rules of conduct. Based on systems theory, Stefan Kühl describes the essential characteristics of an organization in modern times as membership, goals, and hierarchy. People (and the organization or its leaders) decide in favor or against an individual's membership in an organization [26]. Certainly, the other two factors also play an important role in the decision for or against membership in an organization, namely, the question of what objectives the organization is pursuing and to what extent the hierarchical order suits that individual and his or her expectations toward the organization. The alternative to membership in an organization or employment with a company is self-employment. The more attractive a membership in the respective organization appears, the easier it is for the individual to make the decision, and the longer membership is likely to last.

For a person, it can make sense to join an organization if its goals largely coincide with the person's own goals, the social order tends to fit the person's expectations, i.e., it is neither too restrictive nor too loose, and a commitment to the common goal can be achieved [27]. Money certainly plays a role in this as an exchange relationship; as already mentioned above, Generation Z in particular tends to look for a higher purpose, a socially relevant, long-term goal, or an orientation that intrinsically motivates and inspires someone to participate [28]. In addition, the person will also still have an impression of the role they can fill in the given regulatory framework and what the real culture is like. In short, it is about the fit between the expectations of an individual and the organization. This certainly varies from person to person and should in any case be carefully examined before it comes to a commitment.

1.4.2 Awakening Passion for Higher, Individual Performance

What causes someone to self-organize in projects? This might be answered by referring back to scientific studies on cognitive readiness. Cognitive readiness can be defined "as the mental preparation, including skills, knowledge, abilities, motivations, and personal dispositions, needed to establish and sustain outstanding individual and team performance in the complex and rapidly changing environment of project, program, and portfolio management" [29]. It helps project managers and teams to become more focused, responsive, resilient, and adaptive to the changes happening in the context of the project. Furthermore, it´s building on mindfulness as well as emotional and social intelligence.

However, the term "passion" still goes well beyond a person's cognitive readiness. Robert Vallerand describes passion as "a strong inclination toward a specific object, activity, concept or person that one loves (or at least strongly likes), highly values, invests time and energy in on a regular basis, and that is part of one´s identity" [30]. Particular activities cause a particular level of energy in certain people, which exceeds the usual level and lasts longer than ordinary activities. These are mainly activities that are significant for the person and for which the person is willing to spend a considerable amount of time, sometimes even beyond the personal limits. This can be experienced in certain leisure activities such as mountain climbing, people get into a "flow", which makes an experience genuinely satisfying [31]. The main point, then, is to carefully examine the fit between an individual's Cognitive Readiness and the opportunities an organization offer for self-actualization and, based on that, make a decision as to whether it makes sense to enter into a collaboration.

1.4.3 Fostering Project Teams to Achieve High-Performance

The heterarchical form of self-organization builds on diversity of talent, using it for creativity, for organizational learning, and for discussing what is right for the overall organization to evolve [32]. However, achieving high performance depends on the formation of a clear, social order in the project team. Instead of job descriptions that focus on status and narrowly defined authority, role descriptions are agreed upon with a balance of authority and accountabilities in the team. Instead of a boss who controls the team, the project team regulates itself. The team is jointly responsible for achieving the goals, creates the transparency necessary for meeting the goals, and mutually controls each other through peer pressure [10].

In his seminal book "Reinventing Organizations" [33], Frederic Laloux describes the various development stages of an organization on the way to a purposeful form of collaboration. At the highest form, the so-called "Teal Organization", members of an organization do not strive for conformity to external or even internal norms, but trust their inner compass as to what contribution they can make in a given situation to achieving the agreed upon goals. This is primarily about building on strengths, continuously progressing, and, above all, seeking wholeness in relationship to others, to life, and to nature. With these basic principles and a variety of practices, he shows the path to emergence that can lead to high-performance teams.

1.4.4 *Creating Organizations That Are Adaptable and Resilient*

In Sect. 1.2.2, we identified increasing volatility, uncertainty, complexity, and ambiguity as major challenges for most organizations. However, centralized organizations with a strong focus on hierarchy and bureaucracy are rather unsuitable for dealing with these challenges. Self-organizing units, projects, and structures are much more adaptable in this situation and can react ad hoc to the unexpected. Henry Mintzberg describes the key principles to form what he calls an "adhocracy": "To innovate means to break away from established patterns… Of all the configurations, Adhocracy shows the least reverence for the classical principles of management, especially unity of command… different specialists must join forces in multi-disciplinary teams, each formed around a specific project of innovation" [34]. The ability to adapt to a changing environment is only possible through the greatest possible diversity and a self-organizing team of passionate people. Allowing and encouraging this and supporting the team at all times is the key task of leadership [35].

However, it is not enough to adapt to the changing environment from time to time. The frequency and amplitude of change continues to increase, which is why organizations should prepare themselves to be resilient, which basically means absorbing strain and preserving key functions during the presence of adversity as well as developing improved capabilities to cope with future events or crisis [36]. This includes but is not limited to sharpening the senses in the project team for weak signals that often precede an extraordinary event, incorporating diverse perspectives while blocking out "killer phrases", empowering the project team, and incentivizing beyond being compliant [37].

1.5 Conclusions and Outlook

Self-organization is not only increasingly popular among employees, but also offers advantages in many respects when dealing with a VUCA world. First and foremost, it is a matter of leaving behind the traditional principles of organizations of the industrial age, namely, hierarchy, bureaucracy, and subordination to dominant management structures. Projects realized with self-organization help to unleash hidden potentials in the individuals, the project team, and the ambient organization. Potentials such as cognitive readiness, passion, emergence, and performance enhancement on project level as well as adaptability and resilience of the overall organization. However, these effects do not occur on their own, they are rather the result of organizational change and the dedicated efforts of leaders who, above all, have to empower their employees, create space to maneuver, and actively foster a new way of collaborating.

However, self-organization is not a panacea for an organization. Rather, it is a matter of examining in which areas this brings significant benefits. In particular, self-organization will be useful in teams that are closely linked to external markets, in the context of R&D activities, in the delivery of services, etc. In areas that operate in a fairly stable environment, can rely on routine work and repetition, self-organization will not necessarily bring the benefits it shows elsewhere.

A prerequisite for the change toward more self-organization will be a targeted transformation program that checks whether employees are willing to work in self-organized project teams, whether they would like to give this a try, and whether they would prefer to be deployed in such projects after gaining initial experience. Based on a process of organizational learning, the entire organization gradually develops in the direction of more self-organization, like a living organism. This is something that requires further investigation in research.

References

1. Harari YN (2015) Sapiens: a brief history of humankind. HarperCollins Publisher, New York
2. Taylor FW (1989) The principles of scientific management. Reprint. Dover Publications, New York
3. Weber MC (2019) Economy and society. New translation. Harvard University Press, Cambridge
4. Defoe D (1697) An essy upon projects. R.R., London
5. Morris P (2013) Reconstructing project management. Wiley, Chichester
6. Wagner R (2021) The role of projects in the process of transforming automotive industry. In: Cuevas R, Bodea CN., Torres-Lima P (eds) Research on project, programme and portfolio management. Lecture notes in management and industrial engineering. Springer, Cham, pp 127–136
7. Bergmann F (2019) New work. New culture. Work we want and a culture that strengthens us. Zero Books, Winchester
8. Hamel G, Zanini M (2020) Humanocracy: creating organizations as amazing as the people inside them. Harvard Business Review Press, Boston
9. Kotter J (2014) Accelerate. Building strategic agility for a faster-moving world. Harvard Business Review Press, Boston, p vii
10. Pflaeging N (2014) Organize for complexity. How to get life back into work to build the high-performance organization. BetaCodex Publishing, New York
11. Schwab K (2016) The fourth industrial revolution. Crown Business, New York
12. Bodea CN, Ronggui D, Stanciu O, Mitea C (2020) Artificial intelligence impact in project management. IPMA, Nijkerk
13. Savolainen T, Lehmuskoski K, Koskenvaara T (2020) Surfing the digital tsunami. A playbook for continuous business transformation. 4th Wave Business Partnership
14. Von Bertalanffy L (1969) General system theory. Foundations, development, applications, 17th ed. Braziller, New York
15. Haken H, Schiepek G (2010) Synergetik in der Psychologie. Selbstorganisation verstehen und gestalten. 2. Auflage. Hogrefe, Göttingen
16. Oswald A, Köhler J, Schmitt R (2018) Project management at the edge of chaos. Social technologies for complex systems. Springer, Berlin
17. Ryan R, Deci E (2018) Self-determination theory. Basic psychological needs in motivation, development, and wellness. The Guilford Press, New York
18. Robertson B (2016) Holacracy. The revolutionary management system that abolishes hierarchy. Penguin, London
19. Balve P, Schaffitzel N (2019) Agile enterprise structures from a cybernetic perspective. In: Oswald A, Müller W (eds) Management 4.0—handbook for agile practices. Release 3.0. Books on Demand, Norderstedt, pp 111–130

20. Sennet R (2012) Together. The rituals, pleasures and politics of cooperation. Yale University Press, New Haven
21. Tamm J, Luyet R (2019) Radical collaboration. Five essential skills to overcome defensiveness and build successful relationships, 2nd ed
22. Von Foerster H, Poerksen B (2002) Understanding systems. Conversations on Epistemology and Ethics. Kluwer Academic/Plenum Publishers, New York
23. Winter M, Szczepanek T (2009) Images of projects. Gower Publishing, Farnham
24. Pflaeging N (2020) Essays on beta. What's now & next in organizational leadership, transformation and learning, vol 1. BetaCodex Publishing, p 78
25. Gloger B, Rösner D (2017) Selbstorganisation braucht Führung. Die einfachen Geheimnisse Agilen Managements. 2. Auflage. Carl Hanser Verlag, München
26. Kühl S (2013) Organizations. A systems approach. Gower Publishing, Farnham
27. Weick K (2001) Making sense of the organization. Blackwell Publishing, Oxford
28. Fink F, Moeller M (2018) Purpose driven organizations. Sinn-Selbstorganisation-Agilität. Schäffer-Poeschel Verlag, Stuttgart
29. Belack C, Di Filippo D, Di Filippo I (2019) Cognitive readiness in project teams. Reducing project complexity and increasing success in project management. Routledge, New York
30. Vallerand R (2015) The psychology of passion. A dualistic model. Oxford University Press, New York
31. Csikszentmihalyi M (1991) Flow: the psychology of optimal experience. Harper Collins Publisher, New York
32. Fairtlough G (2005) The three ways of getting things done. Hierarchy, heterarchy & responsible autonomy in organizations. Triarchy Press, Axminster
33. Laloux F (2014) Reinventing organizations: a guide to creating organizations inspired by the next stage in human consciousness. Nelson Parker, Brussels
34. Mintzberg H (1993) Structure in fives. Designing effective organizations. Prentice Hall, Upper Saddle River
35. Kotter J (2012) Leading change. Harvard Business Review Press, Boston
36. Arthur J, Moody L (2019) Building resilient organisation. Routledge, Milton Park
37. Kutsch E, Hall M, Turner N (2015) Project resilience. The art of noticing, interpreting, preparing, containing and recovering. Gower, Farnham

Chapter 2
The Whole—More than the Sum of Its Parts! Self-Organization—The Universal Principle!

Alfred Oswald

Abstract The term self-organization has been successfully used for years in management literature, driven by agile organizational concepts in which self-organization plays an outstanding role. Here, self-organization sometimes stands for autonomous teams, time management, agile forms of organization, or new work, to name just a few. Only in the rarest of cases the statement "The whole is more than the sum of its parts" is in the foreground. And this is exactly how we want to understand self-organization. We use the term self-organization as it has been increasingly used in the natural and social sciences since about the mid-1970s. We regard self-organization as a universal principle that creates something new in evolutionary terms and whose basic concepts are applied in nature, in the psychological and social realms, and in technology. Thus, we have a common basic concept that can be applied to individual human beings, teams, organizations, societies, and even to technologies such as artificial intelligence and robotics and their interaction with humans. It helps us to understand, shape, and lead the development and transformation of individuals, teams, organizations, and societies. We outline the characteristics of a project as a teal temporary organization.

Keywords Complexity · Control parameters · Emergence · Governance · Leadership · Order parameters · Self-organization · Setting parameters · Spiral dynamics · Teal project · Teal temporary organization · Value meme

2.1 Motivation—Basic Assumptions and Derived Hypotheses

If you ask project managers whether they have encountered complexity, the vast majority will confirm that they have experienced complexity very often. Very rarely, however, can they explain what they mean by complexity. In very few cases, projects

A. Oswald (✉)
IFST-Institute for Social Technologies GmbH, 52223 Stolberg, Germany
e-mail: alfred.oswald@socialtechnologies.de

R. Ding et al. (eds.), *Research on Project, Programme and Portfolio Management*, Lecture Notes in Management and Industrial Engineering,
https://doi.org/10.1007/978-3-030-86248-0_2

are classified using the Stacey representation or even the Diamond model [1–3]. It is similar with the term "self-organization": it is often put into the context of agile frameworks, but very rarely into the context of complexity.

At the same time, we perceive that the interconnectedness of nature, society and technology, and the dynamics and volatility associated with it are increasing more and more. Our uncertainty is growing at the same time. The effects caused by environmental damage, the collapse of financial systems, positive and negative "social storms" in social media, or the COVID-19 pandemic, make us more and more aware of such interconnectedness and dynamics. In addition, digital transformation is also driving this natural, social, and technical interconnectedness. This is also seen as a driver of innovation to get the harmful effects of networking and dynamics under control.

Projects are the arenas in which innovations are created. Innovations are some of the most essential complexity drivers in these arenas and, very often with a certain time delay, also in the later application areas of these innovations.

The following basic assumptions and derived hypotheses relate to this and the following statements.

The basic assumptions are:

We perceive complexity if a high degree of interconnections in time and/or place is present in a system, small changes have big effects, and unforeseeable system behavior in time and/or space may occur [1]. Complexity is the basis of our being and of life: Therefore, it does not make sense to reduce complexity in every case. Rather, it is necessary to regulate complexity and design it in such a way that it creates added value.

Self-organization thrives in complexity and is the evolutionary mean to regulate complexity. Self-organization creates, under certain conditions and through the interaction of system elements, emergent system properties. We can then state that "The whole is more than the sum of its parts". The consequence of a self-organizing process is an ordered pattern, collective behavior, or a new structure. These emergent system properties cannot be derived from the properties of the system elements and vice versa: if the system properties are known, a decomposition in element properties cannot explain the system properties. Emergence is the process to create emergent system properties [1].

The derived hypotheses are:

Figure 2.1 is a visual translation of the complexity metric proposed in [4]: complexity = emergence * self-organization ($C = \text{constant} * E * S$). This metric expresses that complexity consists in the balance between disorder and order, or in the absorption of completely new information and the processing of known information. It is thus directly related to the approach of "exploring and exploiting" of the so-called ambidextrous organization [5] and it corresponds to the definition of agility = flexibility * speed used in [6]. However, it can also be immediately deduced from this that agility is nothing other than the regulation of complexity, i.e., balancing emergence and self-organization [1, 4].

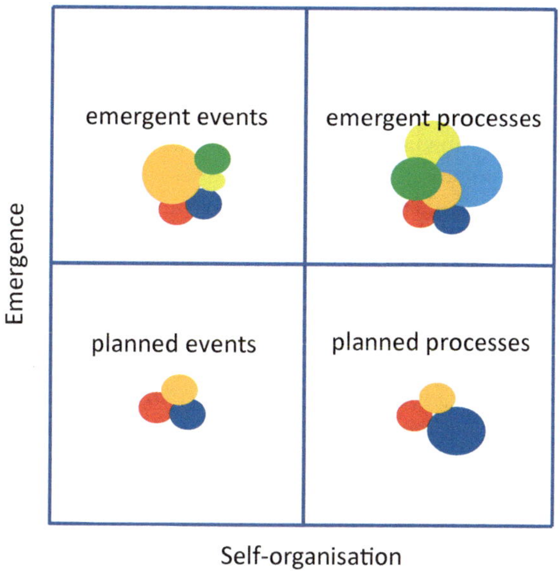

Fig. 2.1 Complexity = Emergence * Self-organization

This metric also expresses that every system organizes itself, only the degree of emergence, i.e., the ability to create something new, is different. The system parameters describing the system as a system are decisive: hierarchical or even autocratic organizations also organize themselves, but the degrees of freedom are so small that probably something new may only arise to a small extent.

Based on Fig. 2.1, plan-based project work and management is in the "planned events" quadrant. Here we use the term "events" to indicate that events or loosely coupled processes dominate. Strongly coupled events and processes of work and management, e.g., production processes, are in the quadrant "planned processes". Because "planned" behavior dominates, these two quadrants show a low or very low emergence. Here, the percentage of stakeholders that show a planning inner attitude (mindset) is very high: New things can only form with difficulty, since the emergence is limited by planning. The importance of the mindset is expressed by the color code of Spiral Dynamics, the development and consciousness model [7]: red means power orientation, blue means control and order orientation, and orange means entrepreneurship and linear scientific thinking orientation. The size of the circles indicates the qualitative significance of the respective orientation.

Separated from the planning quadrants by a very strong jump of the mindset are the emergent quadrants. The quadrant "emergent events" can be roughly associated with the actions that are present in a start-up. The quadrant "emergent processes" shows a high degree of self-organization, with the potential for organizational high performance, and the willingness to adapt quickly to new situations. This is the area of value-creating complexity, which is characterized by a confident balance of self-organization and emergence. This area is characterized by organizational high performance that regulates complexity in a value-creating way. Ideally, this requires a mindset of all key stakeholders that also contains red, blue, and orange value meme components, but is mainly shaped by the transformational value meme components, namely, green (compassion), yellow (non-linear networked system thinking), and teal (holistic-transcendental orientation). The (social) self-organization described in the following section is in this "emergent processes" quadrant and requires the transformational value meme.

This Spiral Dynamics color code introduced by Beck and Cowan to describe levels of social evolution, culture, and mindset development has been applied by Laloux to the classification of organizations [8]. In accordance with these, we speak of a teal (temporary) organization, i.e., a teal project, if this can express teal values memes and these characteristically shape the (temporary) organization. In [6] we referred to this form of color typology as value meme archetypes and applied it to individuals, teams, organizations, or population parts of a society.

The key consequence for (project) management in the "emergent processes" arena is a fundamental re-orientation of the self-understanding of project managers: Project managers are transformational leaders with a teal mindset who have meta-competence to understand complexity, can regulate it and can shape it by means of self-organization [9, 10]. In this way they create agility in people, teams, and organizations and indirectly in society.

2.2 Literature Review—Synergetics and Management 4.0

In the mid-twentieth century, cybernetics, complexity, system theory, and self-organization began to develop more and more as an interdisciplinary theory and practice [11, 12]. Without claiming to be exhaustive, the works that have a direct influence on the understanding presented here are the works of Ashby [13, 14], von Bertalanffy [15], and Bateson [16]. These build the general foundation of the present concept. The works of Jantsch [17] and Kaufman [18, 19] are representative for the statement "self-organization, the universal principle". As a member of the Club of Rome, Vester introduces networked system thinking that is highly relevant to complexity, as well as topics relating to climate catastrophe and the COVID-19 pandemic [20]. The Santa Fe Institute online course provides a necessary in-depth modern understanding of complexity [21]. Luhmann applied the principle of self-organization as a long-range theory to social systems [22]. Baecker supplemented this understanding and applied it to Digital Transformation [23]. This social science understanding coincides to a large extent with the principles of Management 4.0 proposed here [1, 6]. The core of the Management 4.0 concept is essentially based on the theory of self-organization according to Hermann Haken, called Synergetics [24]. Synergetics was first applied to the Laser self-organized technical system. Over the last 50 years, Synergetics has been applied in almost every scientific field resulting in a still growing huge amount of literature [25]. In particular, Synergetics has been applied to psychology and social systems with great success and helps with an understanding of the basic mechanisms of this kind of self-organization [26, 27].

Synergetics differs from other theories of self-organization in that it assigns different time scales to the variables of the theory. A basic assumption of Synergetics is that the variables of large time scales determine system behavior on a macroscopic system level. Due to this ordering character, they are called order parameters. In the following, the names of the variables and parameters known from Synergetics will be used, since we want to retain the connection to the extensive scientific literature.

The core of the present Management 4.0 concept is an adaption of the principles of Synergetics to the (project) management area. It has been developed over the last 10 years in the Agile Management special interest group of the German Association of Project Management [1, 6]. In this book, an additional contribution is connected to this work [28].

2.3 Research Method—Self-Organization in a Nutshell

Before we describe our research method, it seems appropriate to outline some basic ideas of Synergetics using social application areas as examples.

Most people have already experienced how the arrangement of tables and chairs in a meeting room influences how people entering a room behave. If there are several tables, people tend to spread out over several tables. Those people who already know each other will probably gather at the same table. The self-organization of people entering the room, therefore, depends very much on these basic conditions: The arrangement of the furniture, as well as the composition of the people. If there are only standing tables available, people cannot sit down. If there are no tables available, some will simply remain standing in one place, others will move around the room, perhaps looking for contacts. Interior design, here with the example of existing or non-existing tables, is an example of basic conditions, which self-organization designers should consciously arrange. Self-organization designers, however, could also design time; time-boxing in agile techniques is one such possibility. In addition, there are many more options for designing the basic conditions or the setting parameters (SP), according to the theory of self-organization: Team composition is an essential setting parameter of every team self-organization. Currently, for example, the technical possibilities of collaboration tools such as MS Teams or Zoom are also setting parameters of our self-organization in online spaces. Anyone who has ever been part of a team building process knows that it can make a big difference whether the team building process is carried out in rooms in your own company or at a hotel or on a river cruise ship, or maybe somewhere where there is no Internet, or as part of an online collaboration session. All this has an impact on self-organization. As a best-practice collector, you could suggest that we make a list of all potential setting parameters and then, as we will see later, we would already have collected one third of all parameter types. But here, we must disappoint our readers. Personal self-organization begins with a basic understanding of what is being said here and of what is much more essential in a concrete situation, which is the permanent checking of the setting parameters to ensure that the chosen setting parameters lead to the desired effect. If necessary, parameters introduced as "trials" or "tests" must be readjusted based on observations. Parameters are therefore not static but are subject to a passive or active permanent change process. The generalized active leadership process follows the generalized Plan-Do-Check-Act-Cycle of acting in a complex environment [1].

Let us turn to the second third of the system parameter types, the so-called control parameters (CP). The term "control parameter" is, like the other parameter designations, a *terminus technicus* and originates from complexity theory and from the theory of self-organization. Control here means steering, rather than monitoring or surveillance. In Agile Management, the so-called Work-In-Progress (WIP) is one of the most famous control parameters. It plays a dominant role there, especially in the agile frameworks Kanban, Scrum, or Critical Chain Project Management (Theory of Constraints) [1, 6]. It expresses the fact that a person acting in an agile manner processes only one task per time unit. The practice of keeping too many plates spinning should therefore be stopped and instead focused work should be consciously brought to the fore. This considers the fact that focus is a necessary condition for high performance. Meanwhile this basic idea is also explicitly used in smart techniques: MyAnalytics from Microsoft is an office tool that helps you to achieve the desired focus. Here, personal self-organization has a lot to do with self-reflective time management: i.e., not too many meetings per day, scheduling protected time areas for concentrated creative work, sufficient time between meetings to allow for preparation and follow-up work, etc.

In Management 4.0, personality-oriented communication, based on a conscious respectful handling of individual temperaments, values and beliefs, is another very important control parameter [1, 6]. For example, one of the most important tasks of an agile leader, i.e., also of a project leader, is to design this control parameter in such a way that value-creating complexity can develop in a team or an organization. Value-creating complexity means the balance of emergence and self-organization. This balance is enabled and catalyzed by the green (compassion), yellow (non-linear networked system thinking), and teal (holistic-transcendental orientation) transformational value memes.

Individuals, teams, or organizations need a sufficient value-creating complexity of their own to understand and regulate the complexity of the task or environment. This statement stems from Ashby's law and is the engine for innovation [13, 14]. This leads directly to the insight that, for example, highly innovative projects, which drive complexity, require a transformational organizational mindset to develop sustainable solutions. Sustainable solutions are characterized by being based on compassion for nature, animals, and people, anticipating and shaping their systemic impacts in time and space accordingly, and deliver a holistic global sense of purpose. Behrens describes what implications this will have for all social levels for the next few years [29].

The design of a value-creating complexity through the design of control parameters—there may well be several or even many—is a dynamic process. It never ends, because the individual, the team, the organization, or the society are in a permanent flow of contextual change, i.e., subject to self-consistent adaptation of system

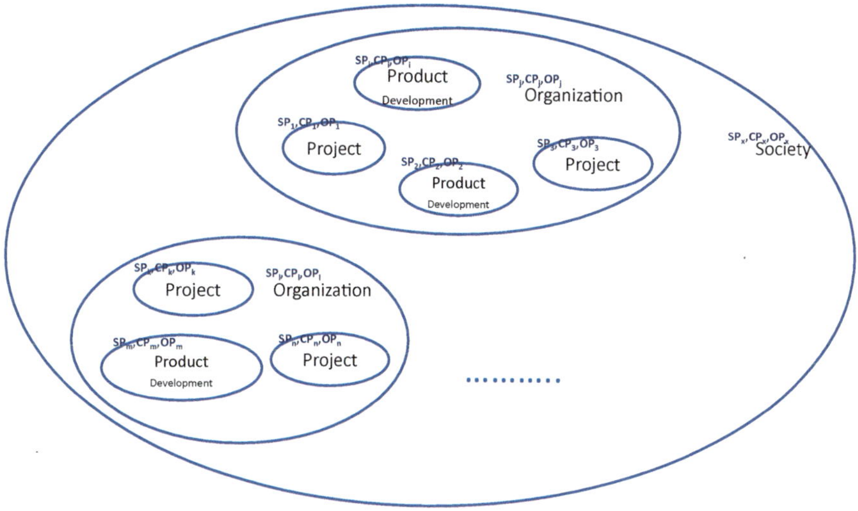

Fig. 2.2 Self-organization on all levels, characterized by system parameters (Setting (SP), Control (CP), and Order Parameters (OP))

parameters. This also demonstrates that the system parameters of self-organization are interdependent in their dynamics.

This corresponds to the approach that every system (S) permanently creates itself by feedback mechanisms. The environment (U) that surrounds the system is constantly included in this self-referential process. In natural science, as well as in modern sociology, this is expressed by the equation $S = f(S, U)$ [22, 30]. This equation represents the core of a systemic perspective. This also means that an organizational structure and its characteristic system features, as shown, e.g., in Fig. 2.2, cannot be understood by breaking down the entire system into its individual parts. Figure 2.2 shows a society, described by self-organized circles (SO-circles), here using two organizations as an example, which contains project and product development SO-circles. These SO-circles are (sub-)systems, which are described by system parameters (SP, CP, OP). According to the statement $S = f(S, U)$ this network of system parameters is self-consistently created by a self-organizing process. Please see Fig. 3 of [28] in this book, too: The figure there illustrates the nested self-consistent self-organization mechanism from a designed and guided self-organization perspective.

Currently, the COVID-19 virus is the dominant setting parameter of our entire society—the virus imposes behavior on us, just like the tables in a meeting room. Politics tries to introduce control parameters through the rules of hygiene and "staying at home" to control the self-organization of society. Unfortunately, not everyone adheres to them and undermines these control parameters—this means, self-organization develops outside the proposed self-organization. This can possibly lead to a tightening of the control parameters by politics; in the worst case, to such an extent, that self-organization no longer deserves its name. So, on the one hand it is an art to design the best (test) system parameters for the respective context and on the other hand the (necessary) system parameters also depend on the behavior of those involved. So, we design the context ourselves, which may then make other system parameters necessary. In the case of COVID-19, this is exactly what we find. Unfortunately, political leaders fall into the usual pattern that is also known from organizational leadership: they try to get a grip on complexity through complicatedness with more and more rules of conduct—which is unfortunately always doomed to failure, because you cannot counter complexity with complicatedness.

This brings us to the third system parameter type of self-organization, the order parameter (OP). As the name suggests, this parameter type creates order in the system. Our market economy is a self-organized system that forms prices under certain setting and control parameters. Prices are the order parameters of the market. But prices are not always (dominant) order parameters; different economic schools use different order parameters, i.e., different long-term variables to describe the market [31]. More generally, it can be said that in a social system, the system elements, i.e., the people, the teams, the organizations, or the entire society, are "ordered" by the order parameters. In a project, a sensemaking collective goal or vision is an order parameter created by all members of a team or an organization. In general, the collective goal or vision is not a simple target or objective but is a target hierarchy [1]. In [6] we have shown that order parameters also play a major role in politics. The statements "Yes we can" by Barak Obama or the "Make America Great Again" are test order parameters, i.e., they are thrown into the social space by respective politicians with the intention to generate an order. This order is visible, if the test is successful, as a strong increase in followers and votes.

If the setting and control parameters are appropriated, the order parameter, i.e., target hierarchy, "enforces" a collective emergent flow state, which we call Collective Mind. It is essential that the Collective Mind of a team or an organization is created by feedback mechanisms in the team or in the organization and not by an external visionary. If there is a Collective Mind, then the orientation toward a target hierarchy creates a social cohesion that is so strong that the team, organization, or society overcomes great difficulties and creates something new. The collective emergent flow can only unfold its effect if it is strongly emotionally charged. SMART objectives (SMART = Specific, Measurable, Accepted, Realistic, Testable), as they have been known in (project) management for many years, do not achieve this. So, SMART objectives are never on the top of a target hierarchy, but at the bottom. If an order parameter is formed in a self-organized manner, this leads to a new macro structure in the entire system, which we call the emergence of the Collective Mind phenomena.

As a necessary condition for the designed and guided (social) self-organization [1], the control parameters must be selected in such a way that the degrees of freedom within the social system team, organization or society are sufficiently large enough to enable self-organization to develop by emergence. The task of WIP limitation or personality-based communication is to specifically enable these degrees of freedom. Degrees of freedom are necessary, but not sufficient. These degrees of freedom must limit themselves to a certain extent by subordinating themselves to a superior goal or vision. Only when this happens is a Collective Mind present. A project team that voluntarily subordinates itself to a collective goal achieves maximum performance and creates innovative project solutions. An organization whose values and beliefs fit and live up to its vision and mission and adaptively align it to its environment will be viable in the long run. A society that demonstrates a degree of freedom in democratic struggle and aligns it with common values and goals shows sustainable development potential.

Agile frameworks are familiar with different forms of a target hierarchy, for example, in the agile framework Scrum, the target hierarchy consists of vision, sprint goal, user story (and task), or in agile portfolio and program management, the story map is another one. The OKR used by google (objective and key results) is a target hierarchy for aligning an entire organization. Sometimes the word "purpose" is used for the top of a target hierarchy. In [6] we have shown that the use of self-organization for large organization requires the synchronized scaling of structure and information: The scaling of structure is built on a team-of-team or project-of-project structure and the scaling of information follows target hierarchy design principles.

The design of the three-parameter types for designed and guided self-organization is called the governance of self-organization. This governance focuses on the alignment of all (sub-) systems of an organization using self-organization to create a self-consistent system-of-systems, i.e., created from top-to-bottom and from bottom-to-top on a regular basis. This understanding of governance is different from that in the ICB 4.0 guidelines [32].

As we have already mentioned, this governance is dynamic. Systems that consist of several subsystems have their own sub-governance for each subsystem, which must be self-consistently coordinated with the higher level governances to enable self-organization in all subsystems and in the entire system. At present, the example of the COVID-19 pandemic shows how difficult it is to achieve a governance of self-organization coordinated self-consistently at all levels in a federal system with many subsystems. According to the current state of knowledge, however, this is the only way to enable self-organization in large systems. In agile management, the usual governance frameworks such as Scrum and Kanban, as well as SAFe and LeSS, are static in nature. Governance of the Holacracy or Sociocracy frameworks are dynamic governance networks for an entire organization with several sub-organizations [6]. Holacracy tries to recapture the resulting complexity (unfortunately) through a complicated network of roles. As we have already indicated, this cannot succeed well, because you cannot master complexity by complicatedness if you want to have self-organization [6].

The design of self-organization through governance is the leadership tool par excellence. As a meta-principle to create a system governance we use the guideline that "any implementation of an SO framework governance should follow the principles of Synergetics and scaling" [6, 28]. If an agile framework cannot fulfill this requirement, that framework is not a framework which supports (scaled) self-organization.

Because self-organization is a universal mechanism, self-organization does not necessarily create "good" structures. "Good" structures mean that the structure is based on a code of ethics. Therefore, good self-organization needs an SO governance which is regulated by a code of ethics and conduct. Just two examples to illustrate this: Mafia-like organizations or social structures are also created through self-organization! Or an AI system trained by means of data, may develop racist macro structures (i.e., decision structures), as a result of consciously or unconsciously selected training data. This is another reason why the social evolution, culture, and mindset development model, Spiral Dynamics, plays a major role in the Management 4.0 concept. If a temporary project has an organizational mindset driven by transformational value memes, this is a critical basis for ethics that guarantee sustainability.

This brings us to our research method:

As outlined, our basic assumption is that self-organization is a universal principle that underlies every system. Self-organization is controlled by the practice of transformational value meme. We combine a deep understanding of the scientific literature of Synergetics with our decades of management experience to adapt the principles of self-organization to the field of management. This is a no mean feat, because in general only the principles are transferable. In doing so, we integrate other scientific theories and models from psychology, neuroscience, sociology, complexity research, system theory, and AI research. We continuously review existing off-the-shelf frameworks to get further ideas for the design of setting, control, and order parameters. In doing this, we learn how to create and apply design principles for self-organized social systems: we create social technologies to build (scaled) self-organized systems according to the team-of-team, project-of-project or system-of-system principle. In addition, insight synergies emerged from the combination of different theories: for example, the value-meme levels of Spiral Dynamics could be related to the neuroscientific consistency theory of basic human needs; or the levels of the Dilts Pyramid are connected to the system parameters. These created a networked theory, which shaped the understanding of social self-organization, helped to explain what complexity regulation, agility, and meta-competence mean and how they are linked [1, 6, 28].

The PDCA-cycle is used as an empirical test framework (see Fig. 2.3): Based on success criteria, key performance indicators, and related success factors in a concrete system context, we build hypotheses, and question why and which social technologies should be used and how. We map the chosen social technologies to the success factors in the practice of (project) work and management, consulting, coaching, and training. In these practices, we check whether the success criteria can be fulfilled with the chosen success factors and related social technologies. If necessary, we adapt these. Clearly, it may also be the case that new major insights make the adaption of success criteria, key performance indicators, and the base theories necessary. This process allows continuous improvement of the self-organized system design and related leadership: the theory network of Management 4.0 is continuously validated by an interplay of theory and practice.

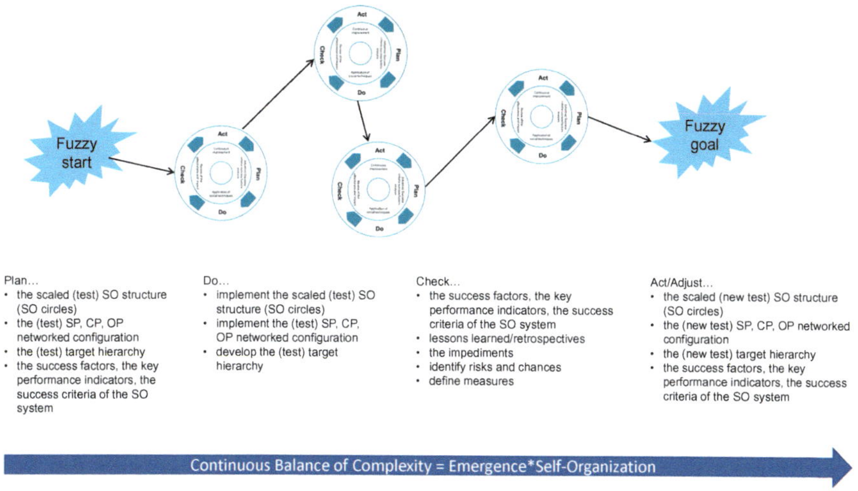

Fig. 2.3 Designed and guided self-organization of a teal project (reproduced from [1] with permission granted by Springer Nature Customer Service Centre GmbH)

2.4 Results and Discussions—The Characteristics of a Teal Project

The qualitative results show that significant performance improvements in efficiency and effectiveness are achieved in project work and at the same time the satisfaction of project staff and stakeholders increases significantly [6].

Based on the assumption that complexity increases in the environment of many (temporary) organizations, this requires, according to Ashby, an increase in the ability of (temporary) organizations to regulate complexity in order to enable value-creating complexity [1, 6]. This regulation of complexity requires a meta-competence characterized by self-reflection, openness and compassion, systemic and networked thinking, and holistic integral sensemaking: a complex or chaotic project (according to Stacey typification) ideally requires a value-meme structure corresponding to that in the "emergent processes" quadrant.

In a teal project, the "limiting" values (red, blue, and orange) are less pronounced and are regulated by the "transforming" values (green, yellow, and teal). Barrett has provided strong evidence for this correlation [33]. This also means that the project manager and, ideally, other stakeholders must have sufficient "transforming" values so that the organizational mindset of the temporary organization "project" shows itself as a teal organizational mindset or teal culture. The composition of the value memes acts as a control parameter, which must be guided by the project manager as teal leader.

Based on the above-described principles of self-organization, we outline the characteristics of a complex or chaotic project which is capable of showing a balance of emergence and self-organization.

- Such a project has a teal organizational mindset. There is a sufficient number of stakeholders (core team and extended team members), and importantly, a project leader who has a teal mindset.
- The stakeholders and in particular, the leaders of the project, show a meta-competence which allows the potential development of the four levels of learning according to Bateson, linked by Dilts to the Dilts Pyramid [1, 6, 16, 34].
- The embedding of the project in its environment is done sustainably, considering long-term consequences for nature, technology, and society. A teal organizational mindset (teal culture) is a necessary condition for this. This may also necessitate premature termination of the project if the solution cannot fulfill sustainability.
- The stakeholders accept uncertainty, also for the entire duration of a project, and accept that the project start, and project goal are not well defined, but are fuzzy.
- Project leaders act as self-organization designers. Leaders design and guide by SO governance, which makes self-organization possible: this is visible in organizational performances with substantially better offset and super scaling factors—i.e., the whole is more than the sum of its parts [1, 6].
- Leaders are capable of switching adaptively between open mode and closed mode, while at the same time increasing focus on the project goal. Open mode means the emphasis is on effectiveness, i.e., emergence. Closed mode means the emphasis is on efficiency, i.e., self-organization. This switching between open mode and closed mode is a basic process characteristic for leading complex projects. This switching process characteristic is typical for self-organized frameworks like Design Thinking and Collective Mind Method [1, 6].
- The systemic embedding of the teal project in its environment is based on hybrid scenarios described in [6, 28].

The Management 4.0 approach has been tested qualitatively through experience in practice. There is no available evidence based on statistical methods. It is also questionable as to whether this is useful. Management 4.0 concepts are based on theories and models that can be implemented using AI. Strong evidence can also be delivered by an AI implementation. An AI implementation can be used to falsify the Management 4.0 theory and measure the offset and scaling factors [6]. The value of implementation using AI is likely to be in all three added value areas—assistance, value focus, and collaboration [35].

References

1. Oswald A, Köhler J, Schmitt R (2018) Project management at the edge of Chaos. Springer, Heidelberg
2. Stacey RD (2015) Strategic management and organizational dynamics: the challenge of complexity, 7th ed. Pearson Education, kindle ed.
3. Shenhar AJ, Dvir D (2007) Reinventing project management: the diamond approach to successful growth & innovation. Mcgraw-Hill Professional, Boston

4. Fernández N, Maldonado C, Gershenson C (2014) information measures of complexity, emergence, self-organization, homeostasis and autopoiesis. In: Propokenko M (ed) Guided self-organization: inception, emergence, complexity and computation, vol 9. Springer, Heidelberg
5. O'Reilly III CA, Tushman ML (2004) The ambidextrous organization. Harvard Business Review. https://hbr.org/2004/04/the-ambidextrous-organization. Accessed 07 Dec 2020
6. Oswald A, Müller W (eds) (2019) Management 4.0—Handbook for agile practices, Release 3, Books on Demand GmbH, Norderstedt
7. Beck DE, Cowan CC (1996) Spiral dynamics: mastering values, leadership and change (Blackwell textbooks in linguistics)
8. Laloux F (2014) Reinventing organizations, kindle ed. Neslon Parker
9. Erpenbeck J et al (2006) Metakompetenzen und Kompetenzentwicklung, QUEM-Report, Heft 95 Teil 1, Berlin
10. Bergmann G et al (2006) Metakompetenzen und Kompetenzentwicklung, Metakompetenzen und Kompetenzentwicklung in systemisch-rationaler Sicht, Selbstorganisationsmodelle und die Wirklichkeit von Organisationen, QUEM-Report, Heft 95 Teil 2, Berlin
11. Wikipedia (2020) Complex systems. https://en.wikipedia.org/wiki/Complex_system. Accessed 04 Dec 2020
12. Wikipedia (2020) Self-organization. https://en.wikipedia.org/wiki/Self-organization. Accessed 04 Dec 2020
13. Ashby WR (1957) An introduction to cybernetics. Chapman & Hall Ltd., London
14. Conant RC, Ashby WR (1970) Every good regulator, must be a model of that system. Int J Syst Sci 1(2):89–97
15. Von Bertalanffy L (2006) General system theory, revised ed. George Brazillier Inc., New York
16. Bateson G (2020) Mind and nature: a necessary unity (Advances in systems theory, complexity, and the human sciences), kindle ed.
17. Jantsch E (1980) The self-organizing universe: scientific and human implications of the emerging paradigm of evolution (Systems science and world order library. Innovations in systems science). Pergamon
18. Kauffman SA (1993) The origins of order: self-organization and selection in evolution. Oxford University Press
19. Kauffman SA (1996) At home in the universe: the search for the laws of self-organization and complexity, kindle ed.
20. Vester F (2002) Die Kunst vernetzt zu denken: Ideen und Werkzeuge für einen neuen Umgang mit Komplexität, Ein Bericht an den Club of Rome, dtv Verlagsgesellschaft
21. Santa Fe Institute (2014) Complexity explorer. Online course introduction to complexity. http://www.complexityexplorer.org/courses. Accessed 20 Sept 2014
22. Luhmann N (1996) Social systems, 1st ed. Stanford University Press
23. Baecker D (2018) 4.0 oder die Lücke die der Rechner lässt. Merve Verlag, Leipzig
24. Haken H (1983) Synergetics: an introduction, 3rd rev. and enlarged ed. Springer series in synergetics. Springer, Heidelberg
25. Kröger B (2015) Hermann Haken: from the laser to synergetics. Springer, Cham
26. Haken H, Schiepek G (2010) Synergetik in der Psychologie: Selbstorganisation verstehen und gestalten. Hogrefe-Verlag, Göttingen
27. Kelso JAS (1995) Dynamic patterns: the self-organization of brain and behavior (Complex adaptive systems). A Bradford Book
28. Tuczek HC, Flore A, Nuhn HFR, Schaffitzel N (2020) A systemic approach to agile management and self-organization for sustainable transformation of organizations, in this book
29. Behrens P (2020) The best of times—the worst of times, futures from the frontiers of climate science. The Indigo Press, London
30. Vogd W (2020) Quantenphysik und Soziologie im Dialog, Betrachtungen zu Zeit, Beobachtung und Verschränkung. Springer Spektrum, Berlin
31. Zhang W (1991) Synergetic economics, time and change in nonlinear economics. Springer, Berlin
32. IPMA (2015) Individual competence baseline, Version 4, ICB 4.0

33. Barrett R (2017) The values driven organization, 2nd new ed, kindle. Routledge
34. Dilts R, DeLozier J (2010) NLP II—the next generation, enriching the study of the structure of subjective experience. Meta Publications, Capitola, CA
35. IPMA, PwC (2020) Artificial intelligence impact in project management, Report

Chapter 3
A Systemic Approach to Agile Management and Self-Organization for a Sustainable Transformation of Organizations

Hubertus C. Tuczek, Agnetha Flore, Helge F. R. Nuhn, and Norbert Schaffitzel

Abstract The understanding of agility and its necessity as a key factor for business success has never been more urgent than today. Many companies are not able to implement truly agile organizations. Why is this so? What are the driving forces behind a sustainable transformation? A holistic approach for the entire company, starting with the management mindset and the governance structure of self-organization, is the way forward for a consistently agile approach. The professional group "Agile Management (AM)" of the GPM developed the Agile Management 4.0 approach, which takes into account various scientific concepts to create a solid foundation for practical relevance in everyday management. AM 4.0 follows a systemic approach based on the theory of synergetics developed by Haken (Synergetics, an introduction: nonequilibrium phase transitions and self-organization in physics, chemistry, and biology. Springer, New York, 1983) and the model of self-organization according to Haken and Schiepeck (Synergetik in der Psychologie. Selbstorganisation verstehen und gestalten. Hogrefe, Göttingen, 2010). We explain Agile Mindset with the help of the neurological layers of the Dilts Pyramid (Dilts in A brief history of logical levels (2014), http://www.nlpu.com/Articles/LevelsSummary.htm), a model frequently used in psychology. Management cybernetics describes the system dynamics of stability and adaptation of organizations. Stafford Beer's

H. C. Tuczek (✉)
Hochschule Landshut, Am Lurzenhof 1, 84036 Landshut, Germany
e-mail: hubertus.tuczek@haw-landshut.de

A. Flore
OFFIS – Institute for Information Technology, Escherweg 2, 26121 Oldenburg, Germany
e-mail: agnetha.flore@offis.de

H. F. R. Nuhn
Wilhelm Büchner Hochschule Darmstadt, Hilpertstrasse 31, 64295 Darmstadt, Germany
e-mail: helge.nuhn@wb-fernstudium.de

N. Schaffitzel
DB Systel GmbH, Jürgen-Ponto-Platz 1, 60329 Frankfurt, Germany
e-mail: norbert.schaffitzel@deutschebahn.com

R. Ding et al. (eds.), *Research on Project, Programme and Portfolio Management*, Lecture Notes in Management and Industrial Engineering,
https://doi.org/10.1007/978-3-030-86248-0_3

Viable System Model (VSM) (Beer in J Oper Res Soc 35(1):7–25, 1984) is applied as a reference for the conditions of self-regulating systems. Finally, we address agile techniques and framework in the context of meta-competencies of a learning organization.

Keywords Agile mindset · Agile techniques · Complexity · Governance · Management cybernetics · Self-organization

3.1 Overview

Today's increasingly complex and dynamic environment requires new management responses to the resulting challenges for companies and institutions. Agility is the key concept to meet the needs in a VUCA world (volatile, uncertain, complex and ambiguous). Mindset and learning as well as meta-competencies and leadership for self-organization characterize the agile approach described in this article. Agile Management creates the necessary framework for agile projects and is therefore the prerequisite for agile practices. The Big Picture (Fig. 3.1) gives an overview of the different elements of the "AM 4.0" concept developed by the GPM professional group "Agile Management".

The upper level shows a transformation model, illustrated according to a systemic understanding of synergetics. The different parameters (control, order, and setting) that influence system properties and performance are presented. On the system level, the *organizational mindset* (mindset of the organization as a whole) is considered the most influential order parameter. On team or individual level (subsystem) the corresponding *team mindset* can be observed, which has a wide variety of variations and may or may not be in line with the organizational mindset. The setting parameters describe the relevant context for the given perspective of the organization or team/individual. The intervention (control parameters) must be chosen in such a way that it causes resonance in the system. The lower level in the graph below shows a more specific and detailed view of the mindset as shown in the Dilts pyramid. The factor 1000 is assigned symbolically to express the high effect on the organizational system. The corresponding governance (symbolic factor 100) stands for a self-organization that leads to an *emergent mindset* (lived culture) of the organization. It is a response to the stimulus of the *lead mindset*. Finally, the different techniques are considered with an assigned factor of 1, as shown in Fig. 3.1. The appropriate interaction of all elements ensures a successful agile management approach.

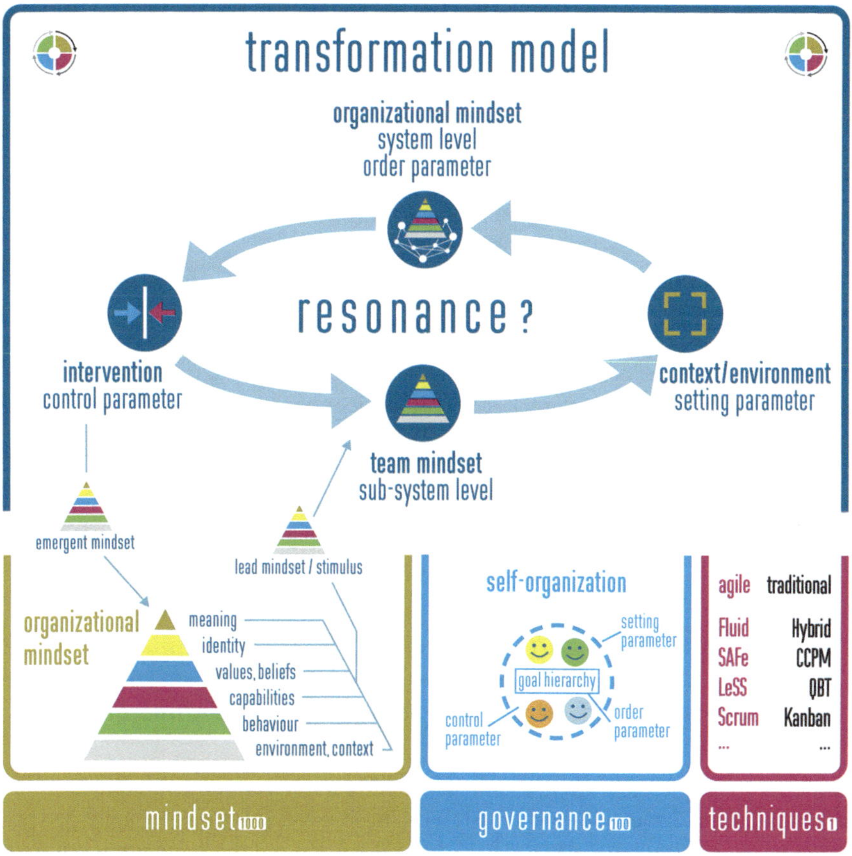

Fig. 3.1 Big picture agile management 4.0 (© Tuczek, Flore, Nuhn, Schaffitzel 2021. All Rights Reserved.)

3.2 Transformation Model

3.2.1 Systemic Approach

Management in today's VUCA world means dealing with organizations in a complex environment and dealing with the challenges involved in an appropriately agile manner. Therefore, leaders must develop a systemic understanding of the interrelationships, as management no longer functions according to simple linear cause-effect intervention patterns, as complex systems do not follow simple cause-effect relationships. The effect of an intervention is essentially non-linear and cannot be predicted to its full and final extent.

Yesterday's management concept of linear projection of trends and figures is no longer up to date. The exponential dynamics of digital transformation is forcing the complex and systemic character of the new business world. Linear thinking and the focus on direct competitors (as known factors) do not generate the right questions for the future development of corporations. The factors that will influence business today are beyond traditional strategic planning and may come from completely different industries or unforeseen global developments. These factors are not connected in a linear and unidirectional way but follow reciprocal and dynamic feedback mechanisms. Therefore, understanding the characteristics of complex systems and thinking in systems and scenarios is essential for today's managers.

At this point, it is important to consider Ashby's principle of cybernetics, "The Law of Requisite Variety": "The greater the variety of actions available to a control system, the greater the variety of perturbations it can compensate for". The well-established paradigm in management for controlling or regulating a system can be expressed by "reducing variety". Managers prefer simplification to complexity. Ashby's law, however, states that in order to manage the complexity of an environment, an equal or greater degree of complexity within an organization is required to achieve control. Managers should therefore welcome complexity in order to regulate environmental complexity. Instead of a leadership style with command and control, new leaders must develop the ability to orchestrate variety with a sense of empowerment and inspiration [1, 5].

3.2.2 Synergetics and the Phenomenon of Self-Organization

The theory of synergetics helps to understand the interaction of the elements in a complex system. Laser theory inspired Hermann Haken to his theory of synergetics—the "science of structure". He described the laser principles as self-organization of non-equilibrium systems. Complex systems have the property of—spontaneously or excited externally—forming synergetic structures (macroscopic patterns) under certain circumstances, a non-linear process. In the case of laser, coherent light is emitted as soon as the light amplitude converges to a stable value above the so-called laser threshold. The emerging systemic pattern forms a new macrostructure by self-organization of the system.

Essential for synergetics is the order-parameter concept, in which only a few specific parameters determine a new macroscopic order. Self-organization takes place, when a system of many non-linear interacting subsystems and external control and setting parameters (energy flows, environment) support self-organization [7, 8].

3.2.3 Systemic Intervention

The principle of self-organization is the answer to the regulation of complexity in social systems and the creation of high-performance organizations. The logic applies that "the whole is more than the sum of its parts".

Therefore, we use the concept of order, control, and setting parameter to describe the system dynamics of an organization and corresponding systemic intervention for our transformation model. We have identified the organizational mindset (culture) as major order parameter for an organization. The neurological layers of the Dilts Pyramid [6], a frequently applied concept of Neuro-Linguistic Programming NLP, serve us as a description and interaction model for the mindset. Robert Dilts developed his model for change processes in the 1980ties. It is related to Bateson's learning theory, which states that learning experience is always context-dependent and can be described as a learning process in four steps. The layers of meaning/purpose, identity, values and beliefs, capabilities, behavior, and environment/context of the Dilts Pyramid are defined in a hierarchical relationship. One layer organizes the information on the next lower layer, which means that changes at one level cause changes at the next level. Usually issues or problems can be solved at the next higher level compared to where they occur. The model gives a holistic overview of all relevant aspects of the goal definition, changes, or problems and can thus make deficits transparent.

This can be widely observed in the introduction of agility in companies. To become agile, most companies start a training program for agile techniques for their employees. In the beginning the employees are motivated, but after a while the effects fade away. Some managers may argue that agility is not the right concept for their company. But what happened? The training of techniques takes place on the levels of capabilities and behavior. However, the higher layers of the Dilts pyramid (values, identity, and meaning) were left in the state they were in before the change and were not adapted. Since we have discussed that the higher layers influence the layers below, the old values and beliefs hinder the changes on the lower layers, and the ambitious undertaking must fail. The problem can be described by: "Doing agile instead of being agile!".

For a successful transformation it must be ensured that the upper layers are part of the change. This concerns the change of the general mindset and underlines the importance of the order parameters. Therefore, mindset has the assigned value 1000, while techniques have the value 1 in relation to its importance for the transformation.

Governance stands for managing the system itself instead of managing within the system. In our context of the transformation model, as shown in Fig. 3.2, this means adjusting the system parameters. In Agile Management 4.0, the system parameters are regulated in such a way that self-organization can be achieved. The intervention can take place at two different levels in an organization: at system level or at subsystem level.

This can be illustrated with the following example. To initiate a change in agility, the head of an organization uses a new lead mindset as a stimulus (control parameter). He would do this at system level by explaining the general meaning and purpose of the change. At the same time, he would exemplify the new values with each individual and each team he encounters in his daily business. As a result of his interventions, a new lived culture emerges in an overall changed mindset. Of course, the leader must ensure that the setting parameters support the change to agility by adjusting the work environment or incentives. In SCRUM, the product owner would be the gatekeeper

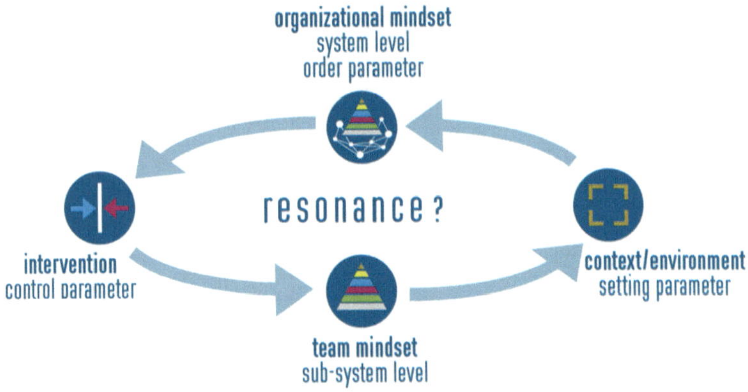

Fig. 3.2 Transformation model based on the theory of synergetics (© Tuczek, Flore, Nuhn, Schaffitzel 2021. All Rights Reserved.)

to control the complexity of the environment for the team. As in laser theory, there is a tipping point in organizations where the new mindset becomes dominant. Usually this point is reached when 30% of the employees support the change.

In complex social systems the outcome of an intervention is not predictable. For this reason, management must regulate the system parameters with the help of constant feedback loops to ensure that the expected result of self-organization in the direction of the desired re-orientation of the company can be achieved (see Fig. 3.3).

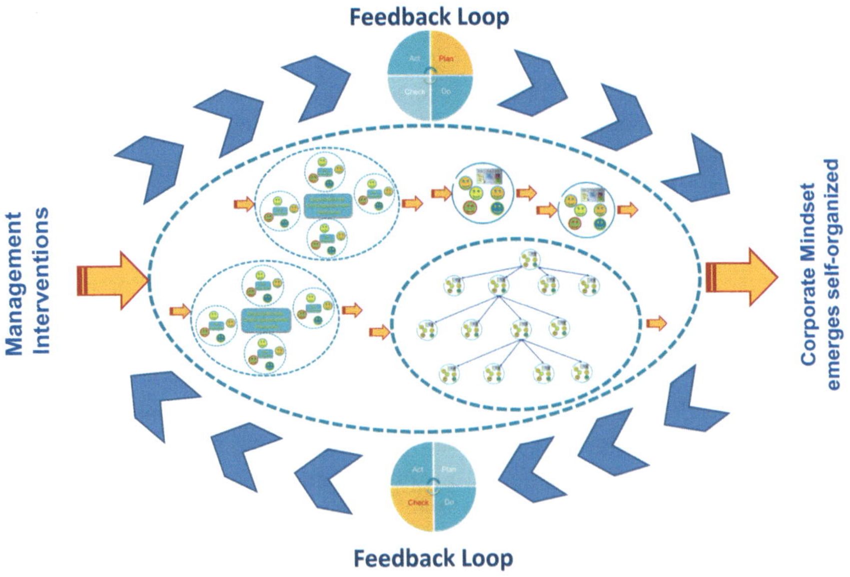

Fig. 3.3 Transformation to self-organization (© Tuczek, Oswald 2021. All Rights Reserved.)

This means that the leaders must continuously adapt the interventions with the logic of Deming's PDCA-cycle (Plan-Do-Check-Act) and a trained intuition based on systemic models. The non-linearity of complexity can lead to a feeling of loss of control. For managers who like to have things under control, this is difficult to accept. However, if the scope of available options for future decisions is to be increased, creativity and a diversity of ideas must be encouraged in the company. This goes hand in hand with a high degree of variety within an organization (see Ashby's Law), which can only be managed by the principles of systemic intervention discussed above, with the ability to act proactively in a complex environment. Furthermore, it is important to explain one's own perception of the context, since the context is not a given thing, but rather an arrangement of relationships and interactions in which personal experience plays an important role. This can lead to misunderstandings, so transparency and feedback on this topic are of central importance.

In the language of Stafford Beer's VSM model [3], the goal of such *Normative Transformation* must be to develop a collective mindset that allows management to address the following key criteria for successful collaboration:

- the values and beliefs of the people involved
- their perceived identity during collaboration
- the knowledge of their common bond
- the decisive driving force for successful collaboration: the deeper meaning, the superior mission and thus the source and inner value of work, which corresponds to the ethical basis of work and profession [14].

As a result, such organizations develop a collective empowerment and mind setting that enables them to adapt quickly to market changes and subsequently contribute to the survival of the entire organization.

For operational aspects and other perspectives on value-based management, see Tuczek [19]. Here you will find the concepts of collaboration and collective intelligence, the requirements for an innovator's mindset, and agile organizational models.

3.3 Mindset

As laid out above, an Agile Mindset is an indispensable prerequisite to successfully establish agility in a company. But what is a mindset anyhow? According to the Cambridge Dictionary, it refers to a person's way of thinking and their opinions [4]. In scientific literature, the term has received little attention to date. Some contributions make use of the term "culture" instead. In business contexts, it is sometimes also referred to as "corporate culture". However, the term "culture" is also not uniformly defined.

We present two definitions and relate them to our concept of mindset. Sackmann [15], for example, understands corporate culture to be a fundamental and collective conviction that significantly influences the thinking, actions, and feelings of managers and employees in the company. This corporate culture is regarded as typical for the company as a whole. Trompenaar [17] on the other hand defines culture as the way in which problems are solved in a group of people or organizations. Similarly, Trompenaar [18] describes culture as layers—like an onion—that one must approach layer by layer. He differentiates between explicit and implicit culture. The explicit culture is represented in the analogy of the onion in the outer layers. These are the things that are first noticeable and observable, such as appearance, language, clothing, and food. Underneath this is an implicit culture that is difficult to recognize and penetrate. Here aspects such as values, norms, and beliefs are hidden.

In the context of the M4.0 approach, we use the term "mindset" and define it as a way of thinking or a set of assumptions, methods, and notations. This mindset can be shared by individuals or groups of individuals and provides a strong incentive to maintain previous behaviors, decisions, or tools. This phenomenon can be described as "group thinking" and it can be difficult for future analysis and decision-making processes to break through and counteract this old pattern of thinking, these old structures.

The Dilts Pyramid, known from NLP (Neuro-linguistic Programming), is a simple but very effective model for representing a way of thinking [6]. It is an extension of the Maslow Pyramid of needs and is used in NLP as a key model for individual and organizational change or transformation processes. Figure 3.4 shows the Dilts Pyramid and consists of six levels: Environment or context, behavior, capabilities, values and beliefs, identity, meaning and arranges these elements in a hierarchical order. This hierarchical order is called the hierarchy of neurological levels [6].

Therefore, the top three levels of the Dilts Pyramid can be described as an implicit culture (cf. Trompenaar), which also includes the way of thinking. The levels capabilities and behavior then correspond to the explicit culture (cf. Trompenaar), which refers to the observable actions.

These top five levels describe the culture or, as we call it in the M4.0 approach, the mindset, which however always goes hand in hand with the context and environment. The context and environment level interacts in a constant interaction with the mindset. Therefore it is always considered in relation to each other.

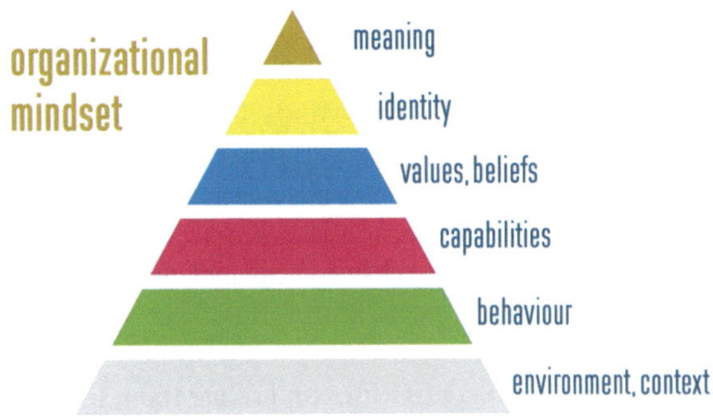

Fig. 3.4 Layers of Dilts Pyramid (© Tuczek, Flore, Nuhn, Schaffitzel 2021. All Rights Reserved.)

To simplify how the levels of the Dilts Pyramid can be further defined, we have compiled sample questions per level. The lowest level of the Dilts Pyramid the "environment or the context" can also be described with the question "Where? And When?" This means that the environment in which individual people or entire organizations move has a strong influence on behavior or processes and methods. The level above the "behavior" can be described by the question "What?" The environment and the context influence the behavior of people and organizations, which is displayed and expressed through processes and methods. The next level is about the "capabilities", which are expressed by the question "How?" Behavior is generated by certain capabilities of people and organizations, i.e., the behavior is controlled and influenced according to the capabilities. Building on this comes the level of "value and beliefs". This level can be explained with the question "For what? And why?" The value and beliefs of people and organizations is a strong influence on the behavior of a person. Ultimately, it is these beliefs and these inner values that determine the next level of "identity" of a person or an organization. This level can be questioned with "Who? How who?" This is about getting an idea of who you are as a person or organization. Or how who—adapting from another role model—one wants to be. Identity can also create a feeling of belonging to a group or organization. The top level of the Dilts Pyramid, the "meaning", is defined by the question "Who else? Why else?" We give orientation to our lives by asking ourselves why we are there and where we want to go. The Dilts Pyramid's concept helps to describe organizational behavior as well as intended transformations of it. Communication blockades are also made visible and dissolved. This brings transparency to management and leadership styles and enables a situational adjustment to act agile [12].

As presented above, the levels of the Dilts Pyramid describe the principles of an individual, team, or organizational mindset. This means that these contents of the levels influence the way of thinking and acting of people, but to different degrees (according to the hierarchical order).

The level of environment and context can be very different, depending on the perspective of a system, subsystem, or an individual. This means that an organization that wants to become agile needs a new, an Agile Mindset. Because this Agile Mindset can serve as a stimulus for self-organization and initiate a transformation process for the entire organization (cf. Fig. 3.1). A few leaders are sufficient to give this impetus with a new Agile Mindset and thus initiate the transformation process [12].

3.4 Governance and Self-Organization

Different approaches to governance set agile organizations apart from non-agile organizations. Non-agile organizations can survive in settings of relative stability, while agile organizations seek to strive in environments that are less predictable, subject to non-linearity, complexity, and ambiguity. These specific environmental parameters cause conventional governance designs to lead to conventional performance levels. Processes may still be functional, but not adaptive to change. They will be efficient and effective in what they are supposed to do, but, because of environmental change, their purposes may not be relevant any more. These days, many practitioners and scholars suggest more decentralized forms of governance or "organization" these days, in order to handle environmental variability.

The concept of self-organization is subject to much confusion in debates about organizational agility, however. Some consider it to describe the processes that allow teams to set up their work environment to their very own requirements and in consequence achieve higher levels of productivity.

We propose to discuss self-organization as a phenomenon that one can observe in nature: self-organized systems are of inherent complexity and display emergent patterns when they reach a certain system state. Examples for this viewpoint are swarm movements of animals, the formation of bubbles or geometric patterns in boiling fluids, different phases of chemical elements, the already mentioned laser beams, and many others.

We employ three types of parameters in order to describe a social system that either displays or does not display agile properties: setting parameters, order parameters, and control parameters (cf. Fig. 3.2). These we derive from [7, 8], and [16], but employ them in parts in specific manners (Fig. 3.5).

First, setting parameters describe the external environment of a system. It will be influential to the system itself, because at its edges, the system will interact with its surroundings. Second, order parameters describe the degree to which a system is aligned and what reference points this alignment is directed toward. Third control parameters describe how the systems are influenced, either from within, or from the outside.

With the help of these defined parameter types, we are able to describe social systems in a helpful way. For example, an agile team is working within an organization (setting parameter) that is being lead to higher levels of agility by an agile-minded

Fig. 3.5 Self-organization and its relevant parameters (© Tuczek, Flore, Nuhn, Schaffitzel 2021. All Rights Reserved.)

leader or scrum master (control parameter) and subsequently displays agile properties—like high responsiveness to changing requirements—because it has aligned its activities according to a common, agile mindset (order parameter).

Another example might be the following: A large corporation's management wants to become more agile. It deems "innovativeness" instead of "cost-effectiveness" as a value that helps the agile paradigm better and changes the setting parameters for the departments and teams they lead by designing a changed incentivization system (control parameter). It should incent activities that benefit "profitable innovations throughput" over "short term profitability" and be proclaimed as the new, relevant incentivization scheme that acts as a setting parameter for all departments. Top management expects it to be a nucleus from which middle managers develop corresponding, aligned goal systems (order parameter), resulting in increased innovation rates and thus quicker market adaptation capabilities of the whole organization.

The following examples will help the reader to understand the concepts of order and control parameters better. In a more practical sense, and in relation to the previous section of this paper, a very strong, and probably the most important order parameter is the shared mindset of a team. As described above, a mindset consists of several layers that add up to the Dilts Pyramid. Therefore, we see that the shared mindset of a team with a strong vision of a team product, a common mission, values and beliefs, as well as compatible skills and competencies and a compatible behavior will produce very low amount of team friction and display high productivity levels. Such teams must be considered as highly aligned toward a common order parameter. The emergent behavior or order parameters that reflect an "agile" order are those in which the activities of the team are aligned with the goal of keeping the structure and organization of the team itself agile. In practice, the team will repetitively reflect upon its self-regulatory processes and common goals. It will also reflect upon the amount of work in progress as well as the tasks that it is processing.

3.5 Viable Systems Model

In order to discuss organizational systems in further detail, we also employ the Viable Systems Model by Stafford Beer [3]. Beer was inspired by Wiener [21] and his works on the theory of cybernetics, which were picked up by other scholars. Ashby, for example, elaborated the Law of Requisite Variety, which we pick up in our reasoning. Also, Beer, in his work of 1959, set out to analyze systems with regard to their functional subsystems. He tried to establish a minimal set of functions that are necessary for the system to function and—from a cybernetic viewpoint—"survive" and are therefore viable. Beer's ideas on VSM are closely linked to the ideas of both (management) cybernetics and complexity theory.

Beer explains five subsystems (cf. Table 3.1). Each subsystem has a different purpose and functionality. All functions together ensure that the system is generating value, in an efficient and productive way that is useful in a specific, strategic

Table 3.1 Systems in the Viable System Model

System #	Purpose
5	Top-management decisions; normative function
4	Analysis and forecasting for future development; strategic management function
3	Optimizing resource usage; command function
(3*	Auditing/monitoring activities/channel)
2	Operational management
1	Operational value generation

context. In addition, the system will display functionalities and interfaces to other systems. Those could be surrounding systems (meta-systems), like enterprises that encompass a team. Alternatively, those could be largely independent, loosely coupled systems, like other teams with similar activities but other purposes, services rendered or products delivered. See also Fig. 3.1, the Big Picture Agile Management 4.0.

The VSM serves as a structural framework to "reflect[…] upon the purposes served by organizations" [9]. Its functions are necessary functions for any organizational system to be viable. Applying the system to an organizational system, one can identify which structural part of it serves what function. In line with Beer, we argue that strongly centralized command-and-control patterns are not optimal in complex and dynamic environments. However, while de-centralization of control can promote efficiency, it can also cause dysfunctions if exaggerated.

The causes for such dysfunctions are, on the one hand, the limited amount of information that a human, or system of humans, can process. On the other hand, the communication bandwidth that limits the amount of information to be conveyed within a system is an additional limiting factor. This pattern can be observed in agile transitions in several organizations; new means of digital communication have vastly increased the amount of information that can be disseminated, but it has not increased the amount of information that can be usefully processed by deciders. We conclude from this observation that conventionally designed systems with centralized command-and-control structures are not able to absorb the external complexity that is given in settings of increased—information-rich—complexity.

Nevertheless, some organizations have strived in unprecedented qualities and velocities, even in environments of high complexity. These organizations are the systems that have learned to structure the layers of their system model differently. The four non-value generating functions of the Viable System Model are implemented differently in agile teams, for example. Nearly all control and organizational management functions reside within the team (hence: self-organization). It is a more promising concept, because an agile-capable team is able to sense the environment directly, along with its ever quicker changing requirements. It is therefore able to react to these changes quicker than other teams, that in order to adapt to changing environments have to transport a lot of information through four different layers, before a directional change can be executed. The different levels of speed with which these different systems can engage with the environment can be equaled with the time differences between execution times of human planned behavior and human reflexes.

The Viable Systems Model can help juxtapose agile and non-agile (sub)systems [9]. In conventional organizations, the organizational layers or subsystems typically reflect the VSM functions. More adaptive organization flexibly assign or attribute VSM functions' responsibilities within the subsystem/team. They also reflect whether those assignments or attributions are still valid and functional and do not let them unchallenged out of fear or laziness.

By mapping the functions to specific elements of each system, one can also identify when an organizational design leaves out functionalities. For example, an agile team may be highly productive, but produce outputs that do not align with the strategy of the embedding system, the company. A look guided by the VSM functions could indicate, why this is the case. For one, the management-related systems could be out of effect in this team. In settings of self-organization, this could further point to lack of professional judgement or decision-relevant information or show that the team's connections with the rest of the organization are dysfunctional or non-existent. However, there may be a plethora of additional hindrances to highly aligned team productivity.

An agile organizational (sub-) system would have the mindset to generate these insights itself. We discussed this in the previous section. It would also be incorporating techniques that help identify these circumstances. This paper's next section will cover these aspects.

In conclusion, the VSM's functions are also helpful in conjunction with the parameter types that describe our concept of self-organization. In conventional organizations with stable processes, higher levels within the VSM may be setting parameters that teams are unable to overcome or effectively influence. Control parameters describe the inputs that may lead to increased self-organization and the subsequent displaying of agile-useful patterns. Order parameters serve as guiding stars to align the processes of continuous challenging and reorganizing VSM layer organizations or function distributions.

3.6 Agile Techniques in the Hybrid Market Model and in the Management 4.0 Approach

The focus of agile management in the sense of Management 4.0 is the creation of business agility by integrating the fundamental conceptual cornerstones mindset and learning, meta-competencies and leadership, self-organization and complex systems as well as fluid organizations and transformation. With the Management 4.0 approach, we created a theoretical foundation of agile principles which help a better understanding of agile practices and its mechanisms of action. It further allows major insights in all principal ingredients what we believe are the basic meta-competencies needed to develop agility in a complex world. They describe skills that everyone should master in the context of an ever faster changing world of work in order to remain successful, self-effective, and healthy.

Our understanding of a comprehensive canon of meta-competencies is contrary to the popular idea, where it is sufficient to integrate individual agile methods and techniques into the existing technical landscape. Usually described as a hybrid approach, it represents only one possible scenario of a "hybrid practice" in the context of Management 4.0. We call it the usual market procedure as a scenario of a company that concentrates on "agile techniques". Within the framework of this approach, an agilisation is to be achieved by building agile team structures and the specific application of agile techniques in the current organization.

It is assumed that the introduction of agile teams and agile techniques automatically ensures the desired business agility. But here one succumbs to a logical error, or as Klaus Leopold puts it: one "escapes into simple derivations" [11, p. 34]. Suddenly, methodological issues and the correct application of the recommended agile techniques are in the foreground in these companies. "But business agility is not only about pure speed. It is about how well you can recognize and respond to changes in the market and how you can take greater and better assessed risks" [10, p. 274].

If companies concentrate too much on agile techniques, the means and objectives of agile efforts are confused. The introduction of agile techniques, methods, and artifacts does not guarantee a transformation to more business agility and self-organization in the company. Just as digitalizing inefficient manual processes by using digital information systems do not necessarily lead to huge productivity gains, improving individual team productivity locally by employing agile techniques will not necessarily increase the overall business outcome (global optimization toward business success). From our perspective, this approach assumes that the mere use of agile working practices leads to a fundamental change in employee attitudes and behavior. But if only the contexts change, employees will quickly adapt their behavior to the new circumstances.[1] Since measures in complex systems very likely cause non-linear effects, it is completely uncertain whether the desired result is effectively realized. It is thus left to the actors themselves to carry out the necessary work on the values and on the changes of their own intrinsic motives.

But for us this is only a kind of hybrid model: Although it introduces agile techniques into companies, it only partially fulfills the transformation work required for the setting conditions and the agile mindset necessary for agile adaptation. In the worst case, it can even be lost out of sight.

We want to illustrate it as follows in the interaction of mindset, governance, and techniques.

In contrast to such a technology-centered concept to achieve entrepreneurial agility, our Management 4.0 approach emphasizes the interaction of a diversity of necessary competencies. At the heart of this approach is a clarification of the mindset of all actors, their location and alignment along a common vision, their initiation and establishment on the basis of coordinated guidelines and governance directives,

[1] Since humans are self-organizing complex systems, this adaptation behavior is logical.

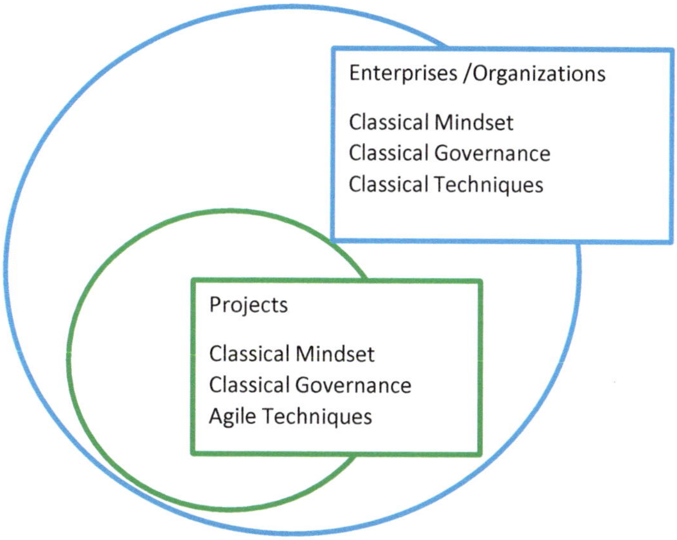

Fig. 3.6 Scenario of a company focused on agile techniques—The hybrid market model (© Tuczek, Flore, Nuhn, Schaffitzel 2021. All Rights Reserved.)

and the coordinated use of both agile and classical (i.e., traditional) management techniques.

Hence, the technical instruments, management methods, and best practices of Management 4.0, which are necessary for the achievement of the agility goal, are drawn from all available and proven management techniques. The Agile Mindset in Management 4.0 consciously integrates the set of traditional practices, because from this perspective, classical/traditional methods and techniques do not per se impede an agilisation. It is rather the mindset of people that prevents agility [12, S. 41]. The following chart illustrates the relationship in contrast to the previous scenario of the hybrid market model (see Fig. 3.6).[2]

It further illustrates how Agile Mindset and Agile Governance can be based on a variety of possible techniques (Fig. 3.7).

Even though Management 4.0 is open to results and variable on the level of the applied methods, we place a certain emphasis on the design of the techniques to be applied especially in product- and project-oriented organizations. Our practical recommendations focus on the principles of Theory of Constraint (TOC) and the application of Critical Chain Project Management (CCPM). The main assumption is that the business agility and output of an organization depends on the workload at the bottleneck of the system. This means that an organization should base its decisions on its ability to exploit its bottleneck to an optimal extent and to eliminate

[2] At this point we would like to mention another hybrid scenario, which we have called "Agile Island". See for this purpose the article "Comparison of hybrid management "in issue 2 (2020) of PROJEKTMANAGEMENT AKTUELL.

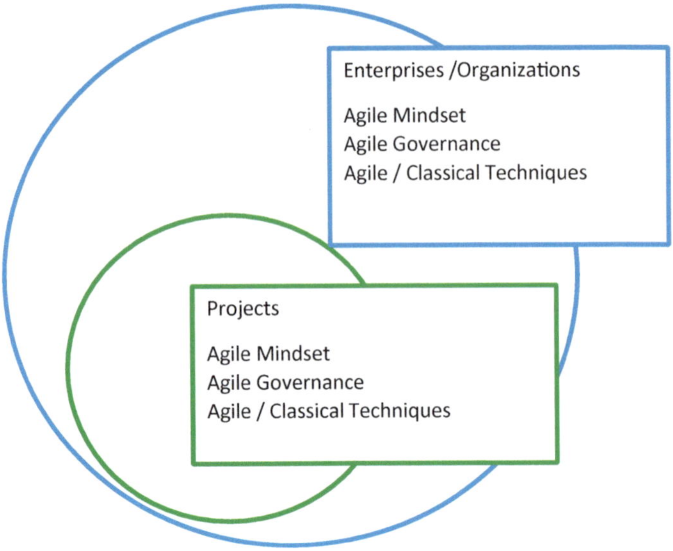

Fig. 3.7 Scenario of an agile organization of Management 4.0 (© Tuczek, Flore, Nuhn, Schaffitzel 2021. All Rights Reserved.)

harmful multitasking as well as overload in the organization. In addition, we assume in Management 4.0 that there is no standard Dilts pyramid and therefore no uniform and ubiquitous mindset for the agile organization. On the contrary: every organization is obliged to follow its own individual path. The frameworks pointed out, which are used in Management 4.0, integrate the necessary meta-competencies into a leadership concept for managing companies in an agile world.

Taking agile project management as an example, this could be illustrated as follows:

- At the macro level, the vision and mission are clearly defined as order parameters and translated into a goal hierarchy and a big picture for the organization.
- Accordingly, at the portfolio level a staggering of the projects and tasks to be carried out is established along the capacity of the bottleneck.
- The environment in which the project operates is clearly defined and the setting parameters are known. The guidelines and the modes of cooperation are reflected in the governance of the project.
- At the micro-level, projects are controlled by so-called fever curves, which reflect project progress and buffer consumption. This ensures that any shortfall can be identified at an early stage and the appropriate remedies can be taken. The control parameter of the system is the maintenance of the workflow as a "one-piece-flow" and the desired control parameter is a Work in Progress (WIP) of 1.
- Finally, at the work level, the workflow and work volume (backlog) within the teams is made visible with buffer and task boards (Scrum or Kanban).

What is illustrated here by the example of a single project can now be extended and scaled to the larger aggregates such as portfolio, program, department, and company. At each level, the relevant specifications must be developed in detail.

In addition, we have gone even further in Management 4.0 and developed an assessment framework to evaluate the scaling effectiveness of the various agile techniques and frameworks. The core of our assessment approach is derived from the three dimensions presented by Geoffrey West in his book "Scale" [20]. In this book, West et al. use these dimensions to estimate the scaling principles of natural physical systems that obey the "fractal-like shapes typical of a complex adaptive system" [20, p. 290]. Therefore, in this book the following three scaling principles are presented, which we apply to social systems according to our scaling requirements:

1. *Networks are self-similar (fractal) and space-filling*
   The principle of self-similarity is based on the occurrence of fractals. This means these are objects that are shaped, designed, or structured in the same way on all levels, from the micro to the macro level.
2. *The ends of the networks are " the same everywhere" (invariant ends)*
   The second principle is directly related to the first principle. It means the ends of a network or in our case the ends of an agile organization, should be invariant or better described as the same everywhere, i.e., based on the same shape of organization.
3. *Principle of freedom from impedances*
   The networks (team of team structures in agile organizations) are designed in such a way that the flow of energy (and information) through the network suffers no impedance barrier.[3] Hence, it is called an impedance match if the flow of energy (information, fluids) is not interrupted.

Management 4.0 extends these evaluation principles by the fourth dimension, which we call the introduction of self-organization at all levels. This fourth principle means that the setting, control, and order parameters are defined and continuously adjusted across all levels according to the respective layer structure [12, p. 218]. The following figure shows the relevant scaling dimensions (Fig. 3.8).

Based on these four principles, concrete estimates and evaluations were made in the book Management 4.0 [12, p. 212 ff.] for the scaling approaches in the Scrum environment, especially for the LeSS- and SAFe frameworks. The main statement about LeSS is that the principle of self-organization is violated in this procedure due to the lack of a target hierarchy and the absence of a scaled self-organization structure [12, p. 228). And although all scaling principles are fulfilled in SAFe, the danger with this approach is that the organizational "setting is so complicated that complexity is transformed to complicatedness. If this happens, the benefits of scaling measures are lost". [12, p. 229].

[3] Impedance is used to detect whether the flow in natural networks stagnates—with the effect of a backflow in the network—or whether the flow is completely interrupted.

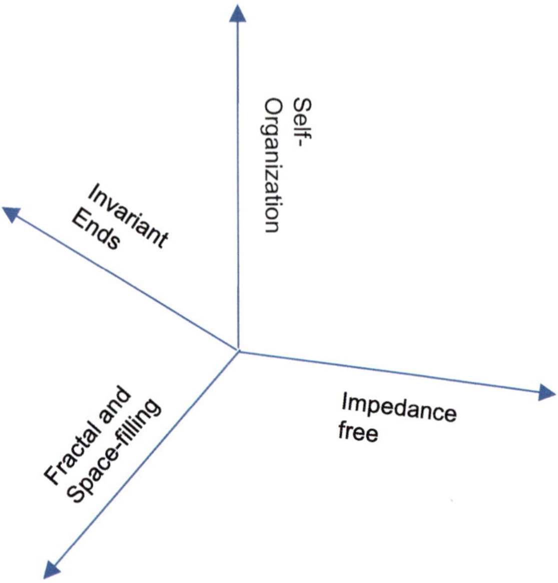

Fig. 3.8 The four dimensions to assess agile scaling

3.7 Conclusion

The main argument in this chapter is that an agile mindset is an indispensable prerequisite for all agile practices. The choice of techniques must be made against the background of higher level meta-competencies. These techniques can consist of agile, classical or hybrid elements, depending on the specific application. The essential means in governance for agility is the ability to induce and develop self-organization.

References

1. Ashby WR (1957) An introduction to cybernetics. Chapman & Hall Ltd., London
2. Beer S (1959) Cybernetics and management. English Universities Press, London
3. Beer S (1984) The viable system model: its provenance, development, methodology and pathology. J Oper Res Soc 35(1):7–25
4. Cambridge Dictionary. https://dictionary.cambridge.org/dictionary/english/mindset (State: 17:04.2020)
5. Conant RC, Ashby WR (1970) Every good regulator must be a model of that system. Int J Syst Sci 1(2):89–97
6. Dilts RB (2014) A brief history of logical levels. http://www.nlpu.com/Articles/LevelsSummary.htm. Accessed 03 April 2020
7. Haken H (1983) Synergetics, an introduction: nonequilibrium phase transitions and self-organization in physics, chemistry, and biology, 3rd rev, enl. Springer, New York
8. Haken H, Schiepek G (2010) Synergetik in der Psychologie. Selbstorganisation verstehen und gestalten. Hogrefe, Göttingen
9. Jackson M (1988) An appreciation of Stafford Beer's „viable system" viewpoint on managerial practice. J Manag Stud 25:557–573. https://doi.org/10.1111/j.1467-6486.1988.tb00047.x

10. Kim G, Behr K, Spafford G (2015) Projekt Phoenix. O'Reilly Verlag GmbH & Co. KG, Heidelberg
11. Leopold K (2018) Agilität neu denken. LEANability GmbH, Wien
12. Oswald A, Müller W (ed) (2019) Management 4.0—handbook for agile practices, Release 3.0. BoD, Norderstedt
13. Oswald A, Tuczek H (2019) Principles of agile leadership 4.0. In: Oswald A, Müller W (eds) Management 4.0—handbook for agile practices, Release 3.0. BoD, Norderstedt
14. Patzelt M (2010) Die Unternehmenspyramide als Vermittlerin zwischen Marketing-, Strategie- und Personalbereich. In: Brinkmann M (Hrsg) Besser mit Business NLP, DVNLP e.V., Berlin
15. Sackmann S (2004) Erfolgsfaktor Unternehmenskultur: Mit kulturbewusstem Management Unternehmensziele erreichen und Identifikation schaffen – 6 Best Practice – Beispiele, 1st edn. Bertelsmannstiftung, Gabler
16. Strunk G, Schiepek G (2006) Systemische Psychologie – Eine Einführung in die komplexen Grundlagen menschlichen Verhaltens. München
17. Trompenaars F (1993) Riding the waves of culture. Nicholas Brealey, London
18. Trompenaars F (1998) Riding the waves of culture. Nicholas Brealey, London
19. Tuczek H (ed) (2017) Management 4.0 und die Generation Y. Landshut Leadership Band 2. Shaker, Aachen
20. West G, Ferguson N, Syed M (2017) Scale. Weidenfeld & Nicolson, London
21. Wiener N (1965) Cybernetics: or control and communication in the animal and the machine

Chapter 4
Value-Orientated Decision-Making in Agile Project Portfolios

Karyne C. S. Ang, Lars Kristian Hansen, and Per Svejvig

Abstract A key characteristic of agile projects is autonomy coupled with self-organization. However, agile practices and decisions do not operate in isolation for projects in a portfolio. Decisions involve different stakeholders at various levels in an organization with different constructs, interpretations, and expectations of value. This chapter explores the role of value in decision-making for agile project portfolios. In consideration of projects as arena for self-organizing, we investigate how agile portfolios function, particularly in decision-making involving multiple stakeholders to maximize value for the portfolio. We present an in-depth single case study of an organization that adopted agile practices in Project Portfolio Management (PPM). The research case is distinct as portfolio decisions and initiatives in the case are driven by considerations of value. The theoretical framework applied in this study draws upon a value spectrum framework that is reinforced by sensemaking principles to explain the agile sequences and decision-making to maximize value in the portfolio. This study traces how an agile portfolio of projects underpinned by value is managed through short iterative cycles. The sequences of the agile cycles and decision-making events could contribute to supporting organizations in considering how value can be embedded in project and portfolio management practices.

Keywords Value · Sensemaking · Agile · Project portfolio · Decision-Making

K. C. S. Ang (✉)
School of Project Management, Faculty of Engineering, University of Sydney, Level 1, 21 Ross St, Forest Lodge, New South Wales, Australia
e-mail: karyne.ang@sydney.edu.au; lkh@mgmt.au.dk; psve@mgmt.au.dk

L. K. Hansen · P. Svejvig
Business and Social Sciences, Department of Management, Aarhus University, Building 2627, Fuglesangs Allé 4, Aarhus V, 8210 Aarhus, Denmark

R. Ding et al. (eds.), *Research on Project, Programme and Portfolio Management*, Lecture Notes in Management and Industrial Engineering,
https://doi.org/10.1007/978-3-030-86248-0_4

4.1 Introduction

This chapter explores the role of value in decision-making for agile project portfolios. Portfolio decisions are typically made to ensure projects in the portfolio have strategic alignment with the organization, future preparedness, and value maximization [1–3]. Specifically, decisions revolve around the evaluation,prioritization, selection, and balancing of various types of projects in the portfolio. Decisions are made about projects to invest in, and subsequently the allocation of scarce resources [1, 2, 4].

Different situations faced by the portfolio may require different decision-making approaches. Whilst formal and rational decision-making is more legitimately accepted compared to informal decisions, decisions made by this basis alone may lead to missed opportunities and an imbalance in project types [5]. Balancing formal and informal approaches in decision-making requires flexibility [5], although, this could lead to potential conflicts in the organization [6].

Where organizations operate in a dynamic environment, the prospects of agile projects with its self-organized teams and portfolios that are adaptive are attractive, but these do not come without challenges in practice. In the project management space, agile projects are well established, especially in the software development and IT space [7, 8]. Some organizations are now considered to be "born agile" or "agile natives", for instance, Spotify and Netflix, whilst others like Amazon have moved from traditional hierarchies to agile enterprises [9] where teams are self-organized and hold more autonomy in decision-making. More recently, interest has moved to studies in the adoption of agile approaches in project portfolios. For instance, organizations are already adopting an "agile at scale" approach. Principles drawn from frameworks and methodologies such as SAFe and Nexus are used although research surrounding agility and decision-making at the portfolio level is scant and still in its infancy [10, 11].

A practical problem identified is that agile practices do not always work well at the PPM level, particularly in the management of value [12]. There is still a lack of research and understanding about how agile portfolios function, particularly in decision-making involving multiple stakeholders to maximize value at the portfolio level. This is important because agile practices do not operate in isolation for projects in a portfolio. Decisions involve different stakeholders at various levels in an organization who may have different constructs, interpretations, and expectations of value [13, 14] that could affect attempts to maximize value in a portfolio.

This chapter presents an in-depth single case study of an organization who has adopted agile practices in PPM. This case was especially selected as the portfolio decisions, actions, and results are distinctively underpinned by value considerations, or value propositions (VP).

The theoretical lens through which we explore the role of value within decision-making in the agile portfolio draws upon elements of a value spectrum framework as part of a typology of multi-stakeholder value perspectives that incorporates sensemaking principles [15–17].

This chapter contributes to the fundamental principles around how value influences the way organizations make decisions in agile project portfolios in order to maximize strategic value. It contributes to a novel theory linking multi-stakeholder sensemaking elements with the role that value plays in agile portfolio environments through an extended application of the value spectrum framework. In practice, the results could potentially extend the way agile portfolio practitioners consider and manage value maximization.

This chapter is structured as follows. First, the concepts of value and its place in decision-making are introduced, followed by a brief primer to sensemaking principles in the context of decision-making. The next section introduces the theoretical framework used in this study: the Value Spectrum Framework [15, 17–19] underpinned by principles of sensemaking [20]. This is followed by the research methodology and a description of the case study, EPSILON. The findings commence with an outline of EPSILON's agile portfolio cycle. Each step in the cycle is interpretively explored to highlight how value and sensemaking are enacted in the cycle. The chapter then examines a key decision-making event more deeply and demonstrates how value underpins the decisions and initiatives in the portfolio. The final section concludes with a discussion on the contributions, implications, and limitations of this study.

4.2 Literature Review

4.2.1 The Concept of Value

Value can be defined as things that are important to human beings in their lives with varying degrees of importance [21]. In this chapter, value is associated with terms including deliverables, benefits, outcomes, and addressing strategic needs and expectations that are important to key stakeholders. We differentiate these terms from the psychology discipline of individual or personal set of "values" which tend to be discussed in the context of individual beliefs, personalities, and motivations and behaviors [22] as well as socio-cultural values in an organization, for instance, values that enable an organization to function as a collective whole through shared values, norms, language, and standards of behavior [23]. While personal and socio-cultural values influence behavioral choices and decision-making [24], and we acknowledge that there is a convergence of the various disciplines, the discussion is not within the scope for this chapter. This could make an interesting research extension capturing the convergence of stakeholder values, strategic value and decision-making in future research.

Value is usually identified as the return of a fair price or exchange by recipients for the benefits received from goods, services, or knowledge that are deemed desirable or useful in both tangible and intangible forms [25]. Tangibles can be identified as financial and other capital-based resources in a firm, while intangibles include relationships and trust, employee knowledge and competencies, group effectiveness, organizational structures, and efficiencies [26].

Project studies today argue that financial or economical value considerations alone are insufficient for holistic and strategic decision-making, particularly in environments requiring more flexibility and agility. Furthermore, different aspects of value are said to be relevant at different project-stages [27] and levels in the organization [13]. Project value can take on dimensions that are economic, social, and environmental [1, 28]. Value also needs to be considered beyond planned or deliberate value, to embrace emergent value that may not have been anticipated upfront. This helps ensure that planning for strategic value in the long term ensures preparedness for the future including opportunities for value that might not have been anticipated upfront [1, 19, 29], and for other non-commercial and intangible aspects of value to be considered [19].

4.2.2 Value and Its Place in Portfolio Decision Making

One of the key purposes of a portfolio is the maximization of value. Value in its role to motivate and influence decision-making is widely accepted by many [21, 30, 31], yet very little is known about how value fits within the decision-making process for agile portfolios. Decisions that need to be made around projects in a portfolio could be influenced by many factors. Cooper et al [32] state that organizations depending on financial decisions alone are less likely to perform well in maximizing the value in their portfolios in alignment to the organization's strategy.

A critical element of the organization's business strategy is the value proposition (VP), that is, what an organization delivers to its customers. Superior value propositions are said to lead to performance success [33]. To support this, organizations embed concepts of VP through the provision and communication of clear VPs in the organization and in its business processes, channels, and service or product features [34].

Beyond the customer, VPs can also be orientated toward other stakeholders. Ballantyne, Frow, Varey and Payne [35] discuss VPs designed for different stakeholders including shareholders, current and potential employees, partners and suppliers, and influencers. Different stakeholders have diverse goals and expectations [36–39], and it could be a challenge to maximize value for portfolios involving multiple stakeholders, particularly for organizations in changing business environments.

4.2.3 Agile Practices in Portfolio Organisations

Organizations are evolving from a command-and-control hierarchical structure into adopting "agile at scale" to adapt to changing business environments [9, 40]. Principles drawn from frameworks and methodologies such as SAFeand Nexus are used. Rigby et al list the characteristics of agile organizations as

- Having small and multi-disciplinary teams.
- Handling large, complex problems by breaking it down into smaller modules, and then using rapid prototyping and tight feedback loops to develop solutions that are then integrated into a coherent whole.
- Focusses on adapting to change rather than following a preset plan of actions.
- Focusses on outcomes (e.g., growth, profitability, and customer loyalty) rather than outputs (e.g., lines of programming code, new products).

Agile practices are said to fit within environments that are complex, where solutions are unclear and requirements for projects are likely to change. Additionally, agile settings mean that closer collaboration with the end users is feasible, and creative self-organizing autonomous teams are viewed to have a performance advantage over command-and-control groups (Rigby et al 2018). As such, closer collaboration and interactions that could potentially maximize value in the portfolio would require a collaborative dialogue among the stakeholders that embraces a common language about value [41]. Value and its language involves the close interaction of stakeholders to make sense of a shared understanding of value, as supported by Cooke-Davis [42] who conclude that humans themselves determine the adequacy of value and project success.

4.3 Theoretical Frameworks Used to Understand Value in Agile Portfolio Decision-Making

The theoretical frameworks drawn upon for this chapter are the Value Spectrum framework (Fig. 4.1) drawn from a typology of Multi-stakeholder Value Perspectives [15, 17–19], underpinned by sensemaking concepts of ongoing interactions, reciprocity of perspectives [20, 43–45], and being retrospective-prospective [20, 46].

The Value Spectrum framework presents how value is perceived as a spectrum or range of stakeholder expressions. The spectrum illustrates how value ranges from perspectives that are, at one end, unknown, unanticipated, unarticulated and hidden; to the known (identified), articulated and qualified realm; to the other opposite end

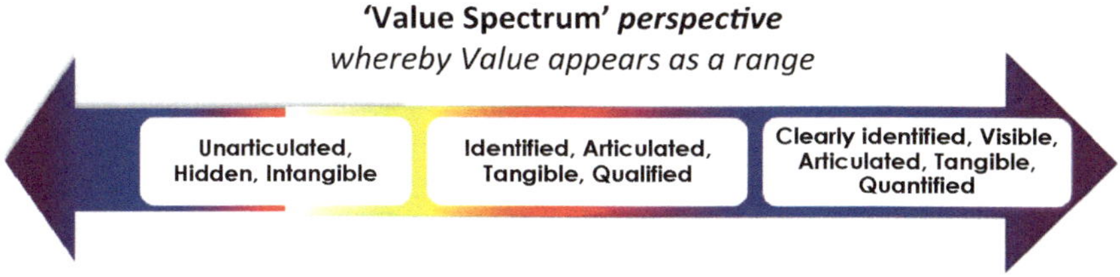

Fig. 4.1 Value spectrum framework [19], adapted from the paper presented by Ang, Killen, and Sankaran, "*Unanticipated value creation: sensemaking and the value spectrum in partnership projects*", with permission granted from convenors of the International Research Network on Organizing by Projects (IRNOP) Conference, London 2015

of the spectrum of the known (clearly identified), articulated, visible and quantified value. Value in the framework takes into account the subjective nuances and dynamics of value as perceived and made sense by different stakeholders associated with agile project portfolios, and is therefore underpinned and complemented by sensemaking concepts.

Sensemaking in organizations is a complex process of forming and re-forming shared understandings built from the ongoing interactions, conversation, and coordinated actions among people [47–50]. These practices are also shaped by language rules, vocabulary, authority relations, work roles, norms, and social structures [20, 48, 51]. Sensemaking comprises an ongoing and dialectic process with a sense of retrospective-prospective in moving backwards and forwards in one's sensemaking practices [43].

4.4 Research Methodology

The overall intention of this study is to understand value in decision-making for agile project portfolios. The overarching research question is therefore:

- What is the role of value in agile project portfolio decision-making?

This study applies a single case study focusing on value and decision practices in an agile portfolio. Guba and Lincoln [52] suggest that case studies could be used to describe what it is like to "experience" a situation. Case studies typically combine data collection methods such as organizational documents, archives, interviews, questionnaires, and observations. This case study applies an interpretive paradigm to provide practice-based narratives surrounding the research question [53] and tests theory in agile portfolio decision-making practice by utilizing the aforementioned Value Spectrum framework underpinned by sensemaking principles in the analysis and discussions. This approach is appropriate in the instances of unique and revelatory cases [54]. Case studies are often utilized in research contexts where precise solutions are difficult to come by. This case can be used to develop background material to explore and discuss particular concrete problems [55]. Single case studies can be used to explain particular phenomenon [56] and to achieve a deeper understanding of the subject matter.

The single case in this study has been codenamed "EPSILON". Data representing the various organizational levels for the project, program, and portfolio, as well as strategic management were collected through semi-structured interviews and multiple organizational artifacts including emails, presentation slides, templates, photographs, and participant's sketches representing their perspectives of the portfolio practices. The data has been kept general and codenamed to protect the anonymity of the organization and individuals. The informants' roles and associated codenames are described in Table 4.1.

In analyzing the data for this case study, the researchers were guided by the research question, and elements drawn from the aforementioned theoretical frameworks of value and sensemaking. The results were also compared to other studies in

Table 4.1 Sample and data types

Data type	Level	Title	Role	Codename
Semi-structured interviews	Management	Head of portfolio & economy/ director of development	Heads the IT change initiatives and development divisions, responsible for planning and budgeting, i.e., finance and portfolio investment; ensures portfolio investments aligned with strategy and finance	HoPE
	Portfolio/Program (PMOs)	PMO/Business development program manager	Controls and coordinates the development programs and portfolio to ensure momentum; supports the program leads and facilitates translation from portfolio decision-making into action with the program leads	BDPM
	Program	Strategy and transformation senior project and program manager	Agile transformation lead	STPP
	Project	Business analyst/Project member	Works on a project as part of an autonomous business design team that reports back weekly to a "squad" on progress and actions/initiatives	BAPM
Documents and images	Illustrations and images	Sketches by participants, photographs of events,		
	Organization documents	Presentation slides, business case templates, value proposition (VP) templates with instructions and examples		

the published literature to understand the findings and to enhance the reliability of the results (Baxter & Jack, 2008; Eisenhardt 1989).

The unit of analysis is the project portfolio. The data was deductively and inductively analyzed for themes and patterns. The data analysis was guided by the Value Spectrum framework underpinned by sensemaking to explore how decisions and value are interpreted and enacted in an agile portfolio setting.

4.5 Case Environment—EPSILON

EPSILON is in the energy industry and has been operating for over 50 years. Their annual turnover is more than 50 Billion Euro. This case study investigates the IT and Business Development portfolio where over 50 employees are involved as part of the agile practices. EPSILON formerly had a traditional hierarchical organization where most members were allocated to permanent tasks. The organization had a goal and

Table 4.2 Summary of the case organization

Organizational summary of EPSILON	
Industry	Energy
Years in business	Created through a merger of two companies that were in business for more than 50 years
Annual company case turnover	Over 50 billion Euro
Type of portfolio	IT and business development
Yearly portfolio cashflow	Over 20 million Euro in the development portfolio understudy
Agile project and portfolio methodologies	Internally developed model, inspired by Spotify
People involved in agile practices	More than 50 employees
Started agile practices	2016, waterfall approach (hierarchical) pre 2016

vision to embrace agility and flexibility to remain competitive in a changing market environment. They are not completely established in their agile practices, but report positive results since adopting agile concepts into the portfolio and organization five years ago (2016).

Table 4.2 summarizes the case organization. EPSILON draws upon a mix of established agile methodologies and is inspired by the Spotify framework [57] that they have adapted to their own contexts.

4.6 Findings and Discussion

The section sets the narrative of EPSILON with its adapted agile project portfolio practices in the organization. This is followed by a description of the cyclical sequence of its agile practices (as events) and how value is positioned in these settings. The sequence of events is discussed using the Value Spectrum framework.

4.6.1 A Shift of Focus: A Portfolio Obsessed with Value

EPSILON shifted from being time and cost focused to reconfiguring the business toward becoming value orientated. "Now we are way more obsessed with, are we delivering the right value? I don't care about costs that much, I need to check it as well, but that is not primarily what I am looking at" (HoPE).

The language used in the organization has been reframed. Projects are framed as "initiatives" and "value propositions" in the new culture and language of agility for EPSILON. At EPSILON, value propositions (VPs) are created based on the strategy and fitted into the investment themes offered in the portfolio. VPs are planned as competitive pitches for resources. These VPs are developed and owned by the business units (BUs) from the bottom up. BUs have their own mechanisms for prioritization to ensure that the decisions at that level also consider value maximization. The final VPs from the BUs are then presented to a wider range of stakeholders and prioritized for delivery. The next section traces how value moves through the portfolio in agile cycles.

4.6.2 Agile Portfolio Cycle Driven by Value Propositions (VPs)

The agile portfolio at EPSILON was described as one that operates in 4-monthly cycles, whilst initiatives implemented at the project level are managed through daily and fortnightly sprints in autonomous teams that self-organize based on the goals provided. Figure 4.2 illustrates Epsilon's agile portfolio cycle through four steps, as described by the informant (STPP). These steps are underpinned by the Preparation, Prioritization, Delivery planning and execution, and Demonstration of value from the VPs.

The steps in the cycle are explained in more detail below.

Step 1: Business units (BUs) review and plan for the VPs to be pitched.

In effect, these are the "business cases", and a specific "business model canvas" template [58] is applied. This is prepared two weeks before a VP Prioritization event.

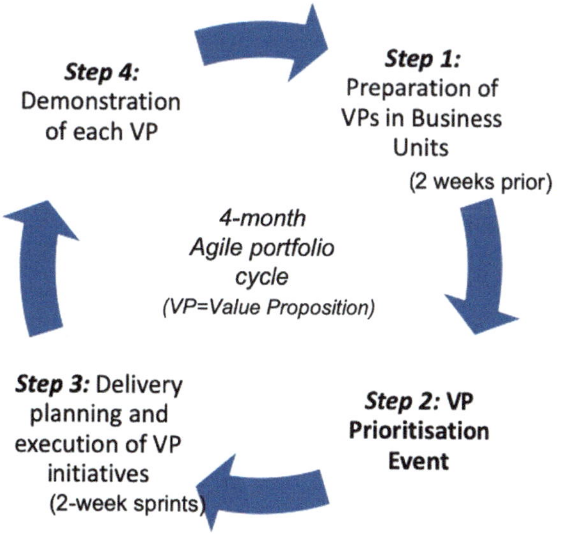

Fig. 4.2 Informant's perspective of the agile portfolio cycle underpinned by value

Step 2: VP Prioritization event.

The VPs are pitched by the VP Owners (VPOs) and prioritized (or reprioritized) by a wide group of stakeholders. Investments, resource allocation and overall capacity requirements are also discussed here. This event is central to EPSILON's startegies and identified as the "heartbeat" to value portfolio management and decision-making.

Step 3: Delivery and capacity planning and execution for the VPs.

This occurs once the VPs have been prioritized and selected for investment. VPs are operationalized by autonomous self-organized teams as delivery initiatives (projects) in fortnightly sprints. Teams have the autonomy to make their own decisions and set their own task agendas as they are aligned with the value expectations required in the execution of the initiatives. Alignment with capacity, backlog groomings (backlogs are prioritized and sized), and Retrospectives (reflections and reviews) are conducted in this step.

Step 4: Results from the planned VP initiatives are demonstrated in actuality.

The data implies that with agile cycles, not all VP initiatives may be completed within four months, and would need to be re-iterated in the next round.

Tables 4.3, 4.4, 4.5 and 4.6 summarizes how the value spectrum framework (Fig. 4.1) is manifested within the decisions and initiatives in the agile cycle of Steps 1–4. Value that is identified and discussed does not travel in a linear direction. Although time is on a continuum, decisions made are not linear. Discussions about value through the VPs appear to shift backwards and forwards across the value spectrum, from the unknown to the known at different points of the cycle, in line with the literature [15, 59–61].

Members engage in sensemaking practices involving reciprocal perspectives identified through the ongoing interactions, identification of facts, speculations, and assumptions, and dialogue about value. Visual cues using score-cards, as well exhibiting previous scores provide a point of focus to which decision-makers discuss,

Table 4.3 Value spectrum manifested in Step 1

Value Spectrum	Step 1: Planning of VPs in BUs
Unknown, unanticipated	VPOs and BUs discuss the business case, new ideas, or what has been achieved in the current VP initiatives VPs could be new, or a further iteration of a previous VP requiring further investments and resources Emergent value potentially be identified through this step
Known, articulated, qualified	VPOs and BUs spend 2 weeks before the VPP event to build the *business cases* (using the Business Model Canvas) and VPs templates. Qualitative and quantitative speculations and assumptions are applied to support the case. Eventually, all VPs presented need to be quantified
Known, articulated, quantified	VPs formally submitted using VP templates for prioritization event VPs quantified through key indicators, metrics, profits, and breakeven numbers

Table 4.4 Value spectrum manifested in Step 2

Value Spectrum	Step 2: VP prioritization event (4-monthly)
Unknown, unanticipated	Any tacit reasons for scoring explained to the group Prioritized VP initiatives may be changed by other senior management not in the event, but through a separate management review based on other considerations beyond the VP event
Known, articulated, qualified	Scores are relative, not absolute Discussions made about previous scores (known) to guide current scores Decisions articulated and explained in the group to get consensus before agreement to invest in the prioritized VPs "All the things we want to do for the coming three months are decided here (at the event)" (BDPM)
Known, articulated, quantified	Previous VP scores (known, quantified, agreed) are presented VPs that have been scored, agreed, and prioritized are made visible to everyone Only VPs with clearly measurable metrics are accepted into the VP event Scores are collectively discussed until a group score for the VP is agreed upon VP prioritization and selection decisions made until the preset budget is depleted

Table 4.5 Value spectrum manifested in Step 3

Value spectrum	Step 3: Agile Delivery Planning, capacity alignment and execution
Unknown, unanticipated	Time is needed by the teams and squads to get to know each other before value initiatives/projects can be rolled out Knowledge accumulation and learning to understand and connect with discussions from the other streams Self-organized autonomous teams conduct 2-weekly agile sprints and weekly meetings to make the unknowns known
Known, articulated, qualified	Monthly planning meetings held for delivery planning. Each VP is distributed to various delivery teams. Initiatives are broken down into further details Known project initiatives are evaluated and mapped to provide a baseline for business needs and resource capacity
Known, articulated, quantified	Transparency is emphasized in communicating the work completed, planned, and decisions made

compare, make sense of, and create shared meanings around the numbers to collectively agree on a VP score, as described by BDPM, "During the discussions, individuals need to negotiate and justify any conflicting scores until a collective agreement is reached". They engage in retrospective-prospective sensemaking in the way they reflect, clarify, justify and negotiate VPs based on past decisions in order to arrive at a shared decision in the present. VP decisions are made relative to each other, and relative to what was decided in the past.

The data also suggests that personal preferences and pet projects in decision-making have been diminished through the collectively accepted new practices and

Table 4.6 Value spectrum manifested in Step 4

Value spectrum	Step 4: Demonstration of VP
Unknown, unanticipated	Emergent/new ideas are planned for translation into initiatives for consideration
Known, articulated, qualified	Value categories identified as Sustainability, Digitalization, Customer satisfaction (happy customers), Business efficiency (e.g., reduced costs), Growth and New solutions
Known, articulated, quantified	VP results are quantified, e.g., lead indicators, VP metrics, profits, and breakeven figures

organizational mindset of "This is for EPSILON's good". Discussions and decision-making about value is influenced by signals from top management in terms of appropriateness of behaviors, expectations, rules, and guidelines stated by the organization, in line with studies of decision-making at portfolio meetings by Christiansen and Varnes [62]. Ultimately, formal decisions, initiatives, and results at EPSILON are expected to be based on known, quantified, and well-articulated perspectives of value, for instance, decision scores, lead indicators, VP metrics, profits, and breakeven figures controlled by strict templates.

While this demonstrates that the decisions made appear to be largely rational, it was also mentioned that senior management could override the group's decisions to include or reprioritize other VP initiatives not selected by the group during the prioritization event. This is because senior management may have obtained additional information unbeknownst to the other stakeholders, that could impact positively on the organization, and they have the power to include particular VPs of their choice. This implies that power and politics at EPSILON could play a role in theprioritization process. Furthermore, these senior management decisions appear to be accepted in good faith, and not perceived as nullifying the group's decision-making efforts. The belief that "this is for Epsilon's good" prevails. Overall, decisions made about value appear to be a social and organizational construct that is articulated rather than a completely rational outcome.

Although EPSILON states that they manage their portfolio in an agile manner, the nature of the sector in which the organization operates is relatively stable and does not face disruptive change and volatility. Managers recognize that ideas require time for consideration and planning before they can be translated into an initiative within the portfolio. They also acknowledge that not all parts related to the portfolio ecosystem are agile, and it is important to be able to combine and balance different internal and external agile and non-agile stakeholders. For example, HoPE exclaims, "They [e.g., shareholders, investors] don't care about agility, they want something else predictability, stability, and plannability. On the other hand, we have several development teams with extremely fast-moving processes, weeks maybe months tops, and those two worlds are my responsibility to combine."

Agile and self-organizing practices could potentially cause tensions between stakeholders driven by different philosophies. Those driven by strategy realization underpinned by planned strategy (predictability) may be in conflict with those operating under an agile banner underpinned by incremental benefits, unpredictability, and swift changes. Both the agile and less agile facets of organizations need to appreciate that strategic direction could be planned or emergent [63]. Boehm and Turner [64] observe that both discipline (planned) and agility approaches are needed in situation-tailored ways. They advocate for a reconciliation and integration of different users' and stakeholders' expectations of value in determining value propositions to ensure a mutually satisfactory win–win result. EPSILON's value-based approach in addressing and prioritizing their stakeholders' evolving value propositions through the cyclical agile portfolio process, providing a management system of sensemaking, communication and decision-making that incorporate realistic expectations is a positive direction toward high-payoffs and remaining competitive. However, Boehm and Turner [64] cautions that any attempts to balance discipline and agility needs to be in consultation with key stakeholders.

4.7 Concluding Remarks

This chapter exemplifies a value-driven PPM process characterized by an organization's ability to reconfigure and reprioritize project initiatives through self-organizing autonomous project teams, shorter iterative decision-making, and prioritization cycles, in order to adapt to changes in the environment to deliver value to stakeholders.

The study demonstrates how the organization has shifted from a cost-focus to value-focus. Value underpins the strategic business model and decision-making process. Value is not an outcome or result that is measured at the end of a project, but is embedded holistically in the strategies and functions of the portfolio and its underlying agile delivery initiatives. Value plays a key role in setting the rules, boundaries, criteria, or thresholds for decision-making. In agile-orientated portfolios, value could act as the driving force or heartbeat of decision-making.

The cycles and pivotal decision-making events presented could potentially support organizations in considering how value might be embedded in their own portfolios to strategically adapt to the demands of a changing environment. In a project portfolio's attempt to be agile and responsive, it is also critical to recognize and balance different agile and less/non-agile stakeholders in the portfolio ecosystem.

While single case studies have their merit, this can also be considered a limitation of this research. There are opportunities to extend the study by either following the single case longitudinally and across more departments or multiple portfolios. The consideration of the political aspects and converging themes of individual and social "values" with value in agile portfolio decision-making was not fully investigated. Further research could also compare the agile portfolio decision-making practices about value with those from other organizations.

4.8 Ethical Statement

We declare that we ensured the objectivity and transparency in our research and that accepted principles of ethical and professional conduct have been followed. Prior informed consent was obtained from individual participants included in the study before the research. No sensitive personal data was accessed. Anonymity of individual participant data is maintained.

The research does not require ethics approval, as it mentioned in the waiver issued by the Aarhus University's Research Ethics Committee under the number 2019–616-000,009, issued on 16 February 2021.

References

1. Martinsuo M, Killen CP (2014) Value management in project portfolios: identifying and assessing strategic value. Proj Manag J 45(5):56–70
2. Cooper RG, Edgett SJ, Kleinschmidt EJ (2014) Benchmarking best NPD practices-II. Res Technol Manag 47(3)
3. Meskendahl S (2010) The influence of business strategy on project portfolio management and its success—a conceptual framework. Int J Proj Manag 28(8):807–817
4. Hollister R, Watkins MD (2018) Too many projects. Harv Bus Rev 96(5):64–71
5. Gutiérrez E, Magnusson M (2014) Dealing with legitimacy: a key challenge for project portfolio management decision makers. Int J Proj Manag 32(1):30–39
6. Bessant J, Von Stamm B, Moeslein KM (2011) Selection strategies for discontinuous innovation. Int J Technol Manage 55(1/2):156–170
7. Serrador P, Pinto JK (2015) Does Agile work?—A quantitative analysis of agile project success. Int J Proj Mana 33(5):1040–1051
8. Conforto EC et al (2016) The agility construct on project management theory. Int J Proj Manag 34(4):660–674
9. Rigby DK, Sutherland J, Noble A (2018) Agile at scale. Harv Bus Rev 2018:88–96
10. Stettina CJ, Hörz J (2015) Agile portfolio management: an empirical perspective on the practice in use. Int J Proj Manag 33(1):140–152
11. Berkani A, Causse D (2019) Explaining the agile transformation through management innovation adoption
12. Sweetman R, Conboy K (2018) Portfolios of agile projects: a complex adaptive systems' agent perspective. Proj Manag J 49(6):18–38
13. Ang K, Biesenthal C (2017) Multilevel value creation in projects, programs, and portfolios. In: Drouin N, Müller R, Sankaran S (eds) Cambridge handbook of organizational project management. Cambridge University Press, Cambridge, pp 295–310
14. van Marrewijk A et al (2016) Clash of the titans: temporal organizing and collaborative dynamics in the Panama Canal megaproject. Organ Stud 37(12):1745–1769
15. Ang K, Killen C (2016) Multi-stakeholder perspectives of value in project portfolios. In: 16th annual conference of the European academy of management (Euram) conference 2016. European Research and Management Conference 2016
16. Ang KCS, Sankaran S, Killen CP (2016) 'Value for whom, by whom': investigating value constructs in non-profit project portfolios. Proj Manage Res Pract 3:5038
17. Eskerod P, Ang K (2017) Stakeholder value constructs in megaprojects: a long-term assessment case study. Proj Manag J 48(6):60–75
18. Ang K, Sankaran S, Killen C (2016) Value for whom, by whom: investigating value constructs in non-profit project portfolios. Proj Manag Res Pract

19. Ang K, Killen C, Sankaran S (2015) Unanticipated value creation: sensemaking and the value spectrum in partnership projects. In: International research network on organizing by projects (IRNOP), London
20. Weick KE (1995) Sensemaking in organizations. vol 3. Sage
21. Schwartz S (2006) A theory of cultural value orientations: Explication and applications. Comp Sociol 5(2–3):137–182
22. Rokeach M (1973) The nature of human values. Free press
23. Campbell A, Yeung S (1991) Creating a sense of mission. Long Range Plan 24(4):10–20
24. Ravlin E, Meglino B (1987) Effect of values on perception and decision making: a study of alternative work values measures. J Appl Psychol 72:666–673
25. Allee V (2000) The value evolution: addressing larger implications of an intellectual capital and intangibles perspective. J Intellect Cap 1(1):17–32
26. Allee V (2008) Value network analysis and value conversion of tangible and intangible assets. J Intellect Cap 9(1):5–24
27. Kopmann J et al (2014) Business case control: The key to project portfolio success or merely a matter of form? In European Academy of Management (EURAM). Valencia, Spain
28. Zhang L, El-Gohary NM (2015) Discovering stakeholder values for axiology-based value analysis of building projects. J Constr Eng Manag 142(4):04015095
29. Rank J, Unger BN, Gemünden HG (2015) Preparedness for the future in project portfolio management: the roles of proactiveness, riskiness and willingness to cannibalize. Int J Proj Manag 33(8):1730–1743
30. Cheng AS, Fleischmann KR (2010) Developing a meta-inventory of human values. In: Proceedings of the 73rd ASIS&T annual meeting on navigating streams in an information ecosystem-volume 47. American Society for Information Science
31. Jahani H, El-Gohary N (2012) Value-sensitive construction: value discovery in building projects. Construction research congress 2012: construction challenges in a Flat World
32. Cooper RG, Edgett SJ, Kleinschmidt EJ (1999) New product portfolio management: practices and performance. J Prod Innov Manag 16(4):333–351
33. Lanning MJ, Michaels EG (1988) A business is a value delivery system. McKinsey staff paper 41
34. Sales, TP et al (2017) An ontological analysis of value propositions. In: 2017 IEEE 21st international enterprise distributed object computing conference (EDOC). IEEE
35. Ballantyne D et al (2011) Value propositions as communication practice: taking a wider view. Ind Mark Manage 40(2):202–210
36. Bentzen E, Christiansen JK, Varnes CJ (2011) What attracts decision makers' attention?: Managerial allocation of time at product development portfolio meetings. Manag Decis 49(3):330–349
37. Beringer C, Jonas D, Kock A (2013) Behavior of internal stakeholders in project portfolio management and its impact on success. Int J Proj Manag 31(6):830–846
38. Bourne L (2009) Stakeholder relationship management: a maturity model for organizational implementation. Farnham
39. Bourne L (2011) Advising upwards: managing the perceptions and expectations of senior management stakeholders. Manag Decis 49(6):1001–1023
40. Ancona D, Backman E, Isaacs K (2019) Nimble Leadership Walking the line between creativity and chaos. Harv Bus Rev 97(4):74–83
41. Thiry M, Deguire M (2007) Recent developments in project-based organisations. Int J Proj Manag 25(7):649–658
42. Cooke-Davies T (2002) The "real" success factors on projects. Int J Proj Manag 20(3):185–190
43. Garfinkel H (1957) Studies in ethnomethodology
44. Allard-Poesi F (2005) The paradox of sensemaking in organizational analysis. Organization 12(2):169–196
45. Taylor JR, Lerner L (1996) Making sense of sensemaking: how managers construct their organisation through their talk. Stud Cult Organ Soc 2(2):257–286
46. Pugh DS, Hickson DJ (2007) Writers on organizations. Penguin, Thousand Oaks

47. Easterby-Smith M, Crossan M, Nicolini D (2000) Organizational learning: debates past, present and future. J Manag Stud 37(6):783–796
48. Weick KE, Sutcliffe KM, Obstfeld D (2005) Organizing and the process of sensemaking. Organ Sci 16(4):409–421
49. Dervin B (1998) Sense-making theory and practice: an overview of user interests in knowledge seeking and use. J Knowl Manag 2(2):36–46
50. Hellgren B, Löwstedt J (2001) Management in the thought-full enterprise: European ideas on organizing. Fagbokforlaget
51. Weick KE (2001) Making sense of the organization. Blackwell, Malden, MA
52. Guba EG, Lincoln YS (1981) Effective evaluation: improving the usefulness of evaluation results through responsive and naturalistic approaches. Jossey-Bass
53. Kidder LH, Fine M (1987) Qualitative and quantitative methods: when stories converge. New Dir Program Eval 1987(35):57–75
54. Yin RK (2014) Case study research design and methods, 5th edn. Sage, USA
55. Gustafsson J (2017) Single case studies vs. multiple case studies: a comparative study
56. Siggelkow N (2007) Persuasion with case studies. Acad Manag J 50(1):20–24
57. Kniberg H, Ivarsson A (2012) Scaling agile @ spotify. [online], UCVOF, ucvox.files.wordpress.com/2012/11/113617905-scaling-Agile-spotify-11.pdf
58. Osterwalder A, Pigneur Y (2010) Business model generation: a handbook for visionaries, game changers, and challengers. John Wiley & Sons
59. Grönroos C, Voima P (2012) Making sense of value and value co-creation in service logic
60. Voima P, Heinonen K, Strandvik T (2010) Exploring customer value formation: a customer dominant logic perspective
61. Helkkula A, Kelleher C, Pihlström M (2012) Characterizing value as an experience: implications for service researchers and managers. J Serv Res 15(1):1094670511426897
62. Christiansen JK, Varnes C (2008) From models to practice: decision making at portfolio meetings. Int J Qual Reliab Manag 25(1):87–101
63. Mintzberg H, Waters JA (1985) Of strategies, deliberate, and emergent
64. Boehm B, Turner R (2003) Using risk to balance agile and plan-driven methods. Computer 36(6):57–66

Chapter 5
New Work—Flexible, Mobile, Project-Driven: Can Increasing Self-Organization Contribute to a New Design of Work?

Eckhard Heidling and Nick Kratzer

Abstract The notion "new work" frequently turns up in current debates about the future of work. New Work is described as a fascinating idea in books and papers. Manuals and websites provide instructions on how to do it. However, the origins and meanings of the term remain rather hazy. Essentially, it stands for new forms of work associated with an important change in the governance and culture of work. Self-organization in work, which is increasingly demanded by companies and requested by employees, is regarded as a major trend in this context. Although the New Work discourse addresses many topics in the development of work that has been subject to extensive research and debate in the sociology of work, those two threads of discussion have been running rather separate up to now. The paper tries to establish connections between them by looking at important topics of the New Work discourse from a sociology-of-work perspective: flexible work in terms of space and time, new spaces of work, (agile) project work and self-organization. The focus is set on the respective potentials, limits, and approaches for the design of our present world of work.

Keywords New Work · Project work · Self-organization · Sociology of work

5.1 Introduction

Technological change, currently accelerated by digitalization, has a great impact on structural changes in the present economy, putting much pressure upon restructuring processes in wide areas of manufacturing and service industries. Fast and volatile changes of market and customer requirements in an internationalized competition

E. Heidling (✉) · N. Kratzer
Institut für Sozialwissenschaftliche Forschung – ISF München, Jakob-Klar-Str. 9, 80796 München, Germany
e-mail: eckhard.heidling@isf-muenchen.de

N. Kratzer
e-mail: nick.kratzer@isf-muenchen.de

R. Ding et al. (eds.), *Research on Project, Programme and Portfolio Management*, Lecture Notes in Management and Industrial Engineering,
https://doi.org/10.1007/978-3-030-86248-0_5

are calling for efforts of the companies: they have to improve their innovation capacities, to boost the flexibility of work processes, and to establish suitable skills and competencies. In this context, digital technologies play a leading part by enabling, facilitating, and accelerating new forms of information creation and exchange of data resp. knowledge. The progress of digitalization rings the bell for a new round of the well-known controversial issues: to what extent will technological progress makes human workforce redundant, to what extent will machine wield their power beyond the sphere of work, determining also the sphere of life, and to what extent will this constitute a big step into a comprehensively technicized world.

Within the debate about these developments of work, most researchers basically agree that the division of work between humans and technology is undergoing a change. Yet there are still open questions concerning the directions and outcomes of this change: whether new technological systems will rather support or rather replace human workers, whether digitized processes will promote new spaces and leeways for working action or else lead to an organizational re-centralization, whether increases in company flexibility are to be achieved through flexible technology or rather through flexible employees. In contrast to previous debates about automation technologies, currently a broad consensus exists that man should take a center stage in the development. However, it remains rather unclear how the division of work and working roles should actually be organizationally shaped on the basis of digital technologies.

In the discussion about future images of work, the term New Work currently plays a very popular part. New Work is described as a "fascinating" idea [15] in books and numerous websites. Whilst there is no shared definition of New Work, there are some crucial characteristics. An increasing flexibility, especially regarding time and space, the self-organizing or self-management of work and a reorganization of work processes toward cooperation and project work. The growing significance of cooperation and coordination for successful projects is an important issue in the debate [64]. In addition to formal knowledge, practical and context-related competencies and corresponding problem-oriented approaches are decisive for project management. In this context, a need for research about the development of self-leadership skills has been stated, including self-regulation, self-control, and self-management [27]. Recently an expansion of self-organized structures within projects has been suggested. It is assumed that such a development would require a marked change of the role concepts of project leaders. In self-organized projects, leadership positions would particularly have to focus on the tasks of coaching and empowering the team members, as well as facilitating their self-leadership [70].

Specific aspects of New Work are being discussed in journals and papers [39]. Manuals and guidebooks are providing instructions on how to do and implement New Work [14, 63]. A "New Work Charter" has been launched on the Internet and endorsed by a number of consultants and enterprises, highlighting five principles (i.e., freedom, self-responsibility, purpose, development, and social responsibility) that should guide a fundamental change of the world of work [49]. Moreover, the business platform Xing (rebranded as "New Work SE" in 2019) has created a "New

Work Award" in 2014, a prize being awarded on an annual basis for the best innovative ideas and implementation practices in the re-design of work.

In this context, New Work is usually introduced as a positive answer to the question in which direction work should develop in a world of work increasingly penetrated by digitalization. However, it often remains rather hazy what this concept precisely means, and there is usually a lack of reference to the research and debates conducted within the sociology of work. Against this backdrop, the paper will first take a closer look into the basic ideas of the founder of this seminal concept, Frithjof Bergmann, and discuss how the current use of the term is related to them (Sect. 5.2). Subsequently, this paper identifies topics where the New Work discourse and the debates within the German sociology of work match, although New Work and sociology-of-work debates have been running rather separate as of now. A remarkable number of topics and subjects are revealed where those debates share essential insights and cognitions. Referring to a selection of these topics, conceptual, and empirical findings from sociology of work will be presented. They will be discussed according to the guideline which advantageous and disadvantageous tendencies can be observed in these New Work areas and which points of leverage can be identified for a labor policy in the sense of a human-oriented design of work (Sect. 5.3). The extent and scope of self-organized design of work by the employees themselves plays a particularly important part within the New Work discourse. This question permeates all topics and areas identified. It will be discussed using the example of recent developments of project work (Sect. 5.4). In conclusion, the main findings will be summarized and the prospects for a human-oriented labor policy will be discussed in the light of these findings (Sect. 5.5).

5.2 New Work—Basics of the Concept and Current Discourse

The Austrian-American philosopher Frithjof Bergmann is seen as the father of the New Work concept. In the course of an economic recession and a thorough automation wave of manufacturing activities, the American automotive firm General Motors was planning to cut the workforce of its factory in the city of Flint and its biggest manufacturing site in the U.S. in half, during the 1980s. In order to avoid the mass lay-offs, Bergmann developed an alternative concept: to cut the working time of the employees in half and to permit them to do New Work for the other half of the time [11, pp. 129–131]. The core of New Work is to make the employees do work "that they really, really want". This formula intends to create a counter-image to classic hired labor characterized by heteronomous definition of objects, kinds, contents, temporal succession, and place of work activities. New Work promises to reverse this condition. Autonomy or "*working for oneself* is the basic building block, the molecule out of which New Work as a system has been piece by piece gradually constructed" [11, p. 116; emphasis Bergmann]. Rather than maintaining a state of heteronomous labor that drains the strength and energy of the workers, the aim of New Work is "to transform work until work will create free human beings" [11, p. 12]. This is emphatically not about "a bit more fun" in work which, according to Bergmann, would still amount to merely a job, "only now a job that wears a mini-skirt" [11, p. 194]. Instead, the "mild illness" of traditional wage work is to be replaced by New Work performed on the basis of a "calling" and fueled by a strong intrinsic motivation to do "meaningful" things [11, p. 161]. Thus, Bergmann sees changes of the system that aim for a re-design of working relations as essential. The idea is not to abolish capitalism but to "develop capitalism further *in a different direction.* [...] The idea is a turning operation, a change of direction, to visualize it: a change of 90 degrees" [11, p. 303]. As instruments to conduct this turning operation, mainly the use of highly developed technologies is envisaged, coupled with a miniaturization of productive equipment enabling local self-production. Thus, a de-centralization of central structures and processes is targeted. As the driving forces of these de-centralization processes, three IT-supported technologies are identified: the Internet, the mobile phone, and the laptop [11, p. 221]. This indicates the functional principle of the turning operation: the very same information technology that is the basis for progressing automation and consequently increasing unemployment should be appropriated by the workers affected by it. And they should make use of it as a tool and instrument for autonomous, self-defined working processes. If workers with a strong intrinsic and autonomous motivation use and combine these progressive technologies, small, efficient, and fast bodies will be able to replace the traditional, big, and slow systems with their hierarchical structures (as enterprises and administrations) [11, pp. 212–223].

This mix of ideas could also be put like that: innovative products and services should be produced particularly through the use of digital technologies, which permits a faster reaction to market requirements and at the same time matches the growing desires of employees for meaningfulness and purpose in their work activities. Thus, essential categories suggested by Bergmann lend themselves to ideas how New Work could actually look like and how it could be designed and shaped. This is reflected in current definitions of New Work, whether they refer to Bergmann's ideas or not.

One definition says that New Work is "a way of working characterized by a high degree of virtualization of work equipment, networking of persons, and flexibilization of places, times, and contents of work" [28, p. 5]. Other suggestions lay the focus on "time sovereignty and high subjective well-being", pointing out potentials for development and governance "through interventions on pedagogical, psychological, technological and political levels" [10, p. 13]. Finally, special value is often placed on the issues of purpose and meaningfulness in work and the idea that employees are increasingly interested in the usefulness of their working activities in order to stabilize their work identity and maintain their capacity to act [23].

To sum it up, the main topics in the New Work discourse include flexibilization of work in temporal and local respect; a growing cooperation and networking beyond organizational divisions and even beyond the boundaries of companies; a tendency of development from hierarchy to self-organization, also in combination with new forms of leadership; and a stronger value-orientation of work, coupled with the expectation of sensemaking and purpose in the concrete work activities [28, 72].

5.3 New Work Seen from a Work Research Perspective

In sociology of work, an essential determinant of work is the distinction as to the *what*, *how*, and *where* of working action. Concerning the *what*, aims, intentions, and contents of working action are relevant. Sociological surveys of working action are often directed toward instrumental and object-related activities. The center of attention is the handling of the objects of work, which was traditionally focused on the dealings with material objects in manufacturing processes. Meanwhile, however, this focus has been greatly extended, now also including service work activities and generally working on and with humans, as in nursing and care activities where the "work object" is a human person. Regarding the *how*, the ways how working action is structured and regulated come into view. Traditionally, sociological research paid most attention to planned rational action, which relates to the human capacity to act according to a pre-conceived plan and directed toward a pre-conceived goal. In the context of service work and especially person-related work, more recent surveys elaborated upon the subject qualities of human work (as empathy, emotions, sensual perceptions, etc.), beyond the characteristics of planned rational action [12]. The *where* of working action includes the work environment with the features of the workplace (light, air, noise, etc.) and the organization of distance and closeness in work, but may also refer to the work object (as in building and designing spaces or overcoming spatial distance) or, in a broader definition, to spatial features as region or "cluster", urban or rural locations, center or periphery. In recent debates, the spatial dimension of work has received growing attention in the context of two opposing propositions: on one hand, the thesis that physical space is losing relevance through internationalization and digitalization of space ("annihilation of space"), on the other hand, the thesis that on the contrary physical and geographical dimensions of space are gaining relevance. For instance, companies spend much money for new office concepts (as "Open Space") but also enable employees to work from other locations, as in Home Office. Cooperation is increasingly independent of location by means of electronic media, but on the other hand, there are efforts to promote creativity and innovation precisely by concentration at one physical location, as in "incubators", "hubs", or "co-working spaces".

In the last three decades of research in sociology of work, there are a lot of findings indicating a tendency to "subjectification of work". Companies have been increasingly interested in employees' subject qualities and performances, not only in complex and complicated working tasks but also in "simple" tasks in manufacturing and service work. Employees should think out of the box, co-create and co-design their work, look beyond their limited workplace or organizational division, which often results in an enrichment of tasks, an enlarged scope of action and disposition, and higher qualification and competency requirements. Employees, for their part, also increasingly introduce their own subjective demands and expectations into their work: they expect meaningful "good work", which refers both to their concrete work contents and to the results and contexts of use of their products [32, 50].

From the view of sociologists of work, it seems hardly possible to identify something like a topical core in the current discourse about New Work. However, it can be demonstrated that there is an area of *intersection* between the great variety of seemingly unrelated topics, trends, and visions within this discourse, and two development tendencies identified by the sociology of work. This area of intersection is constituted, on one hand, by the increasing demand that employees should apply themselves to their work as "whole persons" including their subject qualities, that they should work *both* as goal-directed planners and empathic persons "with all their senses". On the other hand, it is constituted by the tendency to network and cooperate beyond organizational boundaries, which results in re-design and restructuring of working tasks and structures.

The following sections will focus on three topical areas: flexible work in temporal and spatial respect, new spaces of work, (agile) project work. These topics *both* play a prominent part in the New Work discourse *and* represent elements of the common patterns of subjectification and networking in sociological work research. The discussion is structured by two contrary perspectives and tendencies, both occurring and observed simultaneously according to empirical surveys and findings of work research:

- the perspective of a promised or even emerging augmentation and enhancement of individual subjectivities, characterized by greater scope of working action in self-organized work designs and correspondent frameworks;
- the perspective of developments that limit, endanger, or prevent this evolvement and unfolding of individual subjectivities and, consequently, the potentials of self-organized work.

5.3.1 The Working Time of New Work: Flexible, Unbounded, Self-Determined?

The topic of time-flexible work plays an important role in the New Work discourse [28]. The beginning and end of a working day are less rigidly fixed, more and more open to the choice of employees according to individual needs and/or professional requirements, and sometimes there are even "trust-based working hours", meaning that working time isn't recorded at all. Meanwhile, a large percentage of employees has at least a certain influence upon the position, distribution, and duration of their working hours. A recent survey of working time, conducted by the Federal Institute for Occupational Safety and Health (BAuA) in Germany, shows that nearly 40% of all employees have "much influence" upon the beginning or the end of their working hours, even 56% can influence their break times. 45% are able to take some hours off from work in relative self-determination, even 58% some days off. Data about employees with "trust-based working hours" is rather limited, their share is probably rather low. The BAuA working time survey provides a share of 3% of all employees who had "no contractualized working hours" [2].

The flexibilization of working time organization has been observed by the sociology of work for a considerable time, and concepts like flexibilization and "blurring of boundaries" [32] have been developed to describe it. This does not only refer to working time in companies but is also understood as a diagnosis that time patterns in society are changing. The proposition is that the Fordist "normal working hours" have eroded. The Fordist time arrangement had evolved during the first decades after the war as a structuring time order for society. Its essential features include the separation of (wage) working time and private time, a "normal working time" of about 40 h per week with a work-free week-end, and a standardization and heteronomous fixation of working time schedules in the companies.

Since about the 1980s, an increasing flexibilization, de-limitation, and subjectification of working time have been observed. Findings from a number of surveys indicate a growing variety of working time arrangements, a flexibilization particularly of the beginning and the end of the working day in the context of flex time models, a partial erosion of the temporal and spatial boundaries between wage work and private life (home office, permanent accessibility), and an increase of autonomy and self-determination in working time organization. This development is accompanied by two general trends that are only in part directly related to time issues. First, new forms of "indirect" performance control and governance have become prevalent in companies: the performance expectations are not any longer directly bound to the time spent in work but are measured in terms of work results ("result-orientation") [38]. Second, a comprehensive individualization of all spheres of life and a change in the value orientations toward self-actualization, subjective meaningfulness of work, and an equilibrated work-life balance have been stated on a broader level of society [3].

This flexibilization and subjectification of working time offers a greater range of action to the employees to adapt their working time not only to the dynamic needs of the enterprises but also to their individual wants and needs. However, work research studies have also revealed potential drawbacks of this development:

- The planning of working hours is more difficult which may also lead to negative consequences for the relationship between wage work and family life/private life [2].
- Under conditions of high time pressure, flexible and autonomous disposition of breaks may have the consequence that employees (are compelled to) cut down or even cancel their rest periods. Inquiries indicated that this applies to 28% of all employees [18].
- Under conditions of flex time or trust-based working hours, the actual flexibility is oriented rather to company requirements than to individual employee needs and wants. This often leads to overtime work, in the case of trust-based working hours even to unpaid resp. uncompensated working time [56].

This cursory rendering of important work research findings goes to clearly show that the consequences of flexible working time design can only be understood appropriately if the questions of performance control in companies are considered. In subjectified performance control systems, the employees themselves are compelled to balance their temporal resources both against performance expectations (on the part of companies as well as on their own part) and private demands, wants, and needs. This task implies a clash of conflicting logics. From the companies' point of view, time economy has to support the aim to achieve the desired result at the right moment "just in time" with a minimum of time expenditure [34, 35]. The individuals, for their part, are bound to time expenditure and the sequentiality of time both in their work and in their lifeworld. The task to deal with these conflicting requirements in self-organization is very demanding and complicated for the workers, which also goes to explain why the issue of an (improved) balance of working time and "lifetime" ranges high on the agenda. Empirical surveys reveal that, for instance, employees in projects have broad possibilities at their disposal to organize their own working time. In this context, time sovereignty includes flexible disposition of the duration and position of working time as well as working location (including home office). This situation opens up considerable opportunities for a self-determined, autonomous performance of working tasks, as far as the *how* and *where* are concerned. The drawback of this high degree of autonomy is, however, the permanent excess of work beyond the limits of the contractual working time [41], which is closely connected to the time pressure frequently occurring in projects. An important requirement arising from this difficult situation of project workers is to make use of the time sovereignty not only for self-determined working times, but also for the self-regulation of strain and stress which can build upon the different intensity levels of work in the project progression. However, this type of time sovereignty is usually only practiced on an informal level by the project workers and informally tolerated by project leaders, at best. From these informal practices, approaches for coping with stress and strain could be developed, but these solutions should be institutionally supported and safeguarded [55].

5.3.2 New Spaces for New Work: Within and Beyond the Company Facilities

A second important topic in the context of New Work is the question of a "spatialization" of work [54] and its transformation, which is manifested in spatially flexible work concepts. The cultural and organizational change toward self-organization, flexibilization, and self-actualization demands new spaces and finds expression in new spaces [66, 30]. Hence, the "workplace of the future" [31] will consist of an individual and dynamic arrangement of different locations, with work taking place in the office but additionally at several other locations, including co-working spaces, home office, and "mobile work" in transit.

5.3.2.1 Spatial Flexibility Within the Company Facilities: Open Space Office

In the future, the office in the company building will only be one of several locations where work is performed. However, it will definitely continue to play a leading part: as "hub", meaning center of communication, and as "home", meaning a social home base and a physical place to experience togetherness and collegiality within the company. It is in this social function that the office will probably get increasingly relevant with the progress of digitalization. If requirements, structures, and processes become more complex and dynamic, better networking of information, more flexible modes of work, faster decisions, and more stable (trust) relationships will become necessary—and this calls for a location of immediate social exchange: the office at work [37].

However, the office will undergo a functional change in this context. Whereas in its traditional function the office mainly served as a place for concentrated single-person work at his or her desk, it will now become in particular a place of cooperation, knowledge exchange, and social integration. This extension of functions has an effect upon its design, as can already be seen in new terms as "open space", "modern workspace", or "multispace". New concepts as to size and equipment accompany this functional change process. These concepts have two features in common: they pursue the idea of an "activity-based workplace," meaning that different working activities require different working arrangements [58, 59]. And they integrate these working arrangements (in the form of spatial zones and rooms) in an open area. That is the reason for the term "open space."

As to now, the accurate quantitative dissemination of open space offices in the world of work is impossible to determine. Statistics are lacking as to how many people in Germany are working in which kind of workplace and how this change over time. Although several inquiries provide hardly comparable and sometimes inconsistent results, a trend toward open offices is supposed to be relatively certain, owing to the fact that new office concepts are an element of a comprehensive transformation of the working sphere as a whole [31, 36].

But whereas the idea of open space is plausible on a conceptual level and the existing layouts and displays make a good impression, in practice the concept proves to be rather controversial and prone to contradictions and inconsistencies. This can be seen in the assessment of employees working in these new working environments. For sure, a good half of open space office workers are satisfied with their workplace but still a quarter is dissatisfied and almost another quarter is neutral [9].

On one hand, open space offices enable their users to work in a flexible and self-organized way, but on the other hand, flexibility and self-organization turn into a demand that the employees have to comply with. They can but also must choose their place in the office and adapt their way of working to the given circumstances. This allows for subjective strategies and coordination practices but also necessitates them. Interestingly, individual coping strategies often have the result to limit the intended effects of the open space office since it may, for instance, reduce the openness and flexibility. If you isolate yourself to be able to concentrate, you don't take part in communication—if you keep your accustomed place you don't participate in the exchange. But on the other hand, these coping strategies make work in open space offices viable, and hence make open space possible in the first place. Individual coping strategies are a way to appropriate the spatial options and to grapple with the disadvantages of open offices experienced in practice, serving both stress reduction and the organization and warranty of the individual working power and capacity [9, 26].

In this respect, the open space office is an answer to the change of work requirements but it is also a mirror of the present development of work. Particularly regarding the subjective ways to deal with the open office, this concept can be seen as an extension of "subjectified performance policy" [43] to the physical and social work environment. In the context of new forms of performance control and governance, the workers are not only challenged to cope with conflicting entrepreneurial needs and ends by themselves, but also have to take responsibility to organize and maintain the conditions of their own performance [38]. This requires specific competencies, resources, and also autonomy.

5.3.2.2 Spatial Flexibility Beyond the Company Facilities: Mobile Work and Home Office

Another element of the spatial reorganization of work envisioned by the New Work concepts is spatially flexible work outside the company buildings. This includes mobile work, work in the home office, and co-working spaces [28].

Mobile Work

Mobile work is a generic name for several forms of temporally and spatially flexible work performed outside the company site. The term is sometimes equated with "flexible work" in general [1] but a more specific definition emphasizes that mobile work is "linked to spatial mobility" and also a "necessary for the job", hence "part of the work performance owed under the employment contract" [68, p. 7].

Although there are no statistics measuring the extent and development of mobile work on a general scale, existing findings seem to indicate that meanwhile a considerable share of all employees are spatially mobile as part of their jobs. A survey states that mobile work "has become a matter of course for many employees of different skill levels and in nearly all sectors and industries" [46]. However, the temporal extent of job-related travels is very variable, ranging from service engineers traveling almost every day to employees who only spend some days of their annual working time at distant locations.

In the mobile workers' view, business travels, visits, and assignments are an ambivalent matter. On one hand, they value them as providing versatility, variety, and autonomy. On the other hand, there are also disadvantages: time pressure, impaired cooperation, overtime work, ergonomically poor working environments, and last not least problems of reconciliation of work and private life. Consequently, mobile workers are "confronted with both the task to reconciliate job-related mobility with their private lives and the task to solve mobility-induced conflicts of requirements within the job" [51, p. 188].

In this perspective, spatially and temporally flexible work is first of all a matter of necessary and required flexibility which yields autonomy benefits but confronts the employees with the challenge to weigh these benefits against the risks and conflicts of mobile work and find a balance, very often on an individual level. In short: mobile work both enables and demands subjectification performance. The chances for success are better the more the companies lend support to the employees and the more they are open to employee participation [68].

Home Office

Whereas traditionally only some professions, as for instance teachers, had the opportunity and duty to work at their homes, the work from "home office" has spread to many other groups in recent years. The number of companies permitting home office grew from 32% in 2014 to 37% in 2016. The share of employees working at least occasionally at home rose from 19% (2013) to 22% (2017). Until fairly recently, the share of home office workers was growing but was far away from constituting a mass phenomenon. However, in the course of the Corona crisis this share could have risen markedly, and as far as can be said as to now, this might become a continuing tendency.

Until not so long ago, enterprises were rather reticent in matters of home office. Hence home office was mainly discussed as a desire on the part of the employees and as an opportunity to reconciliate job and family. But empirical surveys of home office in practice indicate that the working times of employees (both female and male) are higher in home office than in work at the company facilities. In addition, the findings reveal that the promise of a better reconciliation is by no means automatically redeemed. On the contrary, particularly for mothers, home office may be apt to even aggravate the double burden of job and care work. It has been shown that home office for mothers leads to longer working times and additionally to more time expenditure for care work [41].

Findings from qualitative social research provide certain clues to explain these contradictions and inconsistencies. To be sure, employees who have the opportunity to work in home office are interested to integrate private dates (as with handymen, doctors, or else in the context of care work for family members) into the everyday working life in an uncomplicated manner. But at least the same relevance is attributed to motives to better manage the work itself. For instance, employees tend to invest the time saved by the cessation of commuting into their jobs. In addition, they appreciate the home office particularly as an alternative workplace with less disturbances and interruptions, permitting them to perform certain working activities in better concentration and hence better quality [69]. Quantitative surveys confirm that employees mainly see job-related advantages of home office work. In an inquiry study, employees were asked which were the main advantages of home office. The first two ranks were "better work activity performance" (56%) and "saving of commuting time" (55%). "Reconciliation of job and family" only came third place (52%), and 38% chose the answer "higher working time possible" [22].

Hence, home office is supposed to be an option desired or demanded by many employees but its effects are definitely ambivalent. The opportunity to work (partially) at home does provide a greater range of options to the employees to improve flexibility and reconciliation of job and family. But the potential conflict lines do not disappear: they are relocated and transferred into the self-organization of both their own work and the relationship between job and private life. Hence, home office offers may form an important contribution to a work arrangement and organization better suited also to individual needs and desires, but this is not a "surefire success". It is necessary to do something to avoid social isolation, to organize the own work activities and the cooperation with others, and the problem of "permanent accessibility" needs adequate solutions [52].

5.3.3 Project Work

Within the New Work discourse, project work is discussed in the context of the desires of the employees to spend a larger part of their working time with work content and forms oriented to (self) development. Quantitative surveys indicate that 57% of the employees interviewed desired higher shares of project work for themselves and that the desired extent of project work was 14% higher than in their current work situation [72, p. 131]. Correspondingly, surveys for the German context revealed that projects play an increasing role in the manufacturing and service processes of enterprises. The main reason for this development is supposed to be that projects should help to solve new market-induced complex problems within the companies and also in cooperation across companies [24]. Current studies indicate a growing projectification of German economy. In 2013, 34.7% of the total working time of employees was spent for project work. This constituted a rise versus 2008 (29.3%), and a further rise to 41.3% in 2019 was expected. Project work within the manufacturing industry was mostly found in the form of R&D projects (22%) and customer projects (25%) [71]. The participants in projects are in most cases both integrated in traditionally organized work processes and in project work. They usually belong to different company departments and hierarchical levels, and their occupation in projects is temporary [60]. This dual structure, with a matrix project organization characterized by functional lines and departments on one hand and projects on the other hand, seems to be fairly widespread and predominating also on an international level [19]. It can be expected that the duality of permanent company organization and temporary project organization will continue to prevail in the medium or even longer run in many enterprises.

The difference between projects and permanent organization is that projects set specific targets that have to be achieved in a defined time with limited resources in a specific quality. A characteristic of project work is, however, that the project targets are often not fully determined a priori but are finally specified only in the course of the project progression, which means that the necessary resources and the expenditure of working time can be estimated only approximately in the beginning. This implies a measure of openness and uncertainty which, in combination with the temporariness, makes for a principal difference between projects and the predominant understanding of a permanent organization, which is characterized by continuity and stability and aims at stable and secure structures and processes—prototypically represented by the model of the bureaucratic organization.

Frequently project work is considered as an element of knowledge-intensive work processes which in turn are linked with de-centralization of responsibility, relatively high leeways for the design of work, and self-organized working action. However, this is only partially correct since project work is in wide areas structured by formal planning and control instruments and IT-supported tools [29]. The tools and procedures are often geared to standardization in order to reduce interfaces, uncertainties, and ambiguities. Their intention is to establish transparent and reproducible organizational processes within the company and to effectively steer and control the project work. The predominant idea of project management is therefore characterized by predefined tools and controlling processes. Meanwhile an extensive array of planning and controlling models, tools, and training concepts is available. They include waterfall models, milestone setting procedures, and tools for planning and controlling, supported by information and communication technologies [6]. But in spite of this sophisticated arsenal, the project management results are by no means satisfactory. The optimization of planning and control achieved by these tools is contrasted by the fact that the projects are in practice still faced with events and influences that cannot be fully controlled. Studies on real projects have revealed that considerable delays and cost increases compared to the original planning frequently occur [16, 45]. However, this finding is often not a reason to reassess the usual planning and controlling paths but tends more to result in further optimization of formal management elements [61, 67].

In this situation, actual project work is characterized by a permanent difference between planning and reality. Typical statements of project participants are, "many unexpected events occur and there is not a single project that goes according to plan", or, "actually I have only worked in critical projects up to now". Critical situations, meaning events in the project process that could not be anticipated or predicted in spite of systematical planning, are characteristic for a large majority of real projects. The experts who are temporarily integrated into project work are thus confronted with the necessity to situationally adapt their working action to changing tasks in different organizational settings. They have to master the permanent change between stable and stationary line work and temporary, variable, and cross-domain project work, and they have to find ways to organize and cope with this change in a self-organized manner. In this respect, their working tasks are to process and advance knowledge, information, and experiences within and across different disciplines, actors, and often spatially distributed locations of the manufacturing and service processes [25].

The tensions between a permanent organization in line and matrix structures on one hand and temporary project organization on the other hand are assuming a new form against the backdrop of the progress of digitalization in the production processes of products and services. Currently the way to handle these tensions is more and more frequently the use of agile methods. First established in software development, these methodical solutions move away from planning-oriented project procedures, instead relying on iterative proceedings in small self-organized teams with close participation of the customers. Agile project management methods are devised as variable and flexible and pursue the goal to find innovative solutions that do (more) justice to the customers' needs. In addition, agile project management promises faster and more effective processes [7].

Recent studies about the development of project management in German economy indicate that the majority of companies rely on hybrid models (43%), meaning a mix of agile and traditional models. The following ranks were occupied by a selective application of agile methods (28%), a purely agile proceeding (20%), and a consistent use of classic project management methods (9%) [33]. These findings reveal a widespread co-existence of traditional and agile approaches which in itself is rather tension-filled, with five main areas of tension: insufficient attribution of responsibilities between agile and traditional organization; different opinions concerning the extent and speed of the re-orientation of existing processes and structures; lack of employee participation in the restructuring process and lack of trust on the part of the employees; continued existence of silo mentality in the traditional departments; lack of sufficiently adapted organizational structures for an integrative handling of both co-existent approaches [20]. Findings from other surveys have similar implications: important features of agile structures are not well developed in many enterprises. This assumes the form of sub-average transparency of information, no particular eagerness to experiment, and a relatively underdeveloped degree of knowledge exchange [17].

These areas of tension indicate that frictions, transfer, and translation problems occur at the interfaces between agile and non-agile areas. Sociological work research has found that such problems can occur if agile teams are confronted with (or even domineered by) the prevailing control orientation in a company via hierarchical planning structures. In cooperations beyond the boundaries of a company, such interface problems have been observed if agile teams are permitted to work in self-organization for themselves but are continually within the grasp of the customer who can demand at any time that they have to comply with his or her changing requirements [48].

An important consequence of these tension areas is the emergence of new strains, stress factors, and risks in project work. They cannot primarily be reduced to single influence factors, as is traditionally the case in direct physical strain or negative factors in the work environment (noise, heat, dust). They are rather stress constellations developing from the combination of several factors acting in a certain working situation. An example is time and performance pressure. In project work, this pressure mainly emerges from discrepancies between pre-conceived targets and resources on one hand and unpredictable or hardly predictable changes or critical situations developing in the course of the project, as explained above. Another source of stress is the temporal limitation of projects. If only one project is considered, stages of work intensification and expansion of working time may be seen as transient phases. However, since in practice employees frequently have to work in several projects simultaneously, this rhythm cannot be "lived" and experienced. In addition, recurring work interruptions and quickly changing team compositions often prevent a productive team cooperation. These problems give rise to specific stress constellations engendering particularly mental strain and stress which may even lead to burn-out [40].

Surveys of agile project work have revealed that these stress constellations especially occur if only single elements of agility are implemented. In agile work, they can be mitigated by strengthening of self-organization and self-determination. Hence, an important requirement for leaders and managers in agile organizations is to promote and support the employees' self-responsibility and autonomy [20]. The team planning should be closely linked to the employees' experiences, and the estimation how long the completion of the tasks required will take should rely on the experience of the team members. Another important precondition for a self-responsible team consists of trusting and close cooperation relationships. They create a work climate that permits to understand mistakes not as individual failures attributable to individual workers but as a chance for shared review and learning processes of the whole project team. Finally, a stable project team is a factor that can reduce stress constellations. Stability can be supported if the employees are not deployed to several projects at the same time and if "troubleshooting" detachments to other departments or projects are avoided [4].

5.4 Self-Organization

The previous chapters frequently referred to self-organization as a crucial issue in all the dimensions of the New Work discourse, and indeed the question of extent and scope of self-organization pervades all topics of this discourse. This is especially valid for project work. An essential element of project work is innovative and creative working action. Hence, project work invariably has components of self-organization and autonomy. This feature explains in part the growing attractivity of this form of working. It is therefore understandable that the New Work discourse counts project work among the organization models of the future that are to replace the traditional hierarchical structures. In this discourse, self-organization is predominantly located on the individual level of the single worker, and there are frequent references to the discrepancy between the current state of work and the desires of the employees. Consequently, self-organization is conceptualized as an issue that is frequently demanded by workers but hampered by still domineering hierarchical structures in many enterprises [47].

In order to probe the causes for this discrepancy, it is necessary to transcend the individual level and to discuss the functions and consequences of self-organization on several levels within the company. Besides self-organized action in the design of the individual working tasks, the extent and scope of self-organization in cooperation with other employees has to be included. This leads to the question how team members shape their team structures in self-organization, and how project groups shape their cooperation with other groups and departments within the organization and also beyond it, e.g., with customers, in a self-organized way. Bringing together the observations on all of these levels, the question arises how self-organized individual action is related to the production of an overarching social order and organization. Thus, the interplay between individual self-organized actions of employees and strategic processes of steering and control on the part of the company comes into view. It becomes visible that there are intersections between individual processes of self-organization and strategical targets set by superordinate levels in order to achieve specific goals of the company, combined with the conscious application of control systems. This interplay also implies tensions and leads to the question how these tensions are handled [44].

Against this backdrop, the concept of *emergent self-organization* in management theories is an important idea to answer these questions. The idea is that the actions of the individual actors in the organization combine to create an overarching order without intentionally aspiring to do this [57]. There are different answers to the question if and how the individual actors have to be influenced by external parameters to achieve this order, ranging from the maxim to "respect self-organization" , to the assumption of an "implicit rationality", to the mission to create an adequate organizational culture. "In this perspective, the leader should set impulses, support thriving and flourishing, cultivate what already grows by itself" [5, p. 446]. This amounts to the idea that management's essential task is to create the framework that enables evolutionary self-organization processes. By "symbolic organizing" in this meaning, sense-giving, and self-making processes should be initiated. At the same time, it is assumed that these parameters promote and support flexible thinking and acting of employees, their openness for new developments, and their capacity to learn [42].

This understanding of self-organization, however, is not enough to answer the question of whether and how the demand for individual self-organization of the single employees will yield favorable effects for higher organizational structures, as project teams, departments, divisions, and finally the company as a whole—naturally a crucial question for an enterprise. The concept of *autonomous self-organization* deals with this question, combining the actions of individual actors with the intentional creation of an overarching social order [21]. The basic assumption is that individual actors and their actions are *intentionally* directed both to their individual interests resp. situations and to the emergence and warranty of an overarching order [53]. Usually this is accompanied by another assumption—often remaining implicit—that the aims of actions on the individual level and on the organizational level are not immediately identified. For this reason, an inherent characteristic of this approach may be that it shares the idea of the management setting criteria, norms, and ends from outside, thus creating a framework for the self-organized actions of employees [65]. Regarding the actors of autonomous self-organization, a distinction is made between actors in their role as individual members within the organization and the same actors in their role as conscious and goal-directed producers of the organization. In this latter role, their "organizational role", the actors act like a "management", assuming a point of view "from above" and confronting the actors in their individual roles like a "heteronomous organization" [13]. An important difference to the concept of emergent self-organization is that the employees *intentionally* participate in the creation of the organization. This enables them to introduce and negotiate conflictual and conflicting interests.

The questions of self-organized action on different levels are particularly acute in agile project work. Within a project team, self-organized work means that adaptations of the task distribution can be autonomously organized by the team members in the daily operational business. However, this does not resolve the problem of the relationship between this self-organization and the general rules and procedures of other organizational units, let alone the dynamic changes of external market conditions. The problem is particularly how the junctions and interfaces to non-agile processes within the company (administration, marketing, purchasing, sales) and beyond it (cross-company cooperation, customers) are (to be) organized [4]. This problem may lead to controversial issues in coordination and decision-making and even cause the failure of agile approaches [62]. The conflicting requirements endanger the "boundary regulation of self-organized project work," which is in turn a source for additional stress and strain on the part of the employees [8, p. 138].

Considering these problems, it must ultimately remain an open question if and how the demands for more employee self-organization can at the same time yield favorable effects for the overarching organizational structures (as the project team, the department, the company as a whole). When Bergmann tried to implement his approach in existing enterprises, his experiences pointed to the same problem: Although a number of companies were open for his concept, "the businessmen still wanted the workers to do exactly […] what was good for the company. Predictably this problem could never be really solved" [11, p. 318]. Consequently, self-organization and the range of autonomous action promised or warranted to the employees are always subject to the reservation that they can be constrained or even withdrawn if higher ends are at risk or missed. In these cases, more or less severe restrictions of individual self-organized ranges of autonomous employee action are to be expected. These ranges do exist but their extent, duration, and forms are subject to changes according to the negotiations of the actors involved, they are subject to the "dynamics of heteronomous organized self-organization" [48, p. 42].

5.5 Conclusion

The New Work discourse is predominantly not about utopian ideas of a revolution of the wage work system, but about a change of work within the framework of the existing conditions. The New Work discourse points to important developments in the world of work concerning new and alternative forms of work that also have been investigated by work sociologists, in some cases for a number of years. Both perspectives have in common that they focus both on the demands of the enterprises and the desires of the employees for self-organized design of work, particularly under the conditions of progressing digitalization. In the perspective of sociological work research, the process of an increasing subjectification of work opens up spaces for the employees to self-organize their own work but the very same process can also be a source of new forms of stress. The employees are getting new chances to shape their own working activities and work organization but at the same time they have to take a new responsibility for the results and products of their work. To put it in a simple formula: the employees are required to want as subjects what they shall do as employees, and the managers and leaders have to provide an adequate framework in a suitable fashion.

It is by no means simple to put this idea into practice. It will be necessary to consider the question how the persisting "old world" of pre-conceived bureaucratic and hierarchical organizational structures can be reconciled to the growing requirements of flexibility and fluidity of organizations and to the growing demands of employees for a self-organized and individually meaningful work. Because of the simultaneity of traditional routines and new trends, elements of organizational ambidexterity will be characteristic features for a phase of transition whose duration is hardly predictable as to now. The task is to deal with different organizational principles that exist simultaneously, with different velocities of co-existing company divisions, and, in project management, with the co-existence of traditional and agile methods.

The present process of digitalization offers favorable opportunities to tackle this task. Digital spaces are capable of assimilating a great variety of work arrangements. The relations of technology and work organization, the development and application of competencies, the creation of cooperation forms, all of these developments always constitute a social process. The decisive factors for the qualification (or else de-qualification) of the work of the future are the ways how the division of labor is conceptualized and how work organization is projected and shaped. These factors are also crucial for the question how much autonomy, collaboration, learning, and diversity will be possible in actual working activities. The art of work design will be to bring about a nexus of these factors that is productive both for the employees and the companies.

References

1. Arnold D, Steffes S, Wolter S (2016) Mobiles und entgrenztes Arbeiten. BMAS-Forschungsbericht 460, Berlin
2. Backhaus N, Tisch A, Wöhrmann A (2018) BAuA-Arbeitszeitbefragung. Vergleich 2015–2017. Bundesanstalt für Arbeitsschutz und Arbeitsmedizin, Dortmund, Berlin, Dresden
3. Baethge M (1991) Arbeit, Vergesellschaftung, Identität — Zur zunehmenden normativen Subjektivierung der Arbeit. Soziale Welt 42(1):6–19
4. Bahlow J (2020) Agile Teams – Anspruch und Wirklichkeit einer neuen Selbstorganisation. In: Porschen-Hueck S, Jungtäubl M, Weihrich M (eds) Agilität? Herausforderungen neuer Konzepte der Selbstorganisation. Hampp, München, Augsburg, pp 61–79
5. Bea FX, Göbel E (2008) Organisation. Theorie und Gestaltung. Lucius & Lucius, Stuttgart
6. Bea FX, Scheurer S, Hesselmann S (2008) Projektmanagement. Lucius & Lucius, Stuttgart
7. Beck K et al (2001) Manifesto for agile software development. http://agilemanifesto.org/

8. Becke G (2020) Agile Arbeitskonzepte – Zwischen Rationalisierung und gesundheitssensibler Gestaltung. In: Porschen-Hueck S, Jungtäubl M, Weihrich M (eds) Agilität? Herausforderungen neuer Konzepte der Selbstorganisation. Hampp, München, Augsburg, pp 127–149
9. Becker C, Kratzer N, Lütke Lanfer S (2019) Neue Arbeitswelten. Wahrnehmung und Wirkung von Open Space-Büros. ARBEIT 28(3):263–284
10. Berend B, Brohm-Badry M (2020) New Work: Souveränität im postdigitalen Zeitalter. Springer, Wiesbaden
11. Bergmann F (2017) Neue Arbeit, Neue Kultur. Arbor, Freiburg
12. Böhle F (2018) Arbeit als Handeln. In: Böhle F, Voß GG, Wachtler G (eds) Handbuch Arbeitssoziologie. Band 1: Arbeit, Strukturen und Prozesse. Springer VS, Wiesbaden, pp 171–200
13. Böhle F (2020) Selbstorganisation im Spannungsfeld von individueller Autonomie und übergreifender Ordnung. In: Porschen-Hueck S, Jungtäubl M, Weihrich M (eds) Agilität? Herausforderungen neuer Konzepte der Selbstorganisation. Hampp, München, Augsburg, pp 15–22
14. Breidenbach J, Rollow B (2019) New Work needs Inner Work. Vahlen, München
15. Brommer D, Hockling S, Leopold A (eds) (2019) Faszination New Work. 50 Impulse für die neue Arbeitswelt. Springer Gabler, Wiesbaden
16. Dalcher D (2014) What can project success, or failure, tell us about project management theory? In: Rietiker S, Wagner R (eds) Theory meets practice in projects. GPM Deutsche Gesellschaft für Projektmanagement, Nürnberg, pp 177–191
17. Dettmers S, Jochmann W, Hermann A, Zimmermann T, Knappstein M, Fastenroth L, Pela P (2020) Agile Unternehmen – Zukunftstrend oder Mythos der digitalen Arbeitswelt? Kienbaum & Stepstone, Dortmund, Düsseldorf
18. DGB-Index Gute Arbeit (ed) (2019) Jahresbericht 2019. Themenschwerpunkt Arbeitsintensität. Institut DGB-Index Gute Arbeit, Berlin
19. EBS, GPM (eds) (2010) Global project management survey: cultural, individual and organizational competence in project management. GPM Deutsche Gesellschaft für Projektmanagement, Nürnberg
20. Eilers S, Möckel K, Rump J, Schabel F (2018) HR-Report 2018. Schwerpunkt Agile Organisation auf dem Prüfstand. Hays, Mannheim
21. Göbel E (1998) Theorie und Gestaltung der Selbstorganisation. Duncker & Humblot, Berlin
22. Grunau P, Ruf K, Steffes S, Wolter S (2019) Homeoffice bietet Vorteile, hat aber auch Tücken. IAB Kurzbericht 11. Institut für Arbeitsmarkt- und Berufsforschung (IAB), Nürnberg
23. Hardering F (2017) Die Suche nach dem Sinn: Zur Zukunft der Arbeit. Aus Politik und Zeitgeschichte 67(26):4–10
24. Hays (ed) (2015) Von starren Prozessen zu agilen Projekten. Hays, Mannheim
25. Heidling E (2018) Projektarbeit. In: Böhle F, Voß GG, Wachtler G (eds) Handbuch Arbeitssoziologie. Band 2: Akteure und Institutionen. Springer VS, Wiesbaden, pp 207–236
26. Herbig B, Schneider A, Nowak D (2016) Gesundheit, Kommunikation und Leistung in Großraumbüros – Zusammenhänge mit Personenzahl, spezifischen Umgebungsbedingungen und allgemeinen Tätigkeitsmerkmalen. Wirtschaftspsychologie 18(3):71–81
27. Hoegl M, Muethel M, Gemuenden HG (2012) Leadership and teamwork in dispersed projects. In: Morris PWG, Pinto JK, Söderlund J (eds) The Oxford handbook of project management. Oxford University Press, Oxford, pp 483–499
28. Hofmann J, Piele A, Piel C (2019) New Work. Best Practices und Zukunftsmodelle. Fraunhofer IAO, Stuttgart
29. Kalkowski P, Mickler O (2015) Kooperative Produktentwicklung. Fallstudien aus der Automobilindustrie, dem Maschinenbau und der IT-Industrie. edition sigma, Baden-Baden
30. Kim J, de Dear R (2013) Workspace satisfaction: the privacy-communication trade-off in open-plan offices. J Environ Psychol 36:18–26
31. Klaffke M (ed) (2016) Arbeitsplatz der Zukunft. Gestaltungsansätze und Good-Practice-Beispiele. Springer Gabler, Wiesbaden

32. Kleemann F, Voß GG (2018) Arbeit und Subjekt. In: Böhle F, Voß GG, Wachtler G (eds) Handbuch Arbeitssoziologie. Band 2: Akteure und Institutionen. Springer VS, Wiesbaden, pp 15–57
33. Komus A, Kuberg M, Schmidt S, Rost L, Koch CP, Bartnick S, Graf E, Keller M, Linkenbach F, Pieper C, Weiß L (2020) Studie Status Quo (Scaled) Agile 2019/2020. Hochschule Koblenz, Koblenz
34. Kratzer N (2003) Arbeitskraft in Entgrenzung. Grenzenlose Anforderungen, erweiterte Spielräume, begrenzte Ressourcen. edition Sigma, Berlin
35. Kratzer N (2011) Von der Stechuhr zur Vertrauensarbeitszeit? Betriebliche Arbeitszeitpolitik als Beispiel für den Wandel von Herrschaft. In: Bonß W, Lau C (eds) Macht und Herrschaft in der reflexiven Moderne. Velbrück, Weilerswist, pp 219–243
36. Kratzer N (ed) (2017) Open Space. Oder was? Wandel der Büroarbeit. Eine Trendanalyse im Rahmen des Projekts PRÄGEWELT – Präventionsorientierte Gestaltung neuer (Open Space-) Arbeitswelten. ISF München, München. www.praegewelt.de
37. Kratzer N (ed) (2020) Open Space. Besser machen. ISF München, München
38. Kratzer N, Nies S (2009) Neue Leistungspolitik bei Angestellten, edition sigma, Berlin
39. Kühn T, Niedermeier A, Babic E (2019) New Work und die Bedeutung von Organisationskultur, Team- und Beziehungsarbeit – eine Mitarbeiterbefragung. Organ Superv Coach 26(2):159–172
40. Latniak E (2014) Arbeitsgestaltung bei Projektarbeit – widersprüchliche Anforderungen, Belastungen und Ressourcen. In: Vedder G, Pieck N, Schlichting B, Schubert A, Krause F (eds) Befristete Beziehungen. Menschengerechte Gestaltung von Arbeit in Zeiten der Unverbindlichkeit. Hampp, München, Mering, pp 135–157
41. Lott Y (2019) Weniger Arbeit, mehr Freizeit? Wofür Mütter und Väter flexible Arbeitszeitarrangements nutzen. WSI-Report 47, Düsseldorf
42. Malik F (2006) Strategie des Managements komplexer Systeme. Ein Beitrag zur Management-Kybernetik evolutionärer Systeme. Haupt, Stuttgart
43. Matuschek I (2010) Konfliktfeld Leistung. Eine Literaturstudie zur betrieblichen Leistungspolitik. edition sigma, Berlin
44. Mayntz R (2019) Emergenz in Philosophie und Sozialtheorie. In: Greve J, Schnabel A (eds) Emergenz. Zur Analyse und Erklärung komplexer Strukturen. Suhrkamp, Frankfurt am Main, pp 156–186
45. Merrow EW (2011) Industrial megaprojects. Wiley, Hoboken
46. Monz A, Vogl G (2020) Wandel der ortsgebundenen mobilen Arbeit durch Digitalisierungsprozesse. Zeitschrift für Arbeitswissenschaft 74(3):178–187
47. Munsch JP (2020) Standpunkt: Einführung der Selbstorganisation. In: Wörwag S, Cloots A (eds) Zukunft der Arbeit – Perspektive Mensch. Springer Gabler, Wiesbaden, pp 255–261
48. Neumer J (2020) Selbstorganisation gestern und heute – ein qualitativer Umbruch im Umgang mit Unsicherheit? In: Porschen-Hueck S, Jungtäubl M, Weihrich M (eds) Agilität? Herausforderungen neuer Konzepte der Selbstorganisation. Hampp, München, Augsburg, pp 23–46
49. New Work Charta (2019) https://humanfy.de/new-work-charta
50. Nies S (2015) Nützlichkeit und Nutzung von Arbeit. Beschäftigte im Konflikt zwischen Unternehmenszielen und eigenen Ansprüchen. Nomos, Baden Baden
51. Nies S, Roller K, Vogl G (2017) Mobilität und Leistung. Dienstreisende im Trade-Off zwischen Verdichtung und Entgrenzung. ARBEIT 26(2):173–191
52. Nowak I, Klußmann C, Menz W (2019) Normalität in Bewegung. Beschäftigte zwischen erweiterter Erreichbarkeit und digitaler Konnektivität. ARBEIT 28(3):307–332
53. Oswald A (2019) Self-organization. In: Oswald A, Müller W (eds) Management 4.0. handbook for agile practices. Release 3.0. Books on Demand, Norderstedt, pp 55–58
54. Petendra B (2015) Räumliche Dimensionen der Büroarbeit. Eine Analyse des flexiblen Büros und seiner Akteure. Springer VS, Wiesbaden
55. Peters S, von Garrel J, Düben A, Dienel HL (2016) Arbeit – Zeit – Souveränität. Eine empirische Untersuchung zur selbstbestimmten Projektarbeit. Hampp, München, Mering

56. Piele C, Piele A (2018) Flexible Arbeitszeiten – Arbeitszeitmodelle und Flexibilitätsanforderungen. Fraunhofer IAO, Stuttgart
57. Probst G (1987) Selbst-Organisation. Ordnungsprozesse in sozialen Systemen aus ganzheitlicher Sicht. Paul Parey, Berlin
58. Richter G, Cernavin O (2016) Büro als Treiber gesundheitsförderlicher und produktiver Arbeitsbedingungen. In: Klaffke M (ed) Arbeitsplatz der Zukunft. Gestaltungsansätze und Good-Practice-Beispiele. Springer Gabler, Wiesbaden, pp 82–101
59. Rolfö L, Eklund J, Jahncke H (2017) Perceptions of performance and satisfaction after relocation to an activity-based office. Ergonomics 61(5):644–657
60. Rump J, Schabel F, Alich D, Groh S (2011) Arbeits- und Organisationsstrukturen in Bewegung. Eine empirische Studie des Instituts für Beschäftigung und Employability (IBE) im Auftrag von Hays. Ludwigshafen
61. Sanderson J (2012) Risk, uncertainty and governance in megaprojects: a critical discussion of alternative explanations. Int J Project Manage 30(4):432–443
62. Schaffitzel N, Raue S, Edelkraut F, Lang R (2019) Learnings and guidance for the implementation of agility. In: Oswald A, Müller W (eds) Management 4.0. handbook for agile practices. Release 3.0. Books on Demand, Norderstedt, pp 258–287
63. Schermuly C (2019) New Work – Gute Arbeit gestalten. Psychologisches Empowerment von Mitarbeitern. Haufe, Freiburg, München, Stuttgart
64. Söderlund J (2012) Theoretical foundations of project management. In: Morris PWG, Pint JK, Söderlund J (eds) The Oxford handbook of project management. Oxford University Press, Oxford, pp 37–64
65. Stadelbacher S, Böhle F (2016) Selbstorganisation als sozialer Mechanismus der reflexiv-modernen Herstellung sozialer Ordnung? In: Böhle F, Schneider W (eds) Subjekt – Handeln – Institution. Vergesellschaftung und Subjekt in der Reflexiven Moderne. Velbrück, Weilerswist, pp 324–356
66. Van Steenbergen E, van der Ven C, Peeters M, Taris T (2017) Transitioning towards new ways of working: do job demands, job resources, burnout and engagement change? Psychol Rep 121(4):736–766
67. Thamhain H (2013) Managing risks in complex projects. Proj Manag J 44(2):20–35
68. Vogl G (2019) Präventionsorientierte Gestaltung mobiler Arbeit: Eine Einführung. In: Universität Oldenburg (ed) Mobile Arbeit gesund gestalten – ein Praxishandbuch. Oldenburg
69. Vogl G, Kratzer N (2015) Zuhause – Unterwegs – beim Kunden. Wenn die Arbeit viele Orte hat. In: Kratzer N, Menz W, Pangert B (eds) Work-Life-Balance – eine Frage der Leistungspolitik. Springer VS, Wiesbaden, pp 171–192
70. Wagner R (2019) Self-organization: a new paradigm for project-oriented work. PMR (1), July 26, pp 19–21
71. Wald A, Spanuth T, Schneider C, Futterer F, Schnellbächer B, Schoper Y (2015) Makroökonomische Vermessung der Projekttätigkeit in Deutschland. GPM Deutsche Gesellschaft für Projektmanagement, Nürnberg
72. Wörwag S, Cloots A (2020) Gut zu wissen: Wie soll sich die Arbeit aus Sicht der Mitarbeitenden entwickeln. In: Wörwag S, Cloots A (eds) Zukunft der Arbeit – Perspektive Mensch. Springer Gabler, Wiesbaden, pp 127–162

Part II
Self-Organizing in Projects and the People Competences

Chapter 6
Self-Awareness, Assessment, and Organization with Personal Agility

Raji Sivaraman and Michal Raczka

Abstract In this paper, the question we ask is—"can personal agility be a path to self-organization in projects"? We elaborate the path to self-organization in projects through a Personal Agility Lighthouse Model (PALH™) model that Raji Sivaraman and Michal Raczka have created. This model consists of seven different agilities that are essential for the same. The seven agilities in this model are education agility, change agility, emotional agility, political agility, cerebral agility, learning agility, and outcomes agility. We draw the connections that question whether personal agility is necessary for organizational agility. The synthesis of the diverse concepts within the literature on project management which includes agility and the relationship, based on our PALH™ framework, is used in achieving success with self-organization in projects. Starting from self-awareness, the path then leads to maturity, which will mean honing your Personal Agility. We review ideas and experiences at the nexus of self-assessment through individual observation, how they lead to the view of this dynamism towards the well-being of self-organized teams in projects. Creation of impressions will change the probabilities of the success of projects and teams. Delving deep into oneself enhances self-knowledge, furthering it to departments and ultimately the whole organization.

Keywords Personal agility · Open decision framework · Self-assessment · Self-awareness · Self-organization · TTM

R. Sivaraman (✉)
ASBA LLC, Singapore/USA, Singapore
e-mail: raji.sivaraman@asba-llc.com

Feliciano School of Business, Montclair State University, Montclair, USA

M. Raczka
mBank S.A., Warsaw, Poland
e-mail: michal@raczka.biz

R. Ding et al. (eds.), *Research on Project, Programme and Portfolio Management*, Lecture Notes in Management and Industrial Engineering,
https://doi.org/10.1007/978-3-030-86248-0_6

6.1 Introduction

The premise of this paper is to introduce and inject the Personal Agility Lighthouse Model (PALH™) created by Raji Sivaraman and Michal Raczka into project management looking through various lenses. This model was created after exploring various scenarios by doing workshops around the world which included a self-assessment index [1] as well as exercises to learn more about international industry practices and thoughts on personal agility. The model includes seven agilities. We start with **Education Agility** because it is imperative to learn to feel the pain points of others. Cross-functional areas of collaboration and teamwork are helped with this agility. This means there is going to be changes that an individual needs to deal with. That takes us to **Change Agility**. When change happens, emotions start to play where **Emotional Agility** needs to be dealt with. Emotions are the main ingredient for politics to surface. Therefore, **Political Agility** has to be addressed at this point. The cerebrum which comprises of the brain and mind, the two most agile parts of an individual, needs to be aligned and sharpened to deal with politics and quick responses. Thus, **Cerebral Agility** takes shape here. We need to learn to make ourselves go through continuous improvement strategies to relearn all of the above agilities, which is the **Learning Agility**. Consequently, when these agilities or to the extent of the number of agilities needed are honed will sprout forth **Outcomes Agility** achieving more clarity and taking the bar to the next level.

Several academia and practitioner entities also used these seven agilities and found them to be very useful [2]. That was the deciding factor for us to land on these seven agilities even though there are plenty others. Sailing along towards the lighthouse, we would like to generate an insight into the propelling of the self, to reach the organizational competence with self-determination, and controlled management. This means measuring actual performance, taking corrective actions and to be engaged in a culminated 360-degree environment. Based on these settings, we indulge in emphasizing the principles [3] behind the seven agilities in this paper that can lead the path to self-organized projects. The principles are as follows:

1. We need to constantly keep advancing ourselves to reroute our capabilities—*Education Agility.*
2. We need to relearn ourselves to improve competencies—*Change Agility.*
3. We have to treat others with deference—*Emotional Agility.*
4. We need transparency for organizational growth—*Political Agility.*
5. We need to focus on organizational goals not the impediments of alterations—*Cerebral Agility.*
6. We need to have the courage to say "I don't know"—*Learning Agility.*
7. We need to commit to excel in the outcome that is foreseen—*Outcomes Agility.*

Some of the takeaways from the Royal Norwegian Air Force written by the authors, Antonacopoulou et al. [4] are in the same vein, where responsiveness, configuration, and concentration are vital in the defence arena. As such, self-organizing teams needs to include problem-solving perspectives and behaviors of an individual in a project that can contribute to the team and ultimately the project. To that end, the importance of one's mindset, culture and working atmosphere will be explored in the following sections with practical applications of each of the seven agilities in the model. These prerequisites are the success factors for self-organized teams. Perspectives for today's international challenges in project management and in the agile world with diverse environments and industries are further probed in this paper. We decided to expand our research to explore and find out more about how personal agility helps different verticals, spaces, industries, etc. This is because, to our knowledge, we don't find much research in the area of personal agility tying in with self-awareness, self-assessment, and self-organization.

6.2 Path to Self-Organization Through Personal Agility

"To know thyself is the beginning of wisdom."—Socrates [5]. In Fig. 6.1, the path we take is to make oneself be in an awareness mode at all times with three of the agilities, education, emotional, and political agility. Once these three are honed any situational awareness can help close the gap between the C suite and the individuals below them at any level. This is enabled by the self-assessment that one does with the next three agilities, namely change, cerebral, and learning agility. Relevance of these skill sets is paramount to achieve the outcomes that is envisioned.

Fig. 6.1 Path to self-organization through Personal Agility

6.2.1 Self-Awareness

A concept that was elaborated in a thesis by Amar [6] where he writes about the ambiguities of the situations in non-collocated information technology projects that are dealt with prompt receptiveness to lessen unwanted consequences is an example of self-awareness. He further goes on to explain the deployment of the doctrines of agility to control the processes of projects. Extrapolating this thought, if one can control the controllables, the paradigm shift of the agile concept and the multiple ways of working in a self-organized environment will be smooth.

Therefore, we opine that when one needs to take the step to grow awareness of oneself; the three agilities which in the PALH™ Model that will help achieve this are as follows:

6.2.1.1 *Education agility* which stipulates that when one puts oneself into the shoes of another and feels the pain points, one becomes antifragile.

6.2.1.2 *Emotional agility* is required to do this as one broadens oneself via awareness, coping skills, regulating difficult feelings, killing skepticism, and tolerating challenging situations in setting goals.

6.2.1.3 *Political agility* then silently is a contributing factor. Political awareness and cautiousness is a must, which emerges with distinctive divide may it be, between an individual, other stakeholders, departments, etc.

Being aware of oneself based on the above three agilities helps trigger the actions to make retrospective steps to build the right and guided self-assessment.

6.2.2 Self-Assessment

In a paper published in the Harvard Business Press, Thomas [7] explores the capability to assess oneself to be agile and expand on one's personal strategy to do so. The main objective given here is the starting point of our model where a multidimensional self-assessment is used as a radar to gauge the lay of the overall self. This leads to using the binoculars to further understand the workings of the self and organize the directional and team economy with projects and its transformations.

Once awareness is on its growing path, the following agilities in the model need to take shape to clear the path from diversions, disruptions, and unexpected roadblocks. Here we bring in three more agilities, namely:

6.2.2.1 *Change Agility* means working when we have options. Having options is a luxury, but at some point, you need to take a decision and choose one. In order to continuously hone and maintain this agility you need to communicate what the vision is, what decisions were taken, and to what options we are open to.

6.2.2.2 *Cerebral Agility* creeps into the path now as emotions involve the brain and the mind, two of the most agile parts (as mentioned in Sect. 6.1) that is now seen in the horizon as Cerebral Agility. When your mind and the brain are working in an agile manner, it fetches new viewpoints, shows ease in difficult or obscure situations, coming up with timely guidance. Here is the mantra that combines the ingredients to maturity—know more, read more, be curious—therein lies your adaptive achievement!

6.2.2.3 *Learning Agility* shows the path to accelerate to self-organization by having the courage to admit that one does not know everything. One needs to be brave enough to accept the fact that one can be wrong about one's assumptions. This will open up pathways to new discoveries and self-organizational opportunities. The key concept of always learning more is the Harvard Business Review paper's author, Eurich's [8] argument. The author further extends it to fostering the mindset where self-awareness is a rope that is constant and can never be let go so that the improvement of oneself can always be on a upward direction. This is precisely the premise under which Learning Agility in the model is delved into and created.

The above-mentioned three agilities are actionable and lead to high maturity and adaptability in order to achieve self-organization.

6.2.3 Self-Organization

Maturity is a prerequisite for self-organization. In a world that is so uncertain in 2020 with COVID-19, each individual has to spin oneself in ways that they may never have done before. Therefore, using self-organization as a tool to design, engage, challenge, inspire, and transform professionals is vital. Markus et al. [9], discuss about the various means in which people whose line of work requires one to "think for a living" to self-organize rather than being managed. Twenty years back, this resounding mention from the MIT Sloan Management review could not serve better than these unprecedented times of COVID-19.

Baker and Thomas [10] have examined the inclusion of agility and its various advantages to the workforce and the benefits it brings to the hegemony and headship more than a decade ago. Even though the authors had written this in their paper a while back, this holds true for eternity as we see that change is a norm and being agile is the glue to run projects smoothly. Tying all of the six agilities mentioned in self-awareness and self-assessment, the self-organization in the projects arena leads the path to better oneself. It helps to excel to the next level, to better outcomes, and to the seventh personal agility flavor—Outcomes Agility. It is always good to be enterprising, inspiring, and pushing to excel beyond one's limit. It is ok for you to never reach the lighthouse, but it is not ok to not improve and strive for excellence. It is ok to understand that excellence in outcomes can be treated as an object that moves in a forward direction constantly inspiring you to tread one's paths successfully.

Using the three concepts above, we will describe some applications of the seven agilities in projects, teams, organizations, etc.

6.3 Observations Tying the Seven Agilities and Self-Organization

In this section, we delve into the two different angles of how decision-making plays a part in self-organizing teams and we also probe into how self-organizing in a team takes its root from an individual. These are critical in that, they help the projects as an arena for self-organization to reduce turbulences from setting sail to reaching the lighthouse.

6.3.1 Decision-Making in Self-Organizing Team

According to Wikipedia, [11] Decision-making is the process of identifying and choosing alternatives based on the values, preferences, and beliefs of the decision-maker.

As the idea behind the development of the self-organization with linear and non-linear behaviors is brought forth in a paper written by Laycraft [12] from the University of Calgary, its advent serves ongoing processes and experiences. Her article further elaborates that the idea of self-organization has been introduced to developmental psychology, especially to the emotion-cognition relations, personality development, adolescent development, creativity, brain development, and others. This is directly in tandem with the model.

The premise of decision-making lies in the concepts of

- always wanting to learn more
- relearning competencies
- treating others with deference

All of the above are just a mirror of the seven agilities, education agility, change agility, emotional agility, political agility, cerebral agility, learning agility, and outcomes agility. These seven waves drive a multitude of the personal agility traits in each team member from the initiation of a project to the close of a project. Magali and James [13] in a case study stipulate that pre-eminence of procedures should illuminate self-organization and the way one manages oneself. This research done by the authors triggered us to interview more people (in the practitioner and in the academic world globally) and found that using and honing the seven agilities help achieve high maturity levels in self-organized projects.

This segways to teams that need to have an opportunity to make progress and growth in decision-making. It can be obtained through maturity and decision-making as a skill that needs training. Teams that combine a mix of highly honed and trained Personal Agility traits can easily be self-organized to make better decisions as shown in Fig. 6.2.

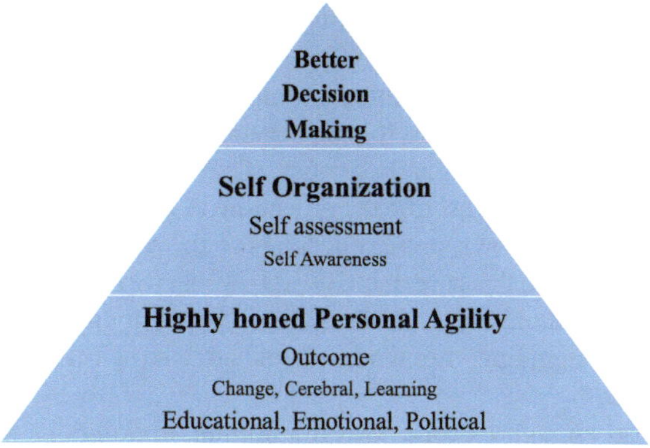

Fig. 6.2 Hierarchy for better decision-making

We will explore a few examples of decision-making tying in to the self-organization and Personal Agility.

While making decisions, there needs to be a clear goal, with all of the encompassing competencies (meaning the seven agilities mentioned before) with external roles and internal roles that self-organizes itself to its best fit. This ensures that people are able to make decisions for the good of the organization, rather than decisions for their own good. Decision-making processes are very complex. This is very pertinently pointed out by Berisha et al. [14] in their paper where they converse about the methods and flair of the treatment of decision-making and how it should be assimilated into the vigorous investigations of outcomes and choices of the decisions. Thus, it is especially important that one looks into the complexities of decisions not just about who decides. We need to understand that collaborative decisions don't apply to every context and self-organization doesn't apply to every framework. Making collaborative decisions in a self-organizing team does not mean that everyone decides.

When under pressure, decision-making processes such as Business Continuity Planning need to have the following embedded in the process like a military setting:

- believing in strong leadership
- clear directions
- directives and orders that needs to be followed to the T
- minimal discussions
- flawless rules

Unfortunately, there is no room or space for self-organization in this scenario. This would mean that *cerebral agility* won't have room here or maybe at a lower level where it may be used by the top level where the orders flow from.

Self-organization applies to the context of creativity, idea generating, and making decisions about which route to choose, and about what hypothesis to check. Teams must stand in, against the biggest challenge which is to overcome the fear of deciding.

Teams with highly honed Personal Agility, especially Learning Agility, are brave enough to fail and learn. When we learn fast, and check our hypothesis fast, we can also fail fast. Failure is just the beginning. Failure creates new opportunities to learn.

Applications of game theory are found everywhere—in business, politics, diplomacy, between family, friends, and enemies. The fundamental ideas of strategic interaction and game theory describe the procedure of decision-making in situations where the outcome of your own strategy depends on the actions and strategies chosen by your "rivals" as well. In this case, it is COVID-19 that we take as an example. Theoretic concepts, such as dominance, Nash equilibrium, credibility, commitment, asymmetric information, signaling, moral hazard, and adverse selection, steers oneself to strategic outcomes.

While it is bad enough to cope with the normal workings in any given arena, where the workforce is scattered in several time zones and geographic areas, adding another layer of multiple skills within a shell makes it harder to have agile inclusion in the mix [15]. This is proven to be a very true statement in a pandemic such as the COVID-19, where self-organized team in a project is the norm. A simple example is when you see large conglomerates have their employees perform their own will without directives from anywhere. They self-organized to come up with ideas to turn their existing tools, methodologies, products, ingredients, etc. into the most required and sought after items. Some of these are sanitizers, ventilators, masks, face shields, etc. It is not a top-down approach in this scenario. The C levels and the upper management did not come up with these designs and concepts. It is the people on the ground who said "we know we have it, we know what can be tweaked and how; so let us". They then pitched it to the upper management. Boeing, Rolls Royce, and many others are examples of this self-awareness, assessment, and organization. Therefore, *cerebral agility* is at its peak here. All of these would not have been possible if education, change, political, emotional, and learning agility were not honed to bring the outcomes agility to fruition.

Another great example of how self-organization and personal agility create better decision-making process is open-source projects. The vast majority of Fortune 500 companies consume open source at some level or another. From the space program, to cryptocurrency, to gaming, open source has taken industries by storm. One interesting statistics highlighted by Liu [16] is that the open-source industry is to reach 30 billion by 2022 from 17 billion in 2017. This statistics points in the direction of how by nature, open-source projects are self-organized and involves thousands of contributors. One of the challenges [17] is how to make decisions in such an environment. The Open Decision Framework that Shah [18] describes is about a process for making transparent, inclusive decisions in organizations that embrace open-source principles. Since, open decision-making is transparent, inclusive, and customer-centric, it guides you to think that it is done in a self-organizing environment.

The Open Decision Framework illustrates an open-source way of working, taking five open-source principles—open exchange, participation, meritocracy, community, and "release early, release often"—and putting them into practice. These five principles can be mapped to the seven principles of the PALH™ model. For example, meritocracy is mapped to the second principle "We need to relearn ourselves to improve competencies – Change Agility". Self-organization in open-source projects needs frames, the same is true for Personal Agility as well. These frames for Personal Agility are the principles of the model. Having principles and honing them makes us more mature, thus our decision can be better, conscious, adaptable, and drive self-organizing teams towards success.

6.3.2 Self-Organizing in a Team Stems from an Individual

In a Research and Development setup, the usual expectation is—"I don't need to be told by someone, what to do, where to look, when to start, stop etc." This is because they are already in a self-organizing environment and they are constantly aware of themselves and are assessing themselves. Objectives will be defined but still ample amount of flexibility surrounds them which is self-organization at play.

On the other hand, an application development team or new employees fresh from college who have not built up enough or no maturity [19] cannot be self-organized. They probably are better off in a setting of an operational (execution of the set process) outfit. This setting is easier for them to manage themselves along with the team. It can also be beneficial for the project by using tools like kaizen, six sigma, so on, and so forth. Guard rails may be erected here and there for finances, human resources, etc. But self-awareness and occasional checks and balances for only the boss to see may be done adhering to protocols. Other than that they do self-organize. Having said this self-assessment and awareness is done a lot in varied ways within each stage, iteration, phase, etc.

Time to Market (TTM) is the duration needed to bring a product from its conception to its availability for sale. McKinsey [20] in their quarterly issue in 2010 emphasized that the agility when at its peak has its advantages to grow profits, have contented stakeholders, better functioning and productivity, and a much quicker time to market. *This is the premise under which the* model brings out the need for honing the seven agilities to reach organizational agility. This means that to satisfy customers, one needs to hone self-awareness of one's strengths and weaknesses. Self-awareness in our minds can be honed if the *education, emotional, and political agility* are honed. Once awareness is peaked, then self-assessment with the aid of *change, cerebral, and learning agility* can improve operational and other efficiencies leading to the *outcomes agility* which is TTM in this example as the outcome.

Self-organization is essentially the direction that organizations lean towards these days. They are not going to disappear anytime soon say Buhse and Stamer [21]. Self-organization succeeds because it emphasizes interdisciplinary peer collaboration that works more effectively than hierarchical management. Companies are turning to self-organization models in order to reduce costs and speed, which in this case shows the direct impact on TTM.

While the number and importance of projects are growing for organizations, there is an increasing pressure on organizations to change their way of working. Self-organizing teams in project arenas being one of them. Globalization, disruptive technologies, and business models are driving this change, requiring organizations to be adaptive, agile, and quickly respond to the challenges at hand. Another perspective is the millennials entering the workplace. They have expectations and viewpoints about work autonomy, development, work–life balance, and much more. These drivers for change will have difficulties to survive in future if not responded to as and when they crop up. Therefore, emergence of the fields of interest in practice and research is growingly seen as self-organizing within and throughout projects.

A number of agile project management approaches and organizational design principles highlight the principles and use of self-organization. One such instance is where Helmel [22] elaborates on the necessity of a robust agile management to ensure the agile corporate design so as to encourage the workforce to bring forth their full potential, capabilities, and know-how.

The ups and downs of the same can be attributed to *education, change, political, emotional, cerebral, learning, and outcomes agility*. The factors that determine these can be

1. Shared goals vs individual goals
2. Team competencies vs individual competencies
3. Leadership styles
4. Motivation meters and gauges
5. Cultural gaps
6. Process pitfalls
7. Strategy alignment.

6.4 Concluding Note

Even though personal agility is the propeller to sail for any ship to move, can this propeller be a forceful injection for it to be included? As we see the COVID-19 situation where agility is not a choice anymore. It is a must. Do we need to wait until that critical moment to make the paradigm shift towards this agility? Certainly not, as the capability and the capacity by honing personal agility would have served the situation much better than the sudden injection.

Ending on the *outcomes agility* note, we feel that *outcomes agility* can be achieved when one is aware of oneself, takes into account the awareness within oneself and the surroundings and environment around oneself to be as organized as possible to achieve the goal and reach one's destination in full form.

To sum up our thoughts, if we ask the following questions as an example among many others, we will be in a great position to pave our path to self-organize ourselves in making our teams and ourselves in a better state for success. Using the seven agilities described in this chapter, the questions one should ponder to reach their desired lighthouses are as follows:

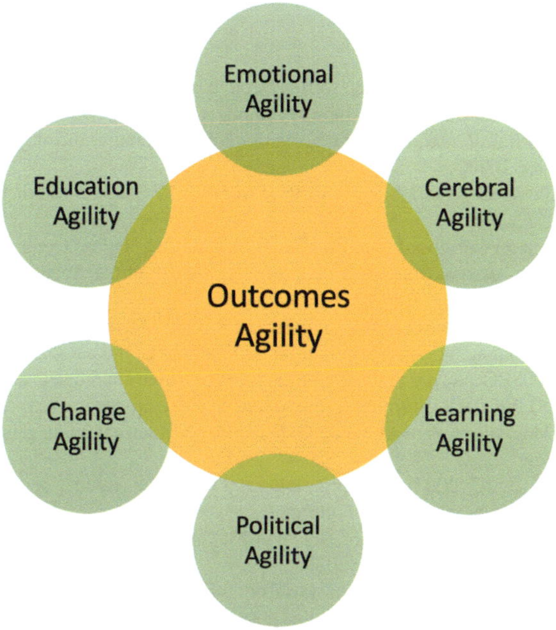

Fig. 6.3 The circle of personal agility

1. Do I have an open mind and am I listening properly? ***Educational Agility***.
2. Am I taking charge of my moods, thoughts, and behaviors? ***Emotional Agility***.
3. Do I comprehend the external and internal scenarios? ***Political Agility***.
4. Am I realistic about what I can achieve with the cards played? ***Change Agility***.
5. Do I respond quickly and aptly to challenges I encounter? ***Cerebral Agility***.
6. What are my capabilities and aptitudes? ***Learning Agility***.
7. Am I producing sustainable outcomes? ***Outcomes Agility***.

As such, the outcomes agility is dependent on the seven agilities as shown in Fig. 6.3, the continuum of the circle of personal agility. The outcomes agility is accentuated as the big circle in the center as the other agilities are the feed into the consequence of honing them, choosing them, and deploying them appropriately. For example, a lighthouse does not have to be tall all the time. It can be short and small as well, as long as its purpose is achieved. A simple example is the Robbins Reef Lighthouse in Bayonne, New Jersey.

Wishing everyone safe sailings towards one's desired lighthouses!!

References

1. Sivaraman R, Raczka M (2020) Discoveries through personal agility
2. Sivaraman R, Raczka M (2017) P, Industry applications. AgilityDiscoveries. http://agilitydiscoveries.com/articles/. http://agilitydiscoveries.com/personal-agility-lighthouse-workshop/. Accessed 29 Dec 2020
3. Principles of Personal Agility Lighthouse™ (PALH™) Model. http://agilitydiscoveries.com/. Accessed 29 Dec 2020
4. Antonacopoulou EP Moldjord C, Steiro TJ, Stokkeland C (2019) The new learning organisation: PART II—lessons from the Royal Norwegian Air Force Academy. Learn Organizat 27(2)
5. Socrates Quote. https://www.azquotes.com/quote/865451. Accessed 29 Dec 2020
6. Amar H (2018) Managing risks in Virtual-Agile IT Projects: the paradigm of responsiveness
7. Thomas RJ (2008) Exploring your capabilities: begin with a candid self-assessment. HBR
8. Eurich T (2018) What self-awareness really is and how to cultivate it. Harvard Business Review
9. Markus ML, Manville B, Agres CE (2000) What makes a virtual organization work? Harvard Business Review
10. Baker SW, Thomas JC (2007) Agile principles as a leadership value system: how agile memes survive and thrive in a corporate IT culture
11. Wikipedia Contributors definition of Decision-Making (2019) https://en.wikipedia.org/wiki/Decision-making. Accessed 29 Dec 2020
12. Laycraft K (2019) Decision-making as a self-organizing process. Ann Cogn Sci
13. Magali S, James L (2019) Self-organizing is not self-managing: a case study about governance challenges in an agile IT unit and its scrum projects. HICSS
14. Berisha G, Pula JS, Krasniqi B (2018) Convergent validity of two decision making style measures, JDDM
15. Theodoridou P (2017) Knowledge integration in self-organizing teams—a practice-oriented perspective. Göteborgs universitets publikationer
16. Liu S (2008) Projected revenue of open source services from 2017 to 2022, Statista
17. Taubert NC (2008) Balancing requirements of decision and action: decision-making and implementation in free/open source software projects. Institut für Wissenschafts- und Technikforschung
18. Shah S (2019) Open Decision Framework for open decision making, Open Practice Library
19. Popovic A, Hackney R, Coelho P, Jaklic J (2012) Towards business intelligence systems success: effects of maturity and culture on analytical decision making, Science Direct
20. Sull D (2010) Competing through organizational agility. Mckinsey quarterly
21. Buhse W, Stamer S (2008) The art of letting go enterprise 2.0. iUniverse
22. Helmel C (2019) An integrated agile organizational design and its impact on a faster response to changing customer needs: the case of INGs one agile way of working, Universitätsbibliothek TU Wien

Chapter 7
Client Experience on Projects

J. R. Turner

Abstract The topic of customer experience is receiving interest in marketing. It is believed vendors who manage customer experience achieve better results than those that do not. Work to date has been done in retail. We consider client experience on projects. Nobody has previously researched this topic, but several authors have written on issues relevant to it. We review that literature. We interviewed several clients, who have experience of interacting with contractors through the project life cycle. We report the results of our interviews. We find clients on projects have similar experiences to customers in retail. But on projects the client controls the interactions whereas in retail it is the vendor who controls the interactions. We suggest on projects we should also consider contractor experience. On projects, maintaining interaction between the client and contractors leads to better performance.

Keywords Client experience · Contractors · Project life cycle

7.1 Introduction

Customer experience is a topic generating interest in the marketing literature [17]. Lemon and Verhoef [17] suggest many firms, including KPMG, Amazon, and Google, have customer experience managers responsible for managing the experiences of their customers. Klaus [13] suggests organizations that manage customer experience achieve better results than those that do not. Pine and Gilmore [23] address the importance of experience in the post-modernist world, and suggest organizations can benefit from creating enduring customer experiences. De Keyser et al. [4] suggest the concept of customer experience can be traced back to Smith [25] and Keynes [12], and identify that customer experience has received attention in the

J. R. Turner (✉)
Europrojex, Wildwood Manor Close, East Horsley, Surrey KT24 6SA, UK
e-mail: rodneyturner@europrojex.co.uk

Department of Civil Engineering, University of Leeds, Woodhouse Ln, Woodhouse, Leeds LS2 9DY, UK

R. Ding et al. (eds.), *Research on Project, Programme and Portfolio Management*, Lecture Notes in Management and Industrial Engineering,
https://doi.org/10.1007/978-3-030-86248-0_7

philosophy, psychology, and sociology literature. Lemon and Verhoef [17] define customer experience as.

> a multidimensional construct focusing on a customer's cognitive, emotional, behavioural, sensorial and social responses to a firm's offerings during a customer's entire purchase journey.

To date, work in customer experience has been in retail. Initially, researchers researched at B2C, looking at experiences individuals have when they purchase goods [17]. More recently, researchers have investigated B2B, suggesting businesses go through similar experiences ([19]; Kuppelweisser and Klaus 2019). Howard and Sheth [10] suggested that customers go through a journey as they purchase an item, consisting of three phases:

- Pre-purchase,
- Purchase, and
- Post-purchase.

In pre-purchase, they recognize their need and search for options; in purchase they make a choice, order, and pay money, and in post-purchase they consume or use the item, and may make service requests. At touchpoints through those three phases, the purchaser makes contact with the vendor, and has cognitive, emotional, behavioral, sensorial, or social responses to the vendor's behavior and offerings. Those responses influence the customer's experience. In retail it is the supplier that controls the interactions. The customer seeks the interactions, but the supplier controls them.

In this paper, we wish to extend this work to investigate customer experiences on projects. On projects, customers are usually called clients, so in the rest of this paper we refer to client experience on projects. To our knowledge, nothing has been written to date on client experience on projects. The customer experience literature in the field of marketing is fairly new, and based on our review of the literature nobody has published a paper in the project management journals which cites and builds on the customer experience literature in marketing. However, people have done research and written papers on topics relevant to client experience. We did a review of papers published in the three journals of project management back to 2015. The three journals are the International Journal of Project Management, the Project Management Journal, and the International Journal of Managing Projects in Business. However, our literature review also draws on results from our interviews. This leads to our first research question:

> RQ1: What has been written in the project management literature about topics relevant to client experience on projects?

On projects it is the client that manages the interaction with the contractors. This is different to retail, where the vendor manages the interaction. When a client has a project to be done, they invite contractors to bid, and have experiences in their interactions with the contractors through the project life cycle. We wish to find what those experiences are. This leads to our second research question:

RQ2: How do project clients experience interaction with contractors at key touchpoints through the project contract life cycle?

In the next section, we consider what has been written in the project management literature about topics relevant to client experience on projects. We then describe the results of interviews we have done with client, project managers, or contract managers about their experiences interfacing with contractors through the project life cycle. We summarize our conclusions in the final section.

7.2 Client Experience on Projects

Many authors in project management have commented on elements of client experience, Table 7.1. Nobody was writing about client experience per se, but the research has covered dimensions of client experience. Turner et al. [29] come the closest. They are talking about the interaction between the client and contractor from the contractor's perspective, but when talking about project marketing they suggest how the contractor can influence the client experience

7.2.1 The Journey

We saw above that in retail [10] suggested a three-stage customer journey. For projects, [16] identified a four-stage cycle:

- Pre-project,
- Tendering,
- Project delivery, and
- Post-project.

Turner et al. [29] confirmed this through their empirical research. Turner et al. [29] were looking from the perspective of the contractor, but clearly showed these four stages for the interaction.

7.2.2 Dimensions

Kuppelweiser and Klaus [15] investigated dimensions of customer experience. Building on a model, called EXQ, to measure customer experience, (Klaus and Aaklan 2013; [13]), they identified two dimensions of customer experience in B2C and three dimensions in B2B. B2B is relevant in projects. The three dimensions are as follows:

- Relationships,

Table 7.1 Authors writing about topics relevant to client experience on projects

Paper	Focus	Topic covered	Touchpoints
Kuppelwieser and Klaus (2019)			
Lecoeuvre and Deshayes [16] Turner et al. [29]	Project marketing	Relationships	Throughout the four stages: pre-project, tender, delivery, post-project
Turner and Müller [30]	Communication between client and contractor	Reliability	Throughout the four stages
Turner et al. [29]	Project marketing	Reliability	Throughout the four stages
Ning et al. [22]	Ambivalence	Reliability	Throughout project delivery
Turner and Zolin [31]	Project success	Offerings	At project completion In the months and years post-project
Value			
Fuentes et al. [8]	Creation of value networks in design	Value	Design during delivery
McColl-Kennedy et al. [19]			
Chih et al. [3]	Knowledge and competence of project team and client enables interaction	Resources	Throughout project delivery
Diegmann et al. [6] Basten et al. [2]	Influence of product and process on customer satisfaction	Activities	Throughout project delivery
Recker et al. [24]	Influence of agile practices on customer responsiveness	Activates	Throughout project delivery and completion
Havermans et al. [9]	Influence of narratives on relationships on projects	Context	Throughout project delivery
Lecoeuvre and Deshayes [16] Turner et al. [29]	Project marketing	Interactions	Throughout the four stages
Söderlund [26] Addyman [1]	Two components of collaboration: cooperation and coordination	Interaction	Throughout the four stages
Yu [35]	Customer participation	Customer role	Throughout project delivery
Floris and Cuganesan [7]	Engagement with stakeholders	Cognitive responses	Project handover
Derakhshan et al. [5]	Legitimacy	Cognitive responses	

(continued)

Table 7.1 (continued)

Paper	Focus	Topic covered	Touchpoints
Williams et al. [34]	Client satisfaction and relationship quality	Happiness (joy)	Project handover
Invernizzi et al. [11]	Information flows	Frustration (sadness)	Transfer of information
Turner and Müller [30]	Communication between client and contractor	Discomfort (fear)	Written and verbal communication

- Reliability
- Outcome.

Relationships

Turner et al. [29] identify the importance of relationships between the client and contractors on projects. Lecoeuvre and Deshayes [16] identified six interactions between clients and contractors, which they called: relationships, communication, trust, collaboration, training, and going-with. Turner et al. [29] called the interaction between client and contractor collaboration and said it had four dimensions: relationships, communication, trust, and going-with. They included training in going with.

Reliability

Turner and Müller [30] and Turner et al. [29] suggest that it is important that the customer should trust the supplier's competence and ethics. Ning et al. [22] investigate how ambivalence can cause the client to trust the contractor in some areas and distrust them in other areas at the same time.

Offering

Turner and Zolin [31] identify three levels of offerings on project:

- The project output: the new asset that is delivered at the end of the project.
- The project outcome: the asset must work to provide the client with new competencies which when operated deliver value. Whether or not this is achieved will be judged in the months after project completion.
- The project goals: with time the client will be able to achieve higher order goals delivering performance improvement. Whether this is achieved will be judged in the years after completion.

Turner and Zolin suggest that during the project, project managers take decisions to finish the project to time, cost, and quality. However, they suggest it is more important that the project managers take decisions to deliver value, that is, the output performs as desired, the outcome is achieved in the months following project completion, and the goals are achieved in the years following project completion.

7.2.3 Value Creation

Value in use is often mentioned as key to customer experience, with the service-dominant school of marketing quoted, [32]. Turner et al. [29] following Turner and Zolin [31] make the point that no value is achieved until the project output works and the outcome is achieved [18]. Fuentes et al. [8] use the service-dominant school of marketing to investigate the creation of value networks to create value from the client's perspective in the design stage of the project. They suggest to involve the client in the design stage so the project on completion will deliver value.

McColl-Kennedy et al. [19] develop a conceptual model showing how five value creation elements can lead to cognitive or discrete emotional responses at touchpoints during customer–vendor interaction. The five value creation elements are: resources, activities, context, interactions, and customer role.

Resources

Chih et al. [3] look at how the knowledge and competence of both project professionals and the client enables interaction between them.

Activities

Diegmann et al. [6] and Basten et al. [2] show that product and process influence customer expectations on projects. Recker et al. [24] look at how agile practices influence customer responsiveness by improving team efficiency and effectiveness.

Context

Havermans et al. [9] look at how narratives can influence relationships. They look at the importance of different groups, the influence of outsiders, and the management of conflicting perspectives.

Interactions

Turner et al. [29] identified four components of collaboration between clients and contractors. Söderlund [26] and Addyman [1] identify two components of collaboration: coordination and cooperation. Cooperation is related to governance, and is about setting common goals, [30], and is not in the four components suggested by [29]. Coordination is about relationships, synchronizing activities, and communicating to jointly manage risk, and is covered by the four components of Turner et al. In this paper, we assume that relationships (above) is coordination and interaction is cooperation.

Customer Role

Yu [35] looks at how customer participation can influence project performance. Knowledge integration is the main contributor.

Cognitive Responses

Floris and Cuganesan [7] discuss the need to make a cognitive engagement with stakeholders, creating a dialogue with the right people at the right time on key issues, and guiding collaborative meaning to align stakeholders. Derakhshan et al. [5] use attribution theory to explain the cognitive responses of external stakeholders, and their assignment of legitimacy to the project organization.

Discrete Emotional Responses

McColl-Kennedy et al . [19] identify five emotions: joy, love, surprise, sadness, and fear. Williams et al. [34] look at what makes the client happy, (joy). Invernizzi et al. [11] show how inefficiencies in communication flow between the contractor and client can lead to frustration, (sadness). Turner and Müller [30] show how the adverse selection and the moral hazard problems [20] can lead to the client experiencing discomfort (fear). The client is never totally certain the contractor is competent or trustworthy. If the client has previous experience of working with the contractor their trust and comfort will be increased. Two key touchpoints, Turner & Müller, discussed are written and verbal reports are made by the contractor. The client trusts written reports to give a valid picture of project progress, but not of looming risks and issues. The client trusts the verbal reports to give a valid picture of looming risks and issues, especially by reading the contractor's body language, but not of project progress.

7.3 Methodology

A qualitative inductive approach was adopted to build on the findings of the literature review. The research broadly followed a radical constructivist approach (von Glaserfeld, 1995). A set of propositions were developed from the literature review, and these were revised based on a series of interviews. The prepositions following the literature review are as follows:

> P1: On a project, the interaction between the client and contractors follows a four-stage life cycle: pre-project, tendering, project delivery, and post-project. Through this life cycle, the client and contractor interact at several touchpoints.
>
> P2: At those touchpoints, the client experiences cognitive, emotional, social, and behavioral responses.
>
> P3: At the touchpoints, the client measures their experience through three dimensions: relationship, reliability, and offering.
>
> P4: The value of the offering is assessed through five value creation elements: activities, resources, context, interactions, and client role.

We interviewed five people, Table 7.2. The interviews were semi-structured. The interviewees were asked to recall a project on which they had been a client representative, such as project manager or contract manager. They were then asked to identify touchpoints with the contractors through the project life cycle as suggested by Lecoeuvre and Deshayes [16]. Table 7.3 shows the interview topic guide. This

Table 7.2 Interviews

Case	Industry	Position	Project	Mode
1	Road	Client project director	Motorway widening	Face to face
2	Rail	Client project manager	New facilities at a station	Face to face
3	Light rail	Client project manager	Station upgrade	Skype
4	Transport	PMO manager	Merger of departments	WhatsApp
5	Energy	Organizational development PMO	Product development	Skype

Table 7.3 Interview topic guide

Interview topics
Customer experience will be explained
Participant asked to recall project on which they were a client representative
Participant asked to identify key touchpoints with the contractor
Participant asked to consider importance of relationship, reliability, and offering at the touchpoints
Participant asked to consider significance of five value creation elements, resources, activities, context, interactions, and customer roles at the touchpoints
Participant asked if these stimulated any emotional responses
Participant asked what cognitive responses they stimulated
Participant asked if they have any other memories of their experience with the contractor

suggests that the interviewees would be specifically asked about the dimensions of customer experience, value creation elements, and emotional and cognitive responses. In the event, the interviewees were allowed to talk freely about the touchpoints, and the interviewer identified when they were talking about the dimensions, value creation elements, and responses. The interviewees also sometimes talked about governance.

7.4 Results

7.4.1 Interview 1

The first interview was with a project director working for an organization managing road infrastructure. The project was the widening of a motorway as part of a major road upgrade. The project was a Design-Build-Finance-Manage (DBFM) project. The interviewee identified four touchpoints, Table 7.4. The client clearly managed the interaction at all four touchpoints. The contractors were invited to the market consultation day, and then to the dialogue meetings at the two stages of tendering.

Table 7.4 Results of Interview 1

Touchpoint	Stage	Event	Actions	Response	Dimension	Value	Governance
Market consultation day	Pre-project	200 people invited to information day	Assess market response	Cognitive			
			Get to know potential bidders	Cognitive	Relationship		
Phase 1 Invitation to tender	Tender	Six sets of contractors invited to two dialogue meetings	Do the they meet our requirements		Reliability		
			Does their proposed solution meet our requirements		Offering Reliability		
			Does their approach to risk management make us feel comfortable	Emotional	Offering Reliability	Interaction	
			Some offered solution would work in southern Europe but not Northern Europe		Offering Reliability		
			Some will design and build Others will manage Prefer the former	Emotional		Value Activities Resources	
			Want openness and transparency	Emotional	Relationship		Transparency Reciprocity
Phase 2 Negotiation	Tender	Three sets of contractors invited to six dialogue meetings	Talk about things in open way to find new solutions		Offering Reliability	Value Activities	Transparency Reciprocity

(continued)

Table 7.4 (continued)

Touchpoint	Stage	Event	Actions	Response	Dimension	Value	Governance
			Choose best value bid, not cheapest bid Choose design and build		Offering	Value	
Project delivery	Projectdelivery	Project control meetings	Formal and informal meetings Legs on table meetings Decisions made at informal meetings Confirmed through contract or at formal meetings		Relationship		Flexibility
			Joint problem-solving Transparency and openness Both parties flexible Decisions made in the best interest of both parties don't strictly conform to contract	Trust	Offering Relationships	Value	Transparency Reciprocity

The client defined the project control requirements, but the meetings themselves were very informal.

7.4.2 Interview 2

The second interview was with a project manager at an organization managing rail infrastructure. The project was effectively an alliance project [28]. The project was to install a new and novel facility at a train station in the country. The same facility was to be installed at the two main station in the country. It was a highly political project performed against a tight timescale. Both the end product and the method of delivery were highly uncertain, so an alliance was the appropriate form of contract [27]. They did not choose the contractors through a bidding process. The project required an engineering (design) contractor, a construction contractor, and a safety contractor. They chose three contractors they had substantial experience of working with at the station concerned, and formed a partnership with them. The interviewee mainly talked about the relationship with the engineering and construction contractors. The interviewee identified four touchpoints, Table 7.5.

7.4.3 Interview 3

The third interview was with a former project manager at an organization running light rail. The project was to increase capacity at station. The nature of the project was described as Innovative Contractor Engagement, and had many of the features of an alliance project [28], but it was not partnering. During the tendering stage, contractors were shown the draft concept design and business plan. They had to suggest innovations to the design which would reduce cost or increase benefit, and revise the business plan accordingly. The contract was awarded to the contractor that made the greatest improvement. The winning contractor improved the benefit to cost ratio from 2.4 in the original business case to 3.5. During project delivery there was a close working relationship between client and contractor based on strong cooperation and coordination. The interviewee identified four touchpoints, Table 7.6. The invitation to tender stage required a 6-month dialogue. During that time, the client formed a working routine with the contractor that would do the work. Addyman [1] describes the value of being able to maintain project capability by transferring routines at temporal transitions on projects. The interviewee said:

> Routines are the foundation of organizational capability… repeatable, recognizable patterns of interdependent action by multiple parties. How can you be in a repeatable, recognizable pattern if you haven't actually spoken to each other?

Turner et al. [29] describe how in the onshore oil and gas industry clients will not let their bidders talk to them during the tender phase. This is for fairness so

Table 7.5 Results of Interview 2

Touchpoint	Stage	Event	Actions	Response	Dimension	Value	Governance
Pre-project familiarity	Pre-project	Client has experience of working with these contractors	Know them well	Cognitive Social	Relationship		
			Trust themcompetence, ethics	Cognitive Emotional	Reliability		
			Have an existing relationship	Social	Relationship		
Initial contact	Tender	Telephone call followed by meeting	Are they willing to work to tight schedule		Offering		
			Explore problems and risks		Reliability Offering		
			Made an alliance agreement with open book				Transparency Reciprocity
			Team spirit	Social	Relationships		
Start-up meeting	Project Delivery	One day with key players	Build relationships	Social	Relationships		

(continued)

Table 7.5 (continued)

Touchpoint	Stage	Event	Actions	Response	Dimension	Value	Governance
			Understand people's characters and behaviors	Emotional Cognitive Social Behavioral	Relationships Reliability		
			Role play, ask people to consider other's position	Behavioral			
			Discuss project		Offering		
Project control design	Project delivery	Control meetings	Construction company involved in design		Relationship Offering	Activity Resource	
			People the same age about 40	Social	Relationship		
Trip to London	Project delivery	Knowledge gathering event	Get to know team Know you can work with them	Emotional Cognitive Social Behavioral	Relationship		
			Learn about project		Offering		

(continued)

Table 7.5 (continued)

Touchpoint	Stage	Event	Actions	Response	Dimension	Value	Governance
	Projectdelivery	Construction company make an offer at end of design phase	Construction company makes an offer based on the design to which they contributed	Cognitive	Offering	Value Activity Resource Client role	Contract
			Client wanted them to make reasonable profit		Reliability	Value Client role	Transparency
			Construction company made good, believable proposal	Cognitive	Reliability	Value Activity Resource	
			Client involved in project, controls finance			Value Client role	
Project control	Project Delivery	Train project went late	Got the ministry to confirm the importance of the schedule	Cognitive Behavioral		Value Context	
		Does this project need to finish on time	Maintain relationship	Social	Relationship		

Table 7.6 Results of Interview 3

Touchpoint	Event	Actions	Response	Value	Dimension	Governance
Market testing		Sharing concept with market and judging response	Cognitive			
Pre-qualification	Issue of notice	Assessing potential contractors against requirements	Cognitive		Reliability	
		Reduce to shortlist of 4			Reliability	
		Ask if interested	Cognitive		Relationship	
Innovative tender	Six months of dialogue	Shared concept design and business case with contractors. The suggest innovations			Offering	
		Tender reissued and contractors rebid	Cognitive	Activities Resources	Offering Reliability	
		Bought innovations off unsuccessful bidders		Activities		
		Got into routine of working together		Interaction	Relationship	
Project delivery	Project control meetings	Informal meetings 10.00 a.m. Thursday Decisions at informal meetings, ratify in contract	Cognitive Emotional Social Behavioral	Interaction Client role	Offering Reliability Relationship	
		Relationship carried forward from tender stage Established routines and ways of working		Interaction Activities Client role	Relationships Reliability	

one contractor does not gain an advantage. But [1] says that this results in a loss of organizational capability. The interviewee also said:

> What I think is unique about what we did is we found that balance between competition and early engagement.

The client is building interaction and relationships, cooperation, and coordination.

7.4.4 Interview 4

Interview 4 was with a manager at a company managing the transport. The project was an organization change project. Several departments in different divisions were doing the same thing, so it was decided to merge them into a new division. This project related to the merger of the departments into one division. A second project would rationalize the systems used by the departments in the one new division. Six touchpoints were described, Table 7.7. This organizational change projects followed a similar cycle to the preceding three construction projects.

7.4.5 Interview 5

Interview 5 was with a manager in an energy company. The company wanted to be able to convert sewerage into fuel capable of being burnt in its furnaces. This was a two-stage product development project. In the first stage, three companies were asked to prototype potential manufacturing solutions to achieve that aim. That stage has been completed and the company is now in negotiation with all three companies to choose one of them to build six plants to perform the duty. The prototyping project went through the four stages identified by Lecoeuvre and Deshayes [16] and Turner et al. [29]. The current negotiations are in the post-project phase of that cycle. But the construction project is in the tendering phase of its four-stage cycle. Four touchpoints were identified, Table 7.8.

A key issue for this form was they wanted to build relationships with the three suppliers, but as an energy company in the public sector they have very strict procurement rules. They suffered the issues experienced by clients in the onshore oil and gas industry [29], but did manage to work around it. As suggested by [1], they see great advantage in carrying forward working relationships through the four stages of both projects. But because of their strict procurement rule, the relationships need to be kept at interaction [19] and cooperation, [26]. However, that still enables them to have good working relationships to carry through the four stages of the two projects. The relationships will break up in the fourth stage of the second project, because the suppliers will switch from construction teams to operation teams, and the client representatives will also switch to operations.

Table 7.7 Results of Interview 4

Touchpoint	Stage	Event	Actions	Response	Dimension	Value
Market research	Pre-project	Researched market for potential consultants	Need to have experience of transportSpeak local languagePrevious experience of working with firmUnderstand peculiarities * Laws, * Technology 5 potential companies identified	Cognitive	Reliability	
First round of tender	Tender	Shortlist from five to three	Showed scope statementAsked for their concept, price, and days Other things	Cognitive	Reliability Offering	Activities Resources
Second round of tender	Tender	Choose consultant to do work	Present concept Present team Do they fit with local culture	Cognitive Emotional Behavioral	Offering Reliability Relationship	Activities Resources Context
Project delivery	Project delivery	Defining how to do the project	Client knew objective Consultants advised on route Consultants advised on stakeholder management	Cognitive Behavioral		
		Progress meetings		Cognitive	Relationship	
Post-project support	Post-project		Help with stakeholder management	Behavioral		

Table 7.8 Results of Interview 5

Touchpoint	Stage/Project 1 Project 2	Event	Actions	Response	Dimension	Value
Specificationand assessment	Tender *Pre-project*	Meetings	Contractors given specification and assessed against criteria	Cognitive	Reliability Offering	
			Contact developed cautiously	Behavioral		Interaction
Prototype	Delivery *Pre-project*	Control meetings	Contractors develop their prototypes	Cognitive	Reliability Offering	
			Contact developed cautiously	Behavioral		Interaction
Prototype	Delivery *Pre-project*	Control meetings	Contractors develop their prototypes	Cognitive	Reliability Offering	
			Contact developed cautiously	Behavioral		Interaction
Delivery	Post-project *Delivery*	Control meetings	Plant construction	Cognitive	Reliability Offering	Value
			Close working relationship can no develop	Behavioral	Relationship	Interaction
Operation	Post-project *Post-project*	Control meetings	Firm will be working with a different people	Cognitive Behavioral		

7.5 Discussion

Our two research questions are as follows:

RQ1: What has been written in the project management literature about topics relevant to client experience on projects?

RQ2: How do project clients experience interaction with contractors at key touchpoints through the project contract life cycle?

7.5.1 *Previous Research*

We found nobody has previously researched client experience on projects. But some authors have written material that is relevant to the discussion, Table 1.

7.5.2 *Nature of Client Experience*

Propositions P1–P4 remain unchanged. But we add two new propositions:

P5: Unlike retail on projects the client manages the interactions with the contractors. In retail, the customer approaches the vendor, but the vendor manages the interactions, and so influences the customer's satisfaction and loyalty. On projects, the client approaches the contractors and so manages the interactions. It is the contractor's satisfaction and loyalty that is at risk.

P6: If the client can build a good relationship and interaction with the contractor during the tender phase, which can create routines in their way of working, which if carried forward to project delivery can improve project performance.

Two features which differentiate projects from retail

The client controls the interaction

In retail the supplier controls the interaction. The purchaser wishes to buy a product and seeks potential vendors, but once the customer makes contact, the vendor controls interactions. Vendors that manage the customer experiences at those touchpoints perform better than those that do not [13]. Some companies employ customer experience managers. On projects it is the clients that manage interactions. Clients invite contractors to participate and manage a series of touchpoints at which they interact with the contractors.

At those touchpoints, clients have cognitive, emotional, social, and behavioral responses, and they manage the relationships, and the contractor's reliability and offering. They also assess value through the activities performed and resources provided by the contractor, interactions with the contractor and their own role. There was less of a focus on context. Interestingly, the contractor will also have experiences at the touchpoints. Perhaps research in project management needs to focus on how the client manages the contractor experiences.

But the contractors also have experiences at those touchpoints, and the way the client manages the touchpoints can influence the contractor's satisfaction and loyalty. It could also influence whether the contractor works according to a principal–agent relationship or a stewardship relationship, Müller [21]. Joslin and Müller [12] suggest a stewardship relationship will often lead to better project performance. It can also influence the nature of the working contractual arrangement. In Interview 2, the management of the touchpoints led to a success alliance arrangement.

The journey

In retail, customers follow a three-stage journey through their purchase, pre-purchase, purchase, and post-purchase (see, for instance, [17]). Following Lecoeuvre and Deshayes [16] and Turner et al. [29], we identified a four-stage journey for projects, pre-project, tendering, project delivery, and post-projects. All of the interviewees followed this four-stage journey. For some the project delivery stage was divided into design and construction. But the nature of the interactions was the same at both those sub-stages.

Interviewee 3 highlighted the importance of carrying forward routines and ways of working from tendering to project delivery. Routines are the foundation of organizational capability and are repeatable, recognizable patterns of interdependent action by multiple parties, [1]. Contractors in the onshore oil and gas industry interviewed by Turner et al. [29] said that their clients would not allow them talk to them during the bid phase, to maintain fairness between the bidders. But all our interviewees reported working closely with their contractors during the tendering phase, and the close working relationships were carried forward to project delivery.

7.5.3 Theoretical Contribution

This paper extends the research on customer experience to projects. However, we see here that the client controls the interactions and so manages their own experiences, and that of the contractors. It raises the question about what the client should be doing to manage the contractor's experiences, to maintain their satisfaction and loyalty.

7.5.4 Practical Contribution

This research has shown how clients can manage their experiences on projects to improve overall project performance.

7.5.5 *Further Research*

This research raises the question, what are the contractor's experiences on projects? How can the client manage those experiences to improve the contractor's satisfaction and loyalty, and achieve appropriate governance and contractual relationships? How does this fit with the concepts of agency theory and stewardship theory [21]?

References

1. Addyman S (2020) Connecting the "demand chain" with the "supply chain": (re)creating organizational routines in life-cycle transitions. In: Pryke S (ed) Successful construction supply chain management: concepts and case studies, 2nd ed. Chichester: Wiley
2. Basten D, Satvrou G, Pankratz O (2016) Closing the stakeholder expectation gap: managing customer expectations toward the process of developing information systems. Proj Manage J 47(5):70–88
3. Chih YY, Zwikael O, Restubog SLD (2019) Enhancing value co-creation in professional service projects: the roles of professionals, clients and their effective interactions. Int J Project Manage 37(5):599–615
4. De Keyser A, Lemon KN, Klaus P, Keiningham TL (2015) A framework for understanding and managing the customer experience. In: Marketing science institute working paper series 2015, Report No 15–121. Marketing Science Institute, Cambridge, MA
5. Derakhshan R, Mancini M, Turner JR (2019) Community's evaluation of organizational legitimacy: formation and reconsideration. Int J Project Manage 37(1):73–86
6. Diegmann P, Basten D, Pankratz O (2017) Influence of communication on client satisfaction in information system projects: a quantitative field study. Proj Manage J 48(1):81–99
7. Floris M, Cuganesan S (2019) Project leaders in transition: Manifestations of cognitive and emotional capacity. Int J Project Manage 37(3):517–532
8. Fuentes M, Smyth JH, Davies A (2019) Co-creation of value outcomes: a client perspective on service provision in projects. Int J Project Manage 37(5):696–715
9. Havermans LA, Keegan AE, Den Hartog DN (2015) Choosing your words carefully: leaders' narratives of complex emergent problem resolution. Int J Project Manage 33:973–984
10. Howard JA, Sheth J (1969) The theory of buyer behaviour. Wiley, New York
11. Invernizzi DC, Locatelli G, Brookes NJ (2018) The need to improve communication about scope changes: frustration as an indicator of operational inefficiencies. Prod Plan Control 299(9):729–742
12. Joslin R, Müller R (2016) The relationship between project governance and project success. Int J Project Manage 34(4):613–626; Keynes JM (2017/1936) The general theory of employment, interest and money. Wordsworth Editions, Ware, Herts
13. Klaus P (2015) Measuring customer experience: how to develop and execute the most profitable customer experience strategies. Palgrave Macmillan, Basingstoke
14. Klaus P, Maklan S (2013) Towards a better measure of customer experience. Int J Mark Res 55(2):227–246
15. Kuppelwiesser V, Klaus P (2020) Measuring customer experience quality: the EXQ scale revisited. J Bus Res
16. Lecoeuvre L, Deshayes P (2006) From marketing to project management. Proj Manag J 37(5):103–112
17. Lemon KN, Verhoef PC (2016) Understanding customer experience throughout the customer journey. J Mark 80(November):69–96
18. Lusch RF, Vargo SL, O'Brien M (2007) Competing through service: insights from service dominant logic. J Retail 83(1):5–18

19. McColl-Kennedy JR, Zaki M, Lemon KN, Urmetzer F, Neely A (2019) Gaining customer experience insights that matter. J Sci Res 22(1):8–26
20. Moe TM (1995) The politics of structural choice: toward a theory of public bureaucracy. In: Williamson OE (ed) Organization theory: from Chester Barnard to the present and beyond. Oxford University Press, New York
21. Müller R (2019) Governance, governmentality and project performance: the role of sovereignty. Int J Inf Syst Proj Manag 7(2):5–17
22. Ning Y, Feng M, Feng J, Liu X (2019) Understanding clients' experience of trust and distrust in dwelling fit-out projects: an ambivalence approach. Eng Constr Archit Manag 26(3):444–461
23. Pine BJ, Gilmore JH (1998) Welcome to the experience economy. Harv Bus Rev 76(July–August):97–105.
24. Recker J, Holton R, Hummel M, Rozencranz C (2017) How agile practices impact customer responsiveness and development success: a field study. Proj Manag J 48(2):99–121
25. Smith A (2012/1776) An inquiry into the nature and causes of the wealth of nations. Wordsworth Editions, Ware, Herts.
26. Söderlund J (2012) Theoretical foundations of project management: suggestions for a pluralistic understanding. In: Morris PWG, Pinto JK, Söderlund J (eds) The oxford handbook if project management. Oxford university Press, Oxford, pp 37–64
27. Turner JR (2004) Farsighted project contract management: incomplete in its entirety. Constr Manag Econ 22(1):75–83
28. Turner JR (2006) Partnering in projects. In: Cleland DI, Garie R (eds) Global project management handbook, 2nd ed, pp 20.1–20.14.S. McGraw-Hill, New York
29. Turner JR, Lecoeuvre L, Sankaran S, Er M (2019) Marketing for the project: project marketing by the contractor. Int J Manag Proj Bus 12(1):211–227
30. Turner JR, Müller R (2004) Communication and cooperation on projects between the project owner as principal and the project manager as agent. Euro Manage J 22(3):327–336
31. Turner JR, Zolin R (2012) Forecasting success on large projects: developing reliable scales to predict multiple perspectives by multiple stakeholders over multiple time frames. Proj Manag J 43(5):87–99
32. Vargo SL, Lusch RF (2004) Evolving to a new dominant logic for marketing. J Mark 68:1–17
33. Von Galsserfeld E (1995) Radical constructivism: a way of knowing and learning. The Falmer Press, London
34. Williams P, Ashill N, Naumann E, Jackson E (2015) Relationship quality and satisfaction: customer perceived success factors for on-time projects. Int J Project Manage 33:1836–1850
35. Yu MC (2017) Customer participation and project performance: a moderated-mediation examination. Proj Manag J 48(4):8–21

Part III
Self-Organizing and the New Technologies

Chapter 8
Projects Organization and Intelligent Technologies

Ronggui Ding and Constanta-Nicoleta Bodea

Abstract Projects are social-technical systems, as networks of interconnected elements comprising groups of people and technologies which are executing together processes for achieving specific objectives. In the actual digital era, the digital transformation of organizations and projects represents an example of how information and communication technologies adoption and usage can change the business. Self-organizing in projects is directly related to the technologies applied in projects. The objective of this chapter is to discuss the potential of some intelligent technologies in transforming the projects towards a self-organizing arena.

Keywords Social-technical systems · Intelligent technologies · Blockchain · Machine learning · Self-organizing

8.1 Projects as Social-Technical Systems

Regardless the level at which we are studying the work systems, meaning: projects, organizations, networks, industries, these ecosystems are integrating technical and social factors. The technical aspects, which are usually managed by the technical engineers, consist of technologies and division of labor. The social aspects are culture, people, competences, motivation, coordination, and control. These aspects are usually performed by managers, HR professionals, sociologists, and psychologists. Considering these duality of technical and social aspects, the work systems are usually called socio-technical systems, for which a specific body of knowledge for designing and operating these systems emerged.

R. Ding (✉)
School of Management, Shandong University, 27 South Shanda Road, Jinan 250100, China
e-mail: ding_rgui@email.sdu.edu.cn

C.-N. Bodea
Bucharest University of Economic Studies, 6 Piata Romana, 1st district, 010374 Bucharest, Romania
e-mail: bodea@ase.ro

R. Ding et al. (eds.), *Research on Project, Programme and Portfolio Management*, Lecture Notes in Management and Industrial Engineering,
https://doi.org/10.1007/978-3-030-86248-0_8

There are strong interactions between technical and social components of work systems influencing the system performance. Due to a higher degree of work mechanization and automation, it is assumed that by including more and more technical elements, this will lead to a higher system performance. But even the first studies of the mechanization in the coal mines [31] and automotive industry [10] revealed that only adding technical elements in the work systems, this does not automatically lead to the increase of work efficiency, mainly because the technical components could have also negative implications. The technology adoption is usually accompanied with the definition of new processes, procedures, and roles for applying the technologies, and all these will lead to the increase of work structure complexity, a high fragmentation of work, lack of communication and trust, weak leadership, and a reduced degree of self-organization. At more general contexts, technologies were considered as being responsible for climate changes, job losses, and alienating work environments. But for the majority of cases, the adoption of new technologies lead to so called "positive deviants" [30], which increase the work systems performance.

The differences between these two categories of impact were mainly explained by how the technical elements were integrated into the work social systems. The research findings underlined the importance of considering the both aspects (technical and social) in designing the work systems. The dichotomy between technological-centric versus human-centric approaches was solved by considering both as being very important in work design, at all levels.

The actual focus of the organization theory on adaptability, innovation, and organizational learning [1, 4, 17] increased the relevance of self-organizing (lean/Six Sigma, employee empowerment, learning organizations, etc.) for the social-technical systems design. In his seminal work [5], Cherns included self-direction and self-organizing as one of the main socio-technical design principles that requires people to have information, power, and accountability in order to behave freely and efficiently at the work place. Several studies were conducted in order to understand how self-organizing can be assured in technological environments [11, 12, 14]. The relevance of self-organizing is even higher for the complex socio-technical systems [2] , where the system design should be undertaken at a higher level that the task, meaning the ecosystem level.

Starting mainly from eighteens, information technologies gained a more important role in the socio-technical systems, due to a higher availability and a higher impact, not only on the production and routine work but also on the administrative and managerial activities. The concepts of Information society and Information Revolution emerged [6, 9, 25]. Designing, implementing, and running an Enterprise Architecture are still challenging for organizations. With the generation of intelligent technologies, the concept of digital transformation emerged, meaning that the impact of Artificial intelligence on the work is considered as transforming completely the work and the way that the work is organized, similar to the impact of electricity [23].

8.2 The Potential of Intelligent Technologies in Transforming Projects in Self-Organizing Arenas

The intelligent technologies are disruptive, in the sense that it produces and amplifies several socio-economic changes. Due to this feature, it is difficult to predict their impact on the workplaces, even if the impact is undeniable. The impact of intelligent technologies depends, on the one hand, on how fast the intelligent technologies are developed and adopted, and, on the other hand, on the policies of organizations and public bodies.

In the short term, the adoption of intelligent technologies can lead to changes in the workforce between sectors through the disappearance of some jobs, transformation of others, and creation of new jobs. In the long term, the number of jobs in the economy is expected to increase due to the adoption of intelligent technologies, especially by increasing the number of jobs in sectors that today do not exist or are not well represented yet. In [29], it is estimated that more than two millions new jobs were created by digitalization in European Union (EU) only in the last decade and 1.75 million new jobs are expected only in IT sector by 2030 in EU.

AI and robotics improve the quality of work, especially due to the fact that they allow workers to focus on creative activities, to the detriment of routine ones, which creates a higher degree of professional satisfaction. But the jobs including routine and repetitive tasks usually are held by low-skilled people. The use of AI could thus accentuate the differences between highly skilled and low-skilled workers.

One of the main challenges for the adoption of intelligent technologies is the requalification of the employees, especially those in the second half of their active life. Their adaptation to the new competency requirements is difficult to be achieved. Also, the speed of the qualification and requalification should be synchronized with the intelligent technologies adoption rate. This is why the educational system has a major role in managing the changes induced by the intelligent technologies in the society, at different levels. Another challenge for the adoption of intelligent technologies is to address the worries related to the negative impact of the digital transformation, respectively, the intrusion in the private life, the risks of discrimination, and socio-economic exclusion. For this reason, it is recommended that the adoption of intelligent technologies to be made transparent and intelligible to everyone.

Several studies were conducted about the artificial intelligence (AI) adoption status in different sectors and types of organizations [8, 13, 19, 27, 34]. In 2020, International Project Management Association (IPMA), in association with PwC, has conducted a study about the impact of AI in project management [3]. More than 52% of the participants considered the AI tools ecosystem to act as intelligent assistants of project managers. 56% of the organizations represented in the study declared that they have a digital transformation strategy that make AI adoption a sustainable process. The participants in this study were asked to characterize the AI-driven project management, as representing the people- or technology-centric approach. Majority of the participants considered AI-driven project management as being people-centric, considering that: "humans are the masters of technology", "the

AI focus is to improve people performance", "it is always about stakeholders and solving business problems", "the business expertise matters more in projects than the technical skills", "AI tools enhance the role of people in projects", and "the use of AI in project management will allow project managers to make smart decisions and effectively manage triple constraints".

When intelligent technologies are adopted in projects, it is important to consider how the project managers will manage the complexity of social-technical components existing in the internal and external project environments. In order to act as integrators of the social and technical components of their projects, project, managers should remove additional planning and reporting layers, empower project team members, and increase the reciprocity in project relationships and communication. All these actions represent triggers for increasing self-organization in projects and for transforming projects in self-organizing arenas.

8.3 Impact of Blockchain Technology in Projects

8.3.1 Introduction to Blockchain Technology

Satoshi Nakamoto proposed the concept of Bitcoin, which can realize the direct transaction between any two parties without the help of a third party [26]. Blockchain technology is the core of the implementation of Bitcoin. Blockchain technology consists of two parts: block and chain. A block refers to all transactions and records, while the chain is a block that is arranged in order of transactions [20]. The essence of blockchain technology is a distributed database system that integrates the asymmetric encryption algorithm, the distributed technologies such as distributed storage, and all factors work together to make the system run normally.

Since the popularity of Bitcoin, the application of blockchain technology has been extensively developed, which can be roughly divided into three stages, namely, Blockchain 1.0 era, Blockchain 2.0 era, and Blockchain 3.0 era [35]. In the first stage, the development of blockchain technology is mainly based on digital currency represented by Bitcoin. In the second stage, blockchain technology is mainly applied to programmable digital assets and smart contracts. In the third stage, blockchain technology began to develop extensively in non-financial fields.

At present, the development of blockchain technology is in the third stage, and as blockchain technology becomes mature, its applications are expected to penetrate more domains, such as public service, management, information security, and Internet of Things.

8.3.2 Blockchain Technology Implementation-Related Factors

The basic factors that underpin the working of blockchain technology can be summarized as follows.

Openness

Blockchain is a distributed ledger, each node on the chain can access a complete audit trail of all transactions of the entire database, so all nodes in the network can see all transactions and data, and they share the same record [16]. Besides, any node in the blockchain can be added or left at will, and their data can get a complete backup. As a result, the information of the entire system remains highly transparent and the integrity of data is easily verifiable.

Traceability

Every block in blockchain has a timestamp which records the time of information generation, so the data of any code can be queried and searched in a chronological order [21]. At the same time, the blockchain encrypts the node information through a unique asymmetric encryption technology. It's difficult to tamper the data of any node as it requires the consensus of majority nodes to complete the modification.

Democratization

The ledger of blockchain stores the information of the node communicating with other nodes and forwards the data to all other nodes [24]. The data stored on the ledger is not controlled by any single party. Each node can verify the records of all transactions across the network without the help of the intermediary or the permission of the third party. Besides, blockchain technology adopts the consensus mechanism, which enables all nodes in the system to store and update data freely and securely.

Token incentive

Token incentive is the core power of blockchain technology. The blockchain system introduces token as an incentive method to encourage all members in the network to participate in the operation of the system and maintain the consensus mechanism [22]. Every member can get corresponding rewards based on his contribution to the system, which effectively promotes the enthusiasm of all members to make contribution to maintain the stability of the system.

8.3.3 The Potential Impact in projects

Project organization is different at each stage of the project life cycle to adapt to the corresponding environment and objective requirements, which is the external manifestation of self-organization. Self-organizing project has lots of advantages. For example, it can fully reflect the demands of stakeholders, give play to their respective professional expertise, and makes the organizational mechanism sufficiently flexible and correctable. However, self-organizing project also has shortcomings. For example, the strategic goals of the self-organization are often not clear, leading to too many internal discussions, and too much time wasted on making decisions. In such an environment, project members need to learn how to deal with various unstable events by cooperation. Successful cooperation involves several key issues, namely, "who is my partner", "who make decisions", and "how to maximize each member's contribution".

The key to solving the first issue is trust within the organization, and the entity that is able to assume the corresponding responsibility to gain the trust of other members is more likely to be selected as a partner. The key to solving the second issue is to conduct a universally accepted dynamic decision-making mechanism based on the common values. The key to solving the third issue is to establish a value recognition mechanism that links the interests of members with the long-term benefits of the project and motivates them to contribute their maximum value. To sum up, the issues of the self-organizing project can be solved with the support of trust, dynamic decision-making, and value recognition mechanisms.

Blockchain technology can provide a technical guarantee for these issues, as shown in Fig. 8.1.

(1) **Improving trust among members**

The trust among members is the determinant of project self-organization. In traditional organizations, power and those who exercise power are set up according to preset scenarios. While self-organization is different. It needs to establish a mechanism to assign powers to different roles according to the scenario changes. Therefore, self-organization can reflect the demands of each role and give them autonomy; meanwhile, it also makes requests for them to assume corresponding responsibilities to

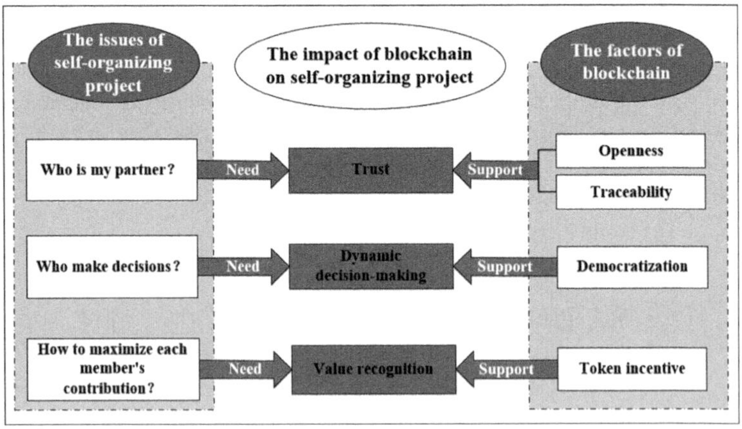

Fig. 8.1 The impact of blockchain on self-organizing project

gain the trust of other parties. Power often comes from people's trust in other's value and competence, so trust is the basis for the existence of self-organization. Establish a trust mechanism is essential to realize the self-organization of the project. The trust establishment needs to be very time efficient, and members should come up with clear evidence quickly to prove that they are competent in their roles and are able to take responsibility for the roles.

The use of blockchain technology has played a key role in trust. Blockchain technology provides a new infrastructure that can ensure the transparency and consistency of interactions between members, thereby making it possible to create an effective trust mechanism for self-organization. Specifically, the data on the chain is open to all members, they can access the data through the public interface. So the team can introduce the mechanism of uploading data to earn points and obtaining data by consuming points to encourage all members of the project to share and exchange data, and realize the efficient collaboration of all members. At the same time, all data storage is attached with a time stamp, which has extremely high traceability, to ensure that the transaction data within the project organization is not easily tampered with and cannot be forged.

The high transparency of the information provided by blockchain technology reduces the mutual suspicion between project members due to information asymmetry, and each member believes in the authenticity of the information provided by other members. Through the realization of traceability, project members do not need to supply clear evidence to prove their competence to undertake responsibilities, and other members can supervise that in real time through the data on the chain. These advantages can provide supporting data for project managers and help project members to select new excellent entrants, thereby promoting the transformation of self-organization.

(2) **Supporting dynamic decision-making**

Blockchain technology can help project members in self-organization to choose trustworthy team partners, but who should make decisions in the team is still one of the problems that needs to be solved. Project self-organization need to determine the activities' priority to complete a certain task and to assign roles and responsibilities through a fairly democratic process in different project phrases. At the same time, the decision-making process of a is dynamic, members in self-organization share or take turns to undertake decision-making tasks in different project phases. All the members have the possibility to become the decision-maker of the phase, which depending on how closely the team members collaborprojectate on the agreed vision and goals. In a self-organizing environment, people need to be navigated through shared values, common vision, and goal frameworks to support the decentralized dynamics decision-making in self-organization.

Decentralized Autonomous Organization (DAO) is a decentralized organizational structure formed on the democratization characteristic of blockchain technology. It plays an important role in achieving decentralized decision-making and improving the enthusiasm of members to participate in decision-making. DAO refers to an organization that operates autonomously without human intervention and management through a series of open and fair rules. Continuously iterative management and operation rules are gradually coded on the blockchain in the form of smart contracts, so that the organization can achieve self-operation and self-evolution according to preset rules through automated algorithms without third-party intervention.

Therefore, the algorithmic foundation of DAO can be useful for the project to achieve an effective self-organization structure, so that all project members can share a large database, a set of value transfer protocols, and target frameworks. Under this framework, smart contracts enable every member to make decisions automatically based on established rules, to realize the automation of large-scale decision-making, and then realize dynamic decision-making in the whole process of the project. To sum up, the DAO can not only make full use of the knowledge and competence of all members, but also guide the development of self-organization in a favorable direction.

(3) **Guaranteeing value recognition**

How to inspire members to contribute their best value is an issue that any organization needs to consider. It's the same with self-organization. Unconventional projects, especially innovation projects, are particularly suitable for self-organization. In an innovation project, the behavior of each member will have a significant impact on the development of the entire innovation project. However, many members focus only on project results and ignore the future benefits of project outcomes, which greatly discourages them from contributing their best value to improving the quality of the outcomes. All members work together to expand the project's revenue instead of gambling in the price of a limited project contract is the correct way to the success of an innovation project.

Therefore, a value recognition mechanism that can combine the long-term benefits of project results with the contributions of members is indispensable to motivate them contributing their biggest value. In order to establish this value recognition mechanism, the concept of project shareholder was proposed. Project shareholder means that members can obtain long-term benefits of the project by investing in the project based on their contribution to the project, and the distribution ratio of their dividends is based on the contribution value of each member to the project. However, it is not easy to determine the contribution value of members to the project. The personnel mobility in the self-organizing project is great, and whether members can continue to obtain benefits after they leave is the key consideration for members to contribute their ability. For example, in phase A of the project, a member puts forward an idea for technological improvement, which can effectively improve the performance of the project, but its value only comes into play in phase D. Can he get the value of their creativity if he leaves in phase B? And how should this value be determined?

The token incentive of blockchain technology can provide support for the realization of *project shareholders*. Specifically, a blockchain community oriented by project shareholders can be built by using the token incentive of the blockchain technology. In this community, each member is given the same status. All members directly invest in the project based on their contributions, forming a long-term community of interests with the project, and achieving a win–win situation. The value recognition rules jointly recognized by each member will be encoded on the smart contract of the blockchain, which can be automatically executed. The workload of each member can be calculated in real time to prove the contribution of each member during the operation of the project, so that the contribution of each member can be matched with its compensation. At the same time, the creative contribution of each member can be recorded and queried through distributed ledger technology, even if the member leaves the project organization, the system can still track the innovation contribution of each innovation contributor. In addition, the accurate positioning of each revenue of the project product through the blockchain technology can realize the fair distribution of revenue among the members.

Blockchain technology has huge potential value in self-organizing projects. Firstly, blockchain technology can improve the trust among members and help self-organization to select suitable members; secondly, it can support the dynamic decision-making of self-organization; finally, it can assist self-organization to establish an effective value recognition mechanism to improve the enthusiasm of members, and realize the maximum benefit of project members. However, due to the immature development of blockchain technology, the technology has not been well promoted and applied. With the continuous development and improvement of blockchain technology, we should actively throw an olive branch to it and use blockchain technology to provide support for self-organizing project.

8.4 Impact of Machine Learning Technologies in projects

8.4.1 Introduction to Machine Learning

This is an era of information explosion. The development of communication and information dissemination technology has overcome the barriers of time and space of traditional information dissemination, making the speed and scale of information collection and dissemination reach an unprecedented level, and realizing information sharing and interaction all over the world. Digitization has brought all kinds of convenience to people's lives, and also brought huge profits to data-intensive companies. It is estimated that Google contributes more than 100 billion US dollars to the US economy every year. Data and information have become indispensable foundations for social development.

However, there exists a large amount of useless content in the information, so that the truly valuable information is submerged in a large amount of spam. Facing with the complex information, no one can accurately screen out all the content that is useful to himself, so in the era of information explosion, the interesting phenomenon of "information scarcity" appears. The reason for this phenomenon is the limited data processing capabilities of human beings at the beginning of the information age. Thanks to the powerful data computing capabilities of computers, machine learning is becoming the main method of analyzing big data and mining potential laws.

Machine learning is used to refer to the behavior of computers using experience to automatically improve the performance of the system. Its origin can be traced back to Alan Turing's automaton model theory [32]. Through the exploration of many scientists in this field, machine learning has developed into an interdisciplinary subject, covering the knowledge of probability theory, statistics, approximate theory, and complex algorithms. As the main tool, the computer can realize the real-time simulation of human learning, and divide the existing knowledge into module structure to improve the learning efficiency. Specifically, under the guidance of algorithms, the computer can automatically learn the data structure and internal laws of a large number of input data samples, thereby the computer can identify new samples intelligently and even realize future predictions. The general process of machine learning is shown in Fig. 8.2. Machine learning is one of the most intelligent and cutting-edge research fields in artificial intelligence, at the same time, as a way to realize artificial intelligence, the capacity of machine learning has become the core competitiveness of technology-based enterprises.

Learning ability is a sign of whether the system is "smart", and improving the self-learning ability of the system has become one of the important research topics in the field of machine learning. The research of machine learning can be divided into two main research directions. The first is the research of traditional machine learning, which mainly focuses on the learning mechanism, and pays attention to simulate the learning mechanism of human. The second is the research of machine learning in the big data environment [33]. This kind of research is mainly about how to use information effectively and focus on obtaining hidden, effective, and understandable knowledge from big data.

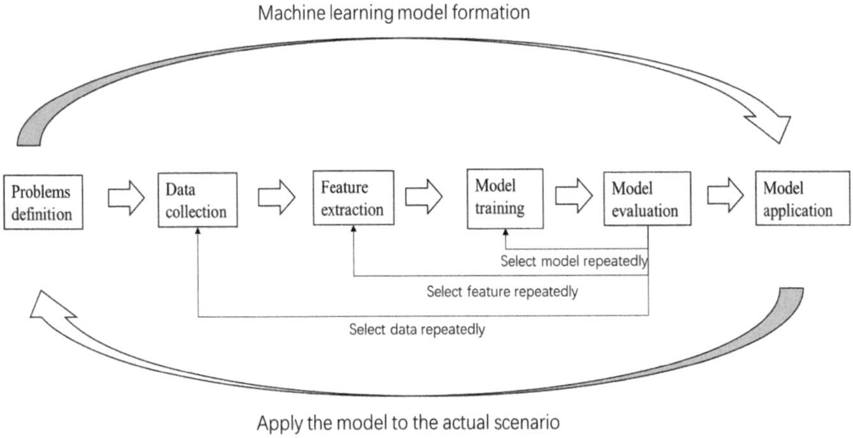

Fig. 8.2 Machine learning workflow

Stimulated by the long-term research foundation of academia and the huge demand of the industry, and supported by powerful computing capabilities provided by the rapid development of computer hardware technology, Hinton, a leader in the field of machine learning, proposed a deep learning model on the published paper "Reducing the Dimensionality of Data with Neural Networks" in 2006 [15]. The deep learning model is the intelligent learning method and cognitive process closest to the human brain. It draws on the multi-layer structure of the human brain, the connection and interaction of neurons, the layer-by-layer analysis and processing mechanism of information, and the powerful parallel information processing capability of self-learning. Deep learning has made breakthroughs in many research fields and achieved great commercial success. One of the most influential events is the smart program AlphaGo, developed by DeepMind Company, a subsidiary of Google, based on deep convolutional neural networks and Monte Carlo tree search algorithm, defeating the world Go champion Lee Sedol with a 4:1 score. It fully demonstrates the powerful learning ability and great development potential of machine learning. The proposal of the deep learning model opened a new era of machine learning, making the application of machine learning a new research field.

8.4.2 *Machine Learning Implementation-Related Factors*

Successful machine learning includes three elements: data, the model of transforming data, and the loss function to measure the quality of the model. These elements are closely related to the basic structure of the project.

Data

For machine learning model, the more data, the better. In fact, data is the core reason that deep learning becomes the mainstream of machine learning, because complex nonlinear models need more data than other machine learning models. The types of data include picture, data, sound, etc. There are a lot of data in the project, including both static data and dynamic data. Sufficient data sources create conditions for the training of machine learning model, which is conducive to the self-learning and self-optimizing of machine learning model.

Model

Model and algorithm are the key of machine learning to produce prediction results based on data [18]. Machine learning benefits from a variety of algorithms that meet different needs. Supervised learning algorithm can learn the mapping function of a classified dataset, while unsupervised learning algorithm can classify unlabeled dataset based on some hidden features. Finally, reinforcement learning can learn decision-making strategies in an uncertain environment by repeatedly exploring the environment. In the self-organizing project, using project parameters to construct the project scenario structure is the key to extract universality from the particularity

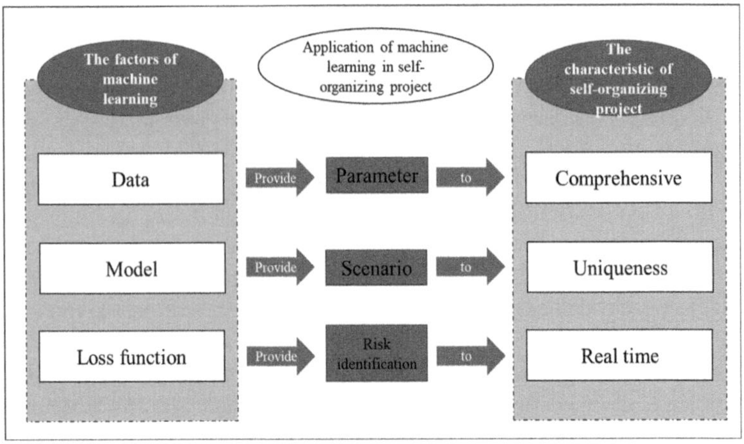

Fig. 8..3 The application of machine learning in self-organizing project

of the project. After finding the universal project parameter structure, we can input different project data to generate targeted management decision-making scheme.

Loss function

The loss function is used to compare the error between the model output and the real value, which is the key to optimize the function parameters and improve the prediction accuracy of the machine learning model [7]. In application, the loss function is usually used as a learning criterion to evaluate the model by minimizing the loss function. In the self-organizing project, the loss function based on the project scenario can be used as the basis for judging the project risk. When there is a big difference between the estimated value generated by the machine learning model fitting the project scenario and the actual value of the project, the project risk can be identified in the project.

The application of machine learning in self-organizing project is shown in Fig. 8.3.

8.4.3 The Potential Impact in Projects

There are two possible applications of machine learning in self-organizaing project.

(1) **Constructing project scenarios**

The project has the characteristics of one-off and uniqueness, so solving the contradiction between the uniqueness of the project and the universality of the law is a major challenge for project management research. On the one hand, empirical research that ignores the uniqueness of the project is difficult to obtain a universal statistical law. On the other hand, it is difficult to apply the statistical law that ignores the characteristics of individual cases to specific project practice. Project analysis over-relying

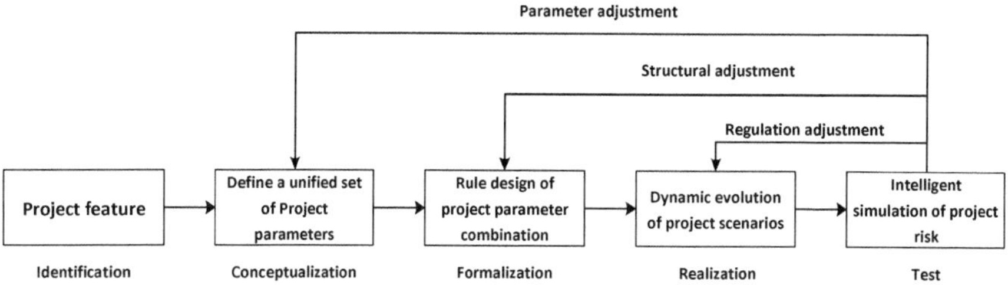

Fig. 8.4 Logical structure of digital project scenario

on personnel experience because it often falls into the dilemma of "One Matter, One Approach". Thereby, it is urgent to use machine learning model to realize the construction of intelligent project system.

The key to building an intelligent project system is to determine the basic modules that can constitute the project, and to describe the universal laws existing in the project in a standardized way (Fig. 8.4). Universality resides in uniqueness. Revealing the essence without being confused by the superficial uniqueness as well as finding the common ground behind the surface can effectively solve the problem of project uniqueness. Projects are unique, but the process of completing the project can be unified. Under a unified process, different inputs will get different output. In 2002, IBM spent 2.1 billion US dollars on acquiring Rational to get the RUP of Rational which is a unified product development process and management tool. Componentization and combination are the best strategies that use the efficiency of industrialization to meet individual needs. Using a unified process to describe different projects, and combining standardized components into personalized products, can not only meet the specific needs of the project, but also achieve the high efficiency and high quality brought by management standardization.

The basic module that can realize the unified expression of the project process and describe the universal law of the project is the project scenario. Project scenario refers to the combination of project organization and project tasks in a time slice. The project scenario is the smallest fragment describing the execution of the project, and the basic unit of project management. According to the needs of digitization and intelligence, the project scenario can be abstracted into a set of parameter structures that express project organization and project tasks. A project can be regarded as a combination of a series of project scenarios, so a unified parameter structure can be used to express its characteristics, thus realizing the description of the universal law of the project. The uniqueness of the project can be seen as the digital expression of different values given on the basis of the unified parameter structure, the different outputs is manifested in the difference in project type and project form.

The biggest limitation of building a project scenario is the lack of data and the insufficient computing power. Lack of data will cause the incompletion of project parameters structure that is describing the project scenario, which will lead to the ineffective transmission of resources and energy within the project scenario, and cannot guarantee the micro-integrity of the project. What's more, the identification of the project scenario will be incomplete as the data shortage, resulting in the lack of corresponding links in the project implementation process, and cannot guarantee the macro-integrity of the project. Insufficient computing power will fail to obtain effective project expression results through calculation under the complex parameter structure, making the constructed complete project scenario parameter structure meaningless. Big data and machine learning, as emerging auxiliary methods, can provide data sources and algorithm guarantees for the construction of project parameter structures. Specifically, using big data can collect sufficient project information, including information of different projects and the information on each time slice of one project. The project information obtained by big data can provide sufficient and effective training data for the machine learning model, and, at the same time, the machine learning model can be optimized based on the comparison of feedback results.

(2) **Optimizing the self-organizing project decision-making**

Because of the expansion of the project scale and the increase of technical requirements of the project, there are numerous uncertain situations that cannot be preset in the project, the role boundary between stakeholders becomes ambiguous, and the requirement for the multiple knowledge domains has been increased. Whereas the self-organizing form of project stakeholders can provide effective production relationships to accomplish the complex task.

Results-oriented special tasks, partners with multiple professional expertise and needs, and unpredictable external environments or opponent strategies require project stakeholders to independently determine their role arrangements and coordination mechanisms based on the project scenario. In traditional project organizations, power and those who exercise power are set according to preset project goals and tasks. Conversely, self-organization requires to establish a mechanism to define power and assign power to specific roles according to changes of project scenario. Since self-organization can conveniently reflect the demands of each role and empower autonomy to each one, it requires each role to undertake corresponding responsibilities so as to gain the trust of other parties. The essential feature of self-organization is to respect the value of each stakeholder and to give full play to the effect of collaboration. In other words, effective self-organization is one in which each person has indispensable value and contributes their value in an appropriate role.

On the basis of this fundamental idea, in 2007, Brian Robertson, the Founder of Ternary Software in the United States, proposed the self-organization model of Holacracy [28]. This new type of organizational management model is completely based on real business operations and management practices, and constantly iterates and evolves a set of empirical management models and methods. However, due to the lack of the support of a complete methodology, some companies that were originally keen to promote the Holacracy gradually abandon this organizational model. But now the emergence of machine learning provides the possibility for the self-organization model to give full play to its value.

The establishment and improvement of the project self-organization mechanism is based on the basic project scenario constructed by the machine learning model.

The project scenario describes the general form of project self-organization and the basic management logic under different project organization forms. The machine learning model will use the parameter structure that is describing the project scenario to analyze the effect of different project implementation under different organizational parameter inputs. It analyzes the possible management effects of different management decisions and possible project risks before the implementation of project decision, and realize the "diagnosis" of the project. By using the machine learning model, the main body of project governance can select the appropriate stakeholders, arrange work tasks, and empower the stakeholders according to the decision-making suggestions based on project scenario. In addition, changes of project parameters will generate new simulation results in the machine learning model in real time and propose corresponding management countermeasures. According to the needs of project scenario, project stakeholders can set up a "tolerance" or "collapse" mechanism to reduce the time from generating project problem to proposing decision-making scheme, thus improving the level of agile project management.

Through the construction of project scenarios and optimization of project decision-making, the machine learning will benefit the self-organizing projects in the following three aspects.

(1) **Improving the efficiency of cooperative partner selection**

For self-organized projects, there is a large amount of information screening and quantifiable evaluation process in the partner selection process. Machine learning has advantages in efficient information collection, processing, and analysis. It can extract key information to describe the company's capacity according to work experience from past projects of different companies, and then match the project's task needs. Compared with traditional bidding methods, it is more efficient and reliable in selecting partners. Machine learning helps to select the suitable partners according to the project scenario, which benefits enterprises to reduce the work flow and saves costs.

Similarly, for the selected companies, the intelligent project partner selection based on machine learning will effectively reduce the company's publicity costs, which is conducive to small-scale and small-influential companies with strong professional capabilities to gain more project opportunities. In addition, automatic matching through machine learning can also reduce the occurrence of companies unable to undertake project tasks to a certain extent.

The machine learning model has self-learning function. The more project scenarios are used and the more entities join, the more accurate the recommendations will be. And it also helps companies to better self-assess and choose projects that are more suitable for themselves.

(2) **Supporting the decision-making process**

In self-organizing projects, project sponsors often need to invest a lot of time in coordinating the relationship between stakeholders. However, due to the different professional nature and work content, it is difficult to form a clear management relationship chain between different stakeholders. The self-centered work style will greatly increase the probability of project conflicts, leading to unfinished self-organizing projects. These conflicts are commonly difficult to avoid through traditional methods using early-stage strategic planning, but they can be dealt with and resolved by improving the agile management level of the project scenario.

The machine learning model can recommend the best decision to the main project decision-making body in real time according to the project scenario, and the stakeholders that have the greatest impact on the project will be recommended to be responsible for the decision-making of a certain stage of the project. Since the main decision-making body will shift as the project scenario changes, the stakeholders will consider the decision-making constraints of other stages when making project decisions, and then choose the decision that gives the most benefit to the project rather than to themselves.

Machine learning can largely replace the daily work of traditional managers, and realize self-management of self-organized projects by selecting the most suitable management body. Besides, it can provide effective suggestions for the management decisions of the project management body according to changes of project scenario, therefore reducing the cost of decision-making and improve the reliability of decision-making.

(3) **Monitoring project in real time and reducing project risk**

Stakeholders of self-organizing projects are connected through project tasks. Their primary goal is to complete individual tasks, so that they lack overall awareness. Therefore, stakeholders in projects often trap in local optimally and damage the overall interests of the project. For example, companies in projects often have to ignore product quality because of the intense project schedule. The introduction of machine learning will monitor the changes of project task parameters in real time. It will automatically generate response decisions and adjust project resource allocation if the project has a local optimal trend.

Machine learning with self-learning function will continuously optimize the reasonable range of project parameters in different project scenarios. It will quickly generate feedback if the changes of parameter exceed the threshold in the project. Meanwhile, the occurrence of each project risk generates historical data records in machine learning, which provides a comparison basis for future risk monitoring and realizes PDCA closed-loop management of project risks.

The introduction of machine learning-based project scenarios will provide huge value to projects in terms of project intelligent decision-making and project risk management, especially in self-organizing projects with a wide variety of resource elements and complex management relationships. The transformation of digital projects triggered by machine learning will greatly liberate project managers' time and energy, and play an important role in improving the efficiency of self-organizing project management under the background of artificial intelligence era.

References

1. Bapuji H, Crossan M (2004) From raising questions to providing answers: reviewing organizational learning research. Manag Learn 35:397–417
2. Mohr BJ, van Amelsvoort P (2016) Co-creating humane and innovative organizations: evolutions in the practice of socio-technical system design. Global STS-D Network Press
3. Bodea CN, Ding R, Stanciu O, Mitea C (2020) Artificial intelligence impact in project management. https://www.ipma.world/researchers/research-projects/
4. Bohm D, Peat FD (2000) Science, order and creativity. Routledge, New York, NY
5. Cherns A (1987) The principles of socio-technical design revisited. Hum Relat 40(3):153–161
6. Cleveland H (1985) The twilight of hierarchy: speculation on the global information society. Public Adm Rev 45:185–196
7. Couso I, Sánchez L (2016) Machine learning models, epistemic set-valued data and generalized loss functions: an encompassing approach. Inf Sci 358–359:129–150
8. Deloitte (2016) Artificial intelligence innovation report. https://www2.deloitte.com/content/dam/Deloitte/at/Documents/human-capital/artificial-intelligence-innovation-report.pdf
9. Duff A (2000) Information society studies. Routledge, London
10. Emery FE, Thorsrud E (1976) Democracy at work. Leiden, Martinus Nijhoff Social Sciences Division
11. Emery FE (1989) Toward real democracy and toward real democracy: further problems. Ontario Quality of Working Life Centre, Toronto
12. Emery FE (1982) New perspectives on the world of work: Sociotechnical foundations for a new social order. Hum Relat 35(12):1095–1122
13. European Union (2020) European enterprise survey on the use of technologies based on artificial intelligence. Euro Comm Direct Gen Commun Netw Content Technol
14. Gustavsen B (1985) Workplace reform and democratic dialogue. Econ Ind Democracy, 6(4):461–479
15. Hinton GE, Salakhutdinov RR (2006) Reducing the dimensionality of data with neural networks. Sci New Ser 313(5786):504–507
16. Hughes L et al (2019) Blockchain research, practice and policy: applications, benefits, limitations, emerging research themes and research agenda. Int J Inf Manage 49:114–129
17. Jantsch E (1980) The self-organizing universe. Pergamon Press, New York, NY
18. Jennings NR, Wooldridge MJ (2012) Foundations of machine learning. MIT Press
19. KPMG (2019) The state of intelligent automation. KPMG International/HFS Research.
20. Li J et al (2019) Big scientific data management. LNCS 11473:239–254
21. Liu L, Xu B (2018) Research on information security technology based on blockchain. In: The 3rd IEEE international conference on cloud computing and big data analysis, pp 380–384
22. Liu S (2019) Research on token incentive mechanism of open-source project—take block chain project as an example. IOP Conf Ser. Earth Environ Sci 252(2): 22029
23. Lynch S (2017) Andrew Ng: why AI Is the new electricity. A computer scientist discusses artificial intelligence's promise, hype, and biggest obstacles. Insights, Stanford Bus. https://www.gsb.stanford.edu/insights/andrew-ng-why-ai-new-electricity
24. Manski S (2017) Building the blockchain world: technological commonwealth or just more of the same? Strateg Chang 26(5):511–522
25. Masuda Y (1981) The information society as post-industrial society. Institute for the Information Society, Tokyo; World Future Society, Bethesda, Md.
26. Nakamoto S (2008) Bitcoin: a peer-to-peer electronic cash system

27. Ransbotham S, Khodabandeh S, Fehling R, LaFountain B, Kiron D (2019) Winning with AI, pioneers combine strategy, organizational behavior, and technology. MIT Sloan Manage Rev. sloanreview.mit.edu.
28. Robertson BJ (2015) Holacracy: the new management system for a rapidly changing world
29. Servoz M (2019) AI. The future of work? work of the future! On how artificial intelligence, robotics and automation are transforming jobs and the economy in Europe. The Commun Netw Content Technol Gen Direct EC.
30. Sternin J, Choo R (2000) The power of positive deviancy. Harward Bus Rev
31. Trist EL, Bamforth KW (1951) Some social and psychological consequences of Longwall method of coal getting. Hum Relat 4:3–38. https://doi.org/10.1177/001872675100400101
32. Turing AM (1950) Computing machinery and intelligence. Mind 59(236):433–460
33. Valsamis A, Tserpes K, Zissis D, Anagnostopoulos D, Varvarigou T (2017) Employing traditional machine learning algorithms for big data streams analysis: the case of object trajectory prediction. J Syst Softw 127:249–257
34. WIPO (2019) Artificial intelligence technology trends. World Intell Prop Organ, Geneva
35. Zhu J, Fu Y (2017) Progress in blockchain application research. Sci Technol Rev 35(13):70–76

Chapter 9
Identifying Organizational Issue for Digital Transformation by an Analysis Based on *Kaizen*

Hiroshi Ohtaka, Motomu Koumura, and Masahiro Isokawa

Abstract Digital transformation requires utilizing information technology (IT) system as well as collaboration between IT vendor and IT user. However, we have been observing disputes between the users and the vendors to compensate for individual loss due to failures of IT projects (IT disputes), which waste tremendous resources and opportunities. Nevertheless, not only root causes of IT disputes, but also why they failed to avoid the disputes, are not clear in most cases. The business risk caused by such IT disputes has been difficult to be visualized sufficiently enough to avoid the same dispute in the future. This paper tries to make it possible for them to manage the business risk of the IT disputes by visualizing the risk. By applying a new method based on *Kaizen* to analyze IT dispute cases of actual IT projects, where recent technologies of package software and agile are introduced for quick response to individual new challenge, we specify individual root cause and visualize a business risk, whose threat has not been understood by organizations. Furthermore, we also discuss development of improved management to cope with the threat of the visualized business risk, from the aspect of organization.

Keywords Business risk · IT dispute · Case analysis · IT system development · Organizational Project Management

9.1 Introduction

Digital transformation or other new challenges may not be achieved quickly and sufficiently, without collaboration between vendor and user of information technology (IT) system. However, we have observed disputes between the users and the vendors

H. Ohtaka (✉)
IT Mieruka Research, 775-14 Hirado-chou, Totsuka-ku, Yokohama, Kanagawa-ken 244-0803, Japan
e-mail: contact@itmieruka.com

M. Koumura · M. Isokawa
IT Mieruka Institute, 3-12-28-903 Shimoochiai, Shinjuku-ku, Tokyo 161-0033, Japan
e-mail: info@it-mieruka.com

R. Ding et al. (eds.), *Research on Project, Programme and Portfolio Management*, Lecture Notes in Management and Industrial Engineering,
https://doi.org/10.1007/978-3-030-86248-0_9

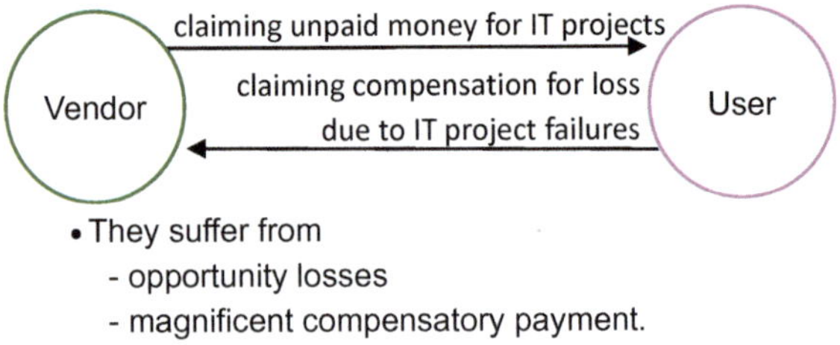

Fig. 9.1 IT dispute due to project failures

to compensate for individual loss due to failures of IT projects (hereafter IT disputes) as illustrated in Fig. 9.1.

When such disputes occur, both user and vendor consume tremendous resources and times to search and explain many evidences for proof, and they suffer from opportunity losses which could be avoided if they did without the IT disputes, even when winning a case. And when losing a case, one suffers from magnificent compensatory payment. Such IT dispute is becoming a major business risk, since it hiders the challenges and sometimes threaten business continuity of company organizations.

Nevertheless, not only root causes of IT disputes, but also why and how their companies failed to avoid the disputes, have not been clear in most cases. Just like IT systems are difficult to be visualized, business risk caused by such IT dispute has been also difficult to be visualized sufficiently. Meanwhile, we have observed similar disputes again and again.

This paper tries to make it possible to manage the business risk of the IT disputes, by visualizing the risk for the user and vendor companies to avoid the risk.

First, by reviewing legacy methods, we clarify that they have failed to specify root causes of the IT disputes and have failed even to show why and how the disputes occur. A new approach is necessary to analyze IT dispute cases. Thus, we present a new method to analyze project cases based on *Kaizen*.

We next apply the method to analyze IT dispute cases which occurred after abort or tremendous cost overrun in recent IT projects, where package software products or agile technology is introduced for quick response to new challenges. We also specify individual root cause and visualize a business risk, whose threat has not been understood and has been overlooked by organizations.

Furthermore, by investigating previous products of societies of project management, we clarify that they have not considered such risk so far. We also discuss development of improved management to cope with the business risk from the aspect of organization.

9.2 Previous Methods for Project Case Analysis

We review the following methods for analyzing project cases, and verify whether they have specified cases of the IT disputes or not.

Quantitative methods have been practiced mainly by academia. For example, Furuyama et al. [2] have many empirical research achievements to prove recommended processes for IT project managers to practice, by statically analyzing disclosed quantitative data of actual development project cases, while the cases themselves are closed. They analyzed data of successful project cases as well as some project cases with minor problems. However, data of serious problem projects (hereafter SPPs) like IT dispute cases are excluded in their statistical analysis, because a book which provided the data source [3] says that if difference between the data and the mean value exceeds the limit, then they are treated as statically singular point (quite rare cases). Quantitative approach by Furuyama and other recent researchers such as Serrador and Pinto [18] have not clarified any root cause nor business risk of IT dispute so far.

On the other hand, qualitative methods have been conducted mainly by practitioners. Smith [19, 20] proposed that troubled IT projects should have originated from 40 causes, by analyzing cases of many troubled projects. However, since he did not disclose the cases, there is no assurance of his proposal. Moreover, there is also no assurance that SPPs including IT dispute are involved in the analyzed cases. Other than above, there have been qualitative case analysis, including Yeo [23] and Sutterfield et al. [22], Nikkei Computer [11], and Standish [21]. However, any of them have the following problems.

- We cannot expect that the root cause of IT dispute may be identified by them, since they have not focused on SPP cases.
- The validity of their proposal lacks proof, unless the analyzed cases are disclosed. They lacked efforts to disclose them by excusing that it may harm personal and corporative privacy.

Moreover, most of suggestions based on such qualitative analysis lack specifics. For example, "the management should be involved in IT project" is often suggested, however the management cannot recognize what specific business risk should be focused on, what actions should be made to avoid the risk, and when the management should participate in IT project. Since the analyzed cases have not been disclosed, they have failed to visualize the risk sufficiently enough to show why and how the risk occurred and to suggest how to avoid the same troubles.

It is IPA (Information-technology Promotion Agency, Japan) that first disclosed problem project cases with which vendor project managers can recognize project risk and understand how the same troubles can be avoided. IPA aggregated 193 raw cases mainly from vendors and disclosed necessary information of them, by concealing harmful privacy information, while preserving facts that indicate what cause made what problem in the individual original case [4, 6–8]. 97 SPP cases, which caused magnificent influence to the management, are included in the 193 cases. By analyzing the SPP cases, research advanced to *Mieruka* (visualizing the risk symptom and to suggest who should and how to avoid the SPPs [12, 13] like Toyota's *Mieruka* activities to make troubles and risks to be manageable) much specifically than legacy methods mentioned before. However, in IT dispute, user and vendor have different idea each other, regarding problem, mistake, and causal relation

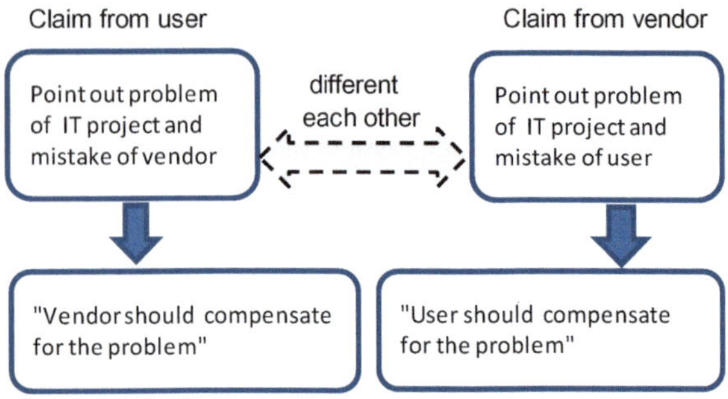

Fig. 9.2 Example of IT dispute case which the previous methods have failed to analyze

between them. Thus, they have different claims regarding cause (who is responsible) as illustrated in Fig. 9.2. Since the IPA cases lack such information, it is difficult to specify which one of user and vendor is true and what is root cause of IT dispute.

9.3 New Methodology

To resolve failures of previous methods, we need a new method that analyze cases objectively by obtaining sufficient information from both of vender and user as shown in the following three steps as illustrated in Fig. 9.3.

(1) Conflicting claims of user and vendor.
(2) Evidences that both user and vendor acknowledged.
(3) Derivation based on the evidences without filtering.

Specifically, causes of IT dispute and process until the dispute occurred are objectively identified, based on all of the fair evidences. In particular, IT Mieruka Institute (ITMI) obtained two cases contributed spontaneously by members of ITMI, who have rich experiences of IT development in vendors and users and also had been members of sectional meeting in IPA. After applying information processing similar

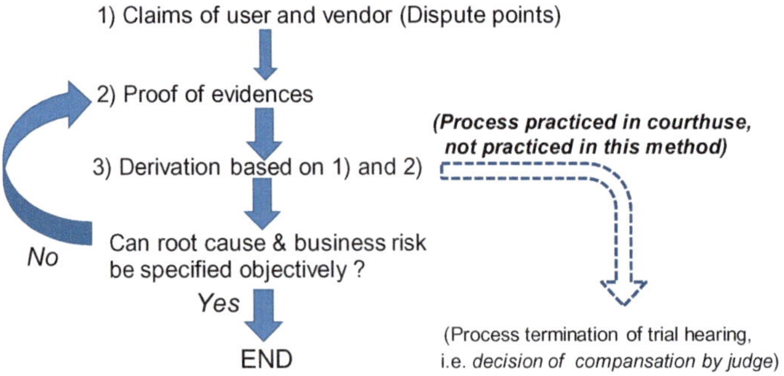

Fig. 9.3 New method for IT dispute case analysis

to IPA's (limiting disclosure by concealing harmful privacy information in the cases to satisfy mandatory privacy requirements from vendors and users) to individual case, we obtained the processed fact information (abstract of the case, (1) dispute points (claims of user and vendor) and (2) acknowledged evidences) as well as (3) derived results from the claims and the evidences.

Note that, although courthouse practices similar process, the process differs from that of the new method, which does not terminate analyzing process until root cause and business risk are specified by introducing a junction point to judge termination of the analysis, as illustrated in Fig. 9.3. On the contrary, in current courthouse, it is probable that process can be terminated when judge believe it even if the root cause is unclear, since mission of courthouse is to make judgement of claimed compensation. This may have been allowing similar IT disputes occur again and again due to the unidentified business risk. On the contrary, in the new method, the proof of evidences is continued until the root cause is identified for preventing similar IT disputes.

The new idea is based on *Kaizen*. *Kaizen* is spontaneous activity self-organized in Toyota, which keeps asking "why?" until root cause is identified. Case analysis based on *Kaizen* does not identify person by name. *Kaizen* discloses only limited evidences necessary to prove what is root cause of accidents/disputes and never disclose any evidences that may identify or even suggest any name of person and company organization. Such *Kaizen* has been practiced also in the following analysis.

9.4 Analysis of IT Dispute Cases

9.4.1 Case 1 (IT Project Which Introduced Package Software Technology)

[Abstract]

User P had a system development plan which provides a new service and presented the system requirement to vendor Vp. Vp made a proposal to develop the system by introducing Vp's own package software and present the development price, based on an assumption that the package software satisfy the requirement and Vp can reduce the development cost. Vp received the order. However, while developing the system, Vp became to be aware of necessity to develop additional software much more than it expected, since the package software does not satisfy P's full demand. Vp claimed to a large amount of pay for the additional cost for developing the additional software to P, however P refused to pay for it.

[Dispute points (claims of user and vendor)]

Claim of Vp: Our proposal clearly states that the system should be developed by introducing our software package, and we received the order on the condition that you acknowledge the proposal. The increased development cost should be paid by P.

Claim of P: We selected Vp as a vender that committed to realize our requirement with the best price. Our requirement does not involve introducing the package software. Since Vp has been ordered by the fixed contract, Vp should pay for the increased cost.

[Acknowledged evidences (Ev1–Ev3)]

Ev1: Vp had a company rule, which any sales person should not present proposal of contract with price to user, unless the sales person and the sales person clarify the price accompanied with corresponding development cost, which is estimated by engineer.

Ev2: All skilled engineers of Vp had been already assigned to other projects for other customers. A responsible upper manager told the sales person to make an approval document for the contract with P, by fulfilling a name of other engineer, who have less skill of the package software, as a responsible person estimating development cost in the document.

Ev3: The engineer practiced Fit&Gap analysis by comparing between functions of the package software and the functions written in the requirement of P, before the sales person make the proposal to P with price estimation. However, since the engineer had less skill, the analysis was not sufficient to clarify difference of business flows provided by the package software and the flows that meets with critical business process, which were not specified in the requirement of P. The difference, which had not been considered by the engineer, caused increased amount of Gap and additional software development, after the contract.

[Derived results from the claims and the evidences]

(1) Ev1 and Ev2 show that Vp have a junction system, based on the company rule for sales person and the approval document, for avoiding troubles by leaving all the decision-making of proposal or contract to sales persons. And Ev3 shows that the sales person and the engineer practiced without provoking the rule of Vp, before getting approval for the contract.
(2) Although skilled engineer could not be assigned for cost estimation, the upper manager did not give up proposing and let the sales person to receive order contract from P. Accompanied with insufficient requirement of P, this also caused increase of development cost and corresponding dispute.

9.4.2 Case2 (IT Project Which Introduced Agile Technology)

[Abstract]

User W, who had wants to create new business and service and had to remake existing IT system, ordered vendor Vw to develop a new system by agile. However, the agile project could not present any achievement expected by W. W decided to abort the project and told Vw to cancel the agreement of the order. Vw claimed the unpaid money to W.

[Dispute point (claims of user and vendor)]

Claim of W: Vw promised that its agile technology can develop a desirable system for W, before the contract. However, systems developed in every iteration in the agile project of Vw had been different from one, which realizes new business and service that W desired. W do not need to pay for Vw, who broke promise.

Claim of Vw: W ordered Vw to develop the system after agreement of semi-delegation contract. Therefore, W should pay Vw as much cost as the resources that Vw consumed for the agile project.

[Acknowledged evidences (EV4–EV7)]

EV4: The management of W requested IS (department of information system) in W to realize new system for new business and services. However, IS has insufficient resources and skill to cope with the request.

EV5: A sales person of Vw proposed IS of W to order system development to Vw, while insisting rich achievements of agile development projects executed by Vw, and explaining "Our agile technology can drastically shorten development period than legacy waterfall-based development, while realizing a system desired by user. Moreover, Vw takes care of everything required to execute agile project, even if user has no experience of agile development. User is just requested to acknowledge deliverables in every iteration, which are provided from team leader of agile project assigned by Vw".

EV6: IS of W, with less experience of agile development, believed quick renewal of existing system, and got approval of the contract from the management of W.

EV7: After the contract agreement, the sales person of Vw resigned and got a new job in another company. However, no history of his sales proposals were preserved in Vw organization. The management of Vw did not take any control of sales activities, although he did as far as technical activities of IT projects are concerned.

[Derived results from the claims and the evidences]

(1) Although agile has potential to shorten development period, by deterring requirement definition more quickly in its iteration process, it is not a magic, which can be applied to every cases unconditionally. It's a mandatory condition that user takes role of requirement definition, which are usually took by product owner (PO) in agile project. If the condition is not satisfied and user leaves all works of PO to vendor, there is no assurance that the agile project achieves its objective. Actually, EV4, EV5, and EV6 show that agile project could not achieve its objective and just repeating iterations, since W left most of works of requirement definition to Vw. It is thought that the agile project aborted because W did not take role of requirement definition itself.

(2) EV5 and EV7 show that the management of Vw just left most of works of getting order from W to the sales person. After he tried to somehow complete the sales mission, "uncontrollable sales" occur, where his proposal become a magic that allowed IS of W to understand that IS has less obligation to define requirement, if the proposed contract is agreed. If the "uncontrollable sales" were controlled by organization of Vw, trouble due to agile project, which is self-organized without involving PO in it, and IT dispute could be avoided.

Table 9.1 Summary of analysis of IT dispute cases

Case	Cause of project abort	Trigger of trouble due to the cause
1	Insufficiently detailed requirement definition by user	Vendor proposed its package solution for user failing to investigate Fit & Gap in user requirement
2	Insufficient skill and resource to define requirement in user	Vendor proposed its agile solution for user to be free from role of requirement definition (product owner)

9.4.3 Summary

Each cause and each trigger of the two trouble cases can be summarized as shown in Table 9.1, based on case analysis mentioned above.

9.5 Visualizing Business Risk Based on the Analysis

Conclusions of previous case analysis have been unclear as far as avoiding serious troubles like IT dispute. For example, "the management should be involved in IT project" has been often concluded by the analysis, however since the management can have less vision of what specific action should be made in actual company, such conclusions have failed moving forward further improvement in actual for many years. To make clear specific action in actual, we first consider the actuality of user and that of vendor, and then visualize the specific business risk based on the analysis accompanied with the actuality.

[Actuality of User]

IPA [5] has been alarming that if user orders to develop IT system with insufficient requirement definition, it is highly probable that troubles such as delay or cost overrun of the IT project. IPA and METI disclosed a principal that buyer (user) should take responsibility of requirement definition [5, 10] to avoid troubles like IT disputes (that is, they made a protective wall to prevent the troubles as illustrated in Fig. 9.4). Thus, user is required to make every effort to observe the principle by taking the responsibility by strengthening its ability to define requirement of IT system.

However, it is not easy matter for user to observe the principle, since most of IS in the user has been allocated limited resources in actual. User, who has insufficient power of requirement definition, are apt to rely upon help of outside vendor.

In such situation, it is easier for sales proposal of vendor to let user misunderstand that there exists a magic which breaks a hole in the protective wall, and let user go forward through the hole without taking sufficient requirement definition obligation. However, the magic is just like "silver bullet", which is denied its existence as far as software are concerned by Brooks [1]. Actually, it made holes in the protective wall, but caused troubles of IT projects and progressing to IT disputes, which are proved by Case 1 and Case 2, as follows.

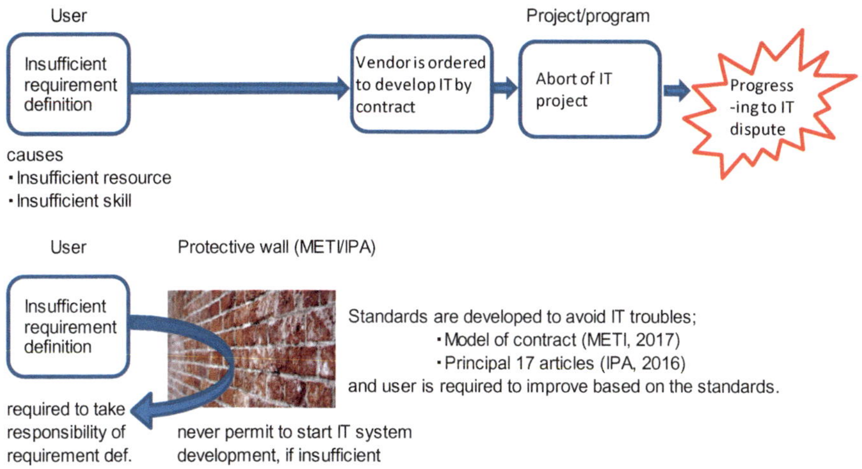

Fig. 9.4 Typical procedure until IT dispute occur and its prevention by METI and IPA

Case 1: User misunderstood that a new (package software) solution proposed by vendor must be "silver bullet", which let user free from obligation of detailed requirement definition.

Case 2: User misunderstood that a new (agile) solution proposed by vendor must be "silver bullet", which let user free from requirement definition (PO) obligation.

[Actuality of Vendor]

Vendor organization generally impose an order quota to sales person. To fulfill the quota, it is usual that the sales person proposes company's solution to user for getting order by insisting its merit. However, the sales person could lead the user to misunderstand the solution as "silver bullet", which don't have any applicable constraint (demerit), whether intentionally or not intentionally as can been seen in the following cases.

(1) Selling "silver bullet" without intention

Generally, solutions have their various applicable constraints. It is probable that sales person, who have less technical skill, may propose solutions without understanding every applicable constraint. In this case, the sales person might sell "silver bullet" without intention.

(2) Selling "silver bullet" with intention

If sales person discloses all applicable constraints to user, risk of failure to receive order increases, since the user may understand the disclosed constraints as demerits and quit order to the sales person. Particular in the organization where the upper manager of the sales person does not accept the failure and pressures the sales person to receive order from the user, the sales person is apt to avoid the risk of the failure by sealing necessary constraints to the user, while another risk of IT trouble becomes greater after receiving order. In this case, the sales person might sell "silver bullet" intentionally.

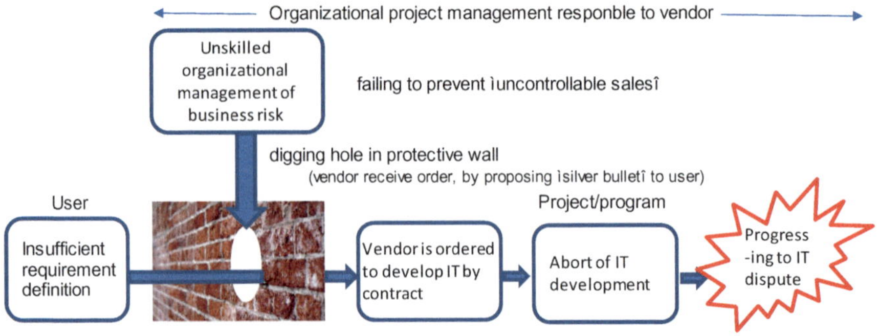

Fig. 9.5 Business risk of IT dispute visualized by the new methodology

[Visualizing Business Risk Related to IT dispute]

It is probable that "uncontrollable sales" may occur in any vendor company, if the company organization fails to control sales person not to sale "silver bullet" solution to user.

Since "uncontrollable sales" allow to dig a hole in the protective wall of Fig. 9.4, the vendor might receive order to develop IT system; however, after agreement of the contract, troubles of IT project/program and successive IT dispute might occur. This must be a business risk, which cannot be managed by project/program manager nor by sales person. No one except the vendor company organization can manage the business risk and take the responsibility to avoid the risk.

If the vendor organization, who usually pressures sales person to receive more orders, fails to prevent "uncontrollable sales", it is a proof that the organization has unskilled management of the business risk. In this sense, the unskilled organizational management of the business risk, which allows "uncontrollable sales", triggers business risk of IT dispute to occur (Fig. 9.5).

The business risk provokes dispute related to IT project/program, particularly when insufficient requirement definition is practiced by user, and have magnificent influences to both of the managements of vendor and user.

9.6 Discussion

The IT disputes, whose cases were disclosed in this paper, could be avoided, if individual user took sufficient role for requirement definition. Efforts of software engineering have been practiced, such as developing requirement engineering body of knowledge (REBOK) [9] and other books for empowering requirement definition skills in users. However, IT disputes have not been prevented for many years in reality. This indicate that there is a limit to prevent IT dispute, if we only rely upon software engineering approach.

This also indicates that it becomes necessary to take one more approach, which copes with the business risk of IT disputes, based on an assumption that incidents of insufficient requirement can be happen in reality. That is, management approach to reduce IT disputes by managing the business risk is also needed to be considered, as well as software engineering approach.

From viewpoint of management approach, the bushiness risk has its root before agreement of the contract to start project/program in Fig. 9.5. Thus, project/program manager cannot take the responsibility. However, it is clearly organizational project management of vendor that should take responsibility of the bushiness risk as illustrated in Fig. 9.5. And if user could know capability of organizational project management of vendors before contract, user could mitigate the risk of IT disputes by selecting a vendor with the best capability and eliminating vendors with poor capability.

Based on the cases analyzed and the background mentioned above, we get three questions to be discussed.

(1) Why not introducing junction system to sales process in vendor's organizational management?

Project management introduces junction system before move forward to next phase (or next iteration in agile) for avoiding trouble risk. If vendor organization allows sales person to propose system development to customer without similar junction system, it may bring serious trouble, which is proved by the actual IT dispute (Case 2). At least, organizational management should introduce junction system before moving forward to sales proposal.

(2) Why not evaluate and continuously improve capability of vendor's organizational capability?

Furthermore, even if the junction system is introduced, the vendor with insufficient capability of organizational project management could cause the actual IT dispute (as proved by Case 1). Vendor organizations are required to evaluate individual level of capability and continuously improve the level. For example, organization should investigate skill of every employees (including not only project managers and engineers, but also senior managers and sales persons) and develop skill inventory. It should use the skill inventory from elementary to Meister level at necessary milestones such as proposal or contract judgement, for avoiding business risk of IT dispute. It should also practice continuous improvement of its capability of organizational project management by evaluating its management and outcomes again and again.

(3) Why not societies related to project management considering avoidance of IT dispute also for project management of user in the future digital transformation era.

Project Management Institute (PMI) developed organizational project management standard, OPM3 [16], to measure and certificate capability of organizational project management for individual company. Although it also specifies portfolio management [14] as well as project/program management [15, 17], we cannot find any specification of risk of IT dispute like Fig. 9.5 nor find management standard for the risk. Thus, user cannot select a proper vendor for avoiding IT dispute, by evaluating individual capability of organizational project management of vendors based on the OPM3.

International Project Management Association (IPMA) and other societies related to project management also do not have scope of the organizational project management of Fig. 9.5, which involves sales activities before starting activities of IT projects/programs in vendor organization. Therefore, we cannot observe any activity to cope with risks caused by problems like "uncontrollable sales" in Fig. 9.5 and any solutions to measure organizational capability for avoiding the risks in the societies.

If the societies related to project management develop a system to certify vendor's organizational capability, and allow project manager of user to select the best vendor with higher organizational capability, IT dispute may be reduced.

9.7 Conclusion

This paper figured out a business risk of IT disputes, which previous methods have failed to visualize and also discussed how to cope with the business risk. However, too many years have been spent for collecting just two cases in this paper, because of heavy obstacles due to security policy of user/vendor companies. We will also try to ask for support from courthouses for moving forward our research of visualization in the future.

References

1. Brooks FP (1987) No silver bullet—essence and accidents of software engineering. IEEE Comput 20(4):10–19
2. Furuyama T et al (2007) Analysis of the factor that affect the performance of software projects. Information Processing 48(8):2608–2619
3. IPA (2006) A white paper of software data. Nikkei BP, Japan
4. IPA (2006) MIERUKA of IT project (lower development phase). Nikkei BP, Japan
5. IPA (2006) Principle 17 articles, pp 87–110 in Ensuring quality of requirement by involving the management 2nd ed. Ohmusha, Japan
6. IPA (2007) MIERUKA (visualization) of IT project (upper development phase). Nikkei BP, Japan
7. IPA (2008) MIERUKA of IT project (middle development phase). Nikkei BP, Japan
8. IPA (2008) MIERUKA of IT project (summary). Nikkei BP, Japan
9. JISA (2011) Requirements engineering body of knowledge (REBOK), 1st edn. Kindai Kagakusya, Japan
10. METI (2007) Model transaction/contract, 1st edn. https://www.meti.go.jp/policy/it_policy/keiyaku/index.html. Accessed 11 Sept 2020
11. Nikkei Computer (2008) 2nd investigation of actual situation of projects. Nikkei BP, Japan
12. Ohtaka H, Fukazawa Y (2010) Managing risk symptom: a method to identify major risks of serious problem projects in si environment using cyclic causal model. Project Manage J 41(1):51–60

13. Ohtaka H, Fukazawa Y (2011) Analysis of causes of serious problem projects focusing on stakeholders. J Soc Project Manag 13(3):19–25
14. PMI (2008) The standard for portfolio management, 2nd edn. PMI, USA
15. PMI (2008) The standard for program management, 2nd edn. PMI, USA
16. PMI (2013) Organizational project management maturity model, 3rd edn. PMI, USA
17. PMI (2016) A guide to the project management of knowledge (PMBOK), 6th edn. PMI, USA
18. Serrador P, Pinto KJ (2015) Does agile work?—a quantitative analysis of agile project success. Int J Project Manage 33(5):1040–1051
19. Smith J (2001) Troubled IT projects—prevention and turnaround. IEE, London, UK
20. Smith J (2002) The 40 root causes of troubled IT projects. Comput Control Eng J 109–112
21. Standish Group (2020) CHAOS report. Standish Group, USA
22. Sutterfield S et al (2006) a case study of project and stakeholder management failures: lessons learned. Project Manage J 37(5):26–35
23. Yeo K (2002) Critical failure factors in information system projects. Int J Project Manage 20(1):241–246

Chapter 10
Identification of Governance Structures for Private–Public Partnership (PPP) Project Through Social Network Analysis

Zhixue Liu, Xinyi Song, Lei Wang, Rui Song, and Itai Lishner

Abstract The governance risks, defined as the likelihood, impact, and manageability of stakeholders' uncertain behavior, vary in different Private—Public Partnership (PPP) projects. While previous research focuses on project management risks in PPP project, such as quality risks, safety risks, cost risks, etc., only few addressed the governance risks that derived from the stakeholders and their governance relationships. The purpose of this study is to find the basic units that constitute the governance risks associated with PPP projects. The paper categorized the project stakeholders into four basic roles and established four basic project governance structures through different role combination. The project governance risks were identified by analyzing the project governance structure through Social Network Analysis (SNA), and this model was validated through a case study of a PPP project in China. This research provides a framework to support the decision-making process for PPP stakeholders by identifying the governance risks and providing guidance on how to mitigate these risks.

Keywords Governance structure · Governance role · Project risk · PPP project

Z. Liu · L. Wang (✉)
Shandong University, 27 Shanda Nanlu, Jinan 250100, China
e-mail: wangklei@sdu.edu.cn

Z. Liu
e-mail: zhixue@mail.sdu.edu.cn

X. Song · R. Song
College of Design, Georgia Institute of Technology, Atlanta, GA 30332, USA
e-mail: xinyi.song@design.gatech.edu

I. Lishner
Technion—Israel Institute of Technology, 3200003 Haifa, Israel

R. Ding et al. (eds.), *Research on Project, Programme and Portfolio Management*, Lecture Notes in Management and Industrial Engineering,
https://doi.org/10.1007/978-3-030-86248-0_10

10.1 Introduction

As a delivery method of public facilities or services, Private—Public Partnership (PPP) model constitutes a systematic change in the fields of infrastructure and public services [2, 10]. It is a long-term cooperative relationship between the government and private investor. PPP project model can solve the government's financial funding gap [3]; improve the quality, quantity, and efficiency of public facilities and services [22]; and is widely used in the fields of economic and social infrastructure [12]. The PPP model has made tremendous development in China in recent years [17]. According to statistical data from the Ministry of Finance of China, as of December 31, 2019, the Ministry of Finance has a total of 12,341 PPP projects on the statistics list, with a total investment of 251 billion dollars.

A project is a platform for stakeholders to realize their needs. The government hopes to ease the financial pressure caused by the construction of public facilities through PPP projects [1], while private investors wish to increase their profits through these projects [4]. The other project stakeholders, such as the suppliers and the consulting companies, also hope to benefit from the project. The project stakeholders start from the irregular organized and self-organized according to the needs of the project, and finally form a stable project structure. Project governance defines the responsibilities, processes, and systems of the project to promote its implementation, to ensure that it achieves project goals, and ultimately to meet the needs and interests of the stakeholders [13]. Therefore, as the basis of project governance, a reasonable set of project stakeholders' self-organizing methods are required in order to regulate and restrict the behavior of each party, which is the project governance structure. The project governance structure is formed by the key stakeholders and their relationships in the project [20], including their roles and responsibilities.

There are multiple analyses of PPP projects' risks in the present research. On the one hand, some of these analyses focus on the external environment of the project, including political risks [23], economic risks [2], etc.; on the other hand, other analyses focus on project's process parameters [15], including the project duration [21], quality, cost risk [7], etc. The environmental risks of the project are not caused by the project itself, and the implement of the project does not affect the generation of such risks. Therefore, this research will not discuss these risks. As for the risks involved in the project's process, i.e., those that are related to the project duration/quality, can be solved through corresponding management methods or procedure by project manager, existing research only focuses on the solution of them, but ignored the cause of them, for example, the inability of incompetent project stakeholders to contribute to the project [9], and, more importantly, unreasonable relationships between the stakeholders might result in insufficient willingness or resources to complete the task [11]. These risks are common in the governance structure formed by project self-organizing. Therefore, it is crucial to analyze the source of these risks existing in the project governance structure.

PPP project governance has more difficulties than the traditional construction project governance [5]. The stakeholder's composition in a PPP project is complex, and it involves the government, private investors, and other numerous stakeholders [27]. The risk of PPP project governance lies in complexity of the status of the government and private investors in the project, resulting in the complex role positioning of other stakeholders in the project. Various project governance models have formed different project governance structures. In order to maximize the realization of their own needs, various stakeholders must adopt different strategies. Effective identification of the roles of various stakeholders in project governance, as well as the establishment of a matching project governance structure is critical to the effectiveness of project governance and project success.

In order to express the relationship in the self-organizing structure between stakeholders more clearly, this study established the governance structure in network form through SNA method. SNA is a commonly used method for network analysis. Stakeholders and their relationships can be expressed using network nodes indicators and structural indicators [25]. SNA methods can effectively analyze the problems of governance roles and structures.

10.2 Identification of the Project Governance Roles

The project role refers to the combination of the responsibilities and rights of the project stakeholders given the project tasks. "Role" refers to a person or an organization, and some specific responsibilities and rights attached to it. The network formed by project stakeholders is, in fact, a network of responsibilities and interests. Stakeholders have both shared needs and responsibilities. Their responsibilities are related to the implementation strategy, aimed to meet the needs. Focusing on the needs of project's stakeholders and key success factors of the project, from the perspective of project governance, the project roles can be separated into four basic governance roles: planning, operation, maintenance, and monitoring [6]. These four roles could realize the fundamental purposes of project governance, i.e., direction (i.e., setting the goals), control (i.e., ensuring projects achieve said goals), and holding project members accountable (i.e., creating accountability for the outputs, outcomes, and benefits initially promised) [14].

The "Planning" role in this research refers to the department and personnel who propose the project's requirements and the project's implementation plans. The project is a highly uncertain system and during the completion of the project, its goals and implementation methods will change as it progresses. Some stakeholders, such as government and the company supervising the project, should take responsibility for clarifying the project's overall goals and implementation method. Specifically, the planning role specifies a series of processes, and the project goal is to be achieved through this series of processes. At the same time, the planning role regulates each party in the process and clarifies the tasks and responsibilities of these parties. The planning role not only clarifies the goals of the relevant parties, but also points out some means to accomplish these goals. Stakeholders who generally assume a planning role are the key elements of project governance, such as the government, who clarify the participants and implementation methods of PPP projects.

The "Operation" role in the present research refers to the department and personnel who perform specific tasks in order to meet the needs of the planning role. Stakeholders who play an operation role in a PPP project are mainly responsible for completing the various plans set by the planning role and completing the tasks of each phase in the project according to the requirements of the planning role. In a PPP project, an operation role is carried out by a stakeholder responsible for a specific job, or a stakeholder who is ultimately responsible for a project task. The performance of those who carry out the operation role will directly determine the nature of the project's completion. Common stakeholders who carry operation roles in PPP projects, such as construction units, assume specific tasks for project construction.

The "Maintenance" role in the current research refers to the departments and personnel who provide resources and tools to other roles. Project stakeholders should be supported by sufficient external environmental resources in order to complete project tasks. The stakeholders responsible for providing various types of information and resource to support other stakeholders in the project carry out the maintenance role of the project. On the one hand, the maintenance role can involve the providing of physical support, such as human resources, materials, and mechanical equipment to the stakeholders carrying out the operation role. On the other hand, it can also provide support in the form of soft knowledge, such as methods and technologies for the planning or operation role stakeholders. Although some maintenance roles also directly undertake the project tasks of the planning role, they are essentially different from the operation roles. The maintenance role is generally targeted at some key links in the project execution process that involves specific resources and technical methods. Management consulting units, suppliers, etc. often play a maintenance role and provide various types of support for the project execution.

The "Monitoring" role in the present research refers to the departments and personnel who supervise, manage, and evaluate the actions of the planning, operation, and maintenance roles. Lack of effective constraints and restrictions of the project implementation process will lead to uncontrolled behavior and will put the project at risk. In the project governance, stakeholders who take responsibility for the supervision and evaluation assume the monitoring role. On the one hand, such monitoring role can constrain the stakeholders by evaluating their behavior. On the other hand, the monitoring role also motivates stakeholders based on the project's efficacy, thereby increasing their motivation to complete the project. There are two main types of monitoring roles in PPP projects: one is the internal stakeholders of the project, such as the supervision company; the other is the external stakeholders of the project, such as the public company.

After determining the governance roles, in order to complete the project, it is necessary to establish the relationships between the governance roles. The relationships between PPP project governance roles can be regarded as a social network-based regulatory system. It is important to establish relationships between project governance roles so that project stakeholders can truly form a reliable value alliance in order to effectively mitigate future project risks. The relationship between the stakeholders in a PPP project is not single dimensional, but rather a complex multi-dimensional interaction relationship, which mainly involves multiple aspects such as resource exchange, contract, and information exchange between project stakeholders. Defining the role relationships between project stakeholders is an important part of establishing a project governance role relationship. Together with the identification of project governance roles, it constitutes the establishment of a project governance structure.

10.3 Classification of Project Governance Network (PGN) Structure

Social Network Analysis (SNA) provides a basic method for analyzing the network structure of project governance. The theoretical foundation of SNA is based on graph theory and sociological theory [25]. SNA abstracts organizational members into the nodes of the network and simplifies the relationship between them into lines between nodes, so as to establish a complete organizational network which is the research foundation. SNA can describe the relationships between stakeholders through constructing conceptual network structural model and represent the characteristics of project network structure quantitatively by giving a variety of network indicators [16, 28]. SNA has shown strong abilities in the following project areas: (1) accurate representation of project structures and process method and (2) interdependence in network-based project organizations [28]. This study uses the indicators of the SNA as the basis for the project governance structural classification and proposes project governance schemes under different project governance structures from the perspective of adjusting the PGN structure.

The overall effect of project governance depends on whether the project governance structure encourages the interaction of information, the transfer of resources, and whether it can achieve effective incentives and constraints. The greater the network density, the better the transmission of information and resources in the project [8]. Tight network relationships can form strong constraints in the network. Network centrality reflects the status of stakeholders in the project governance structure, and it reflects the source of information and resources, and the control center in the project [18]. Therefore, the combination of network density and network centralization can reflect the overall status of project governance.

Network density is a common metric used to describe a given network structure in social network analysis. Generally, network density is used to measure the ratio of the number of actual relationships that exist in the network to the maximum number of relationships that the network may have [19]. In a higher network density project organization structure, there is a closer relationship between various stakeholders. Research shows that in organizations with high network density, the enthusiasm of all stakeholders to carry out work will be higher, and accordingly, the work efficiency of completing tasks will be higher [24]. Firstly, with the increase of network density, the relationship between nodes increases, so that there are enough channels for information communication and resource transfer between different nodes, and the efficiency of interaction is improved. Secondly, under a dense network structure, all parties are more likely to reach agreement on the project goals and the implementation methods because of convergence. Therefore, in a project organization with a high network density, it is easier for stakeholders to build trust, thereby facilitating the expression of their own needs and discussing possible risks in the project.

The calculation formula for the network density is Eq. 10.1:

$$Den = \frac{m}{[n(n-1)/2]}, \tag{10.1}$$

where m is the actual number of edges in the GN and n is the number of nodes in the GN.

Network centralization is another common indicator describing the overall properties of the network, which expresses the degree of concentration of the PGN to a certain node [26]. Degree centrality, closeness centrality, and betweenness centrality are three commonly used indicators to express network centralization, and some research shows that betweenness centrality can better express the ability of core nodes in the network to control the overall network information and resources [24]. On the one hand, these nodes can facilitate the interaction between other actors, while, on the other hand, they can also take advantage of their core position to manipulate information. For the nodes that have small betweenness centrality in the network, their behavior will be greatly influenced by the core stakeholders. Therefore, when the core stakeholder is at the center of the PGN, it can influence behavior expectations and manage information flow to take governance action. The calculation formula for the betweenness centrality of a certain node is given by Eq. 10.2:

$$C_{Bi} = \sum_{j}^{n}\sum_{k}^{n} b_{jk}(i) = \sum_{j}^{n}\sum_{k}^{n} \frac{g_{jk}(i)}{g_{jk}}, \; j \neq k \neq i, \; j < k, \tag{10.2}$$

where $b_{jk}(i)$ means the ability that node i controls the communication between node k and j.$g_{jk}(i)$ means the number of the shortest path between node k and j that go through node i,g_{jk} means the number of the shortest path between node k and j.

The calculation formula for the network centralization is given in Eq. 10.3:

$$C_B = \frac{2\sum_{i=1}^{n}(C_{Bmax} - C_{Bi})}{n^3 - 4n^2 + 5n - 2}. \tag{10.3}$$

According to the indicators of the social network and the basic structure of the PGN, the types of PPP project governance can be divided into the following four categories (shown in Fig. 10.1): strong relationship-strong center which is the alliance

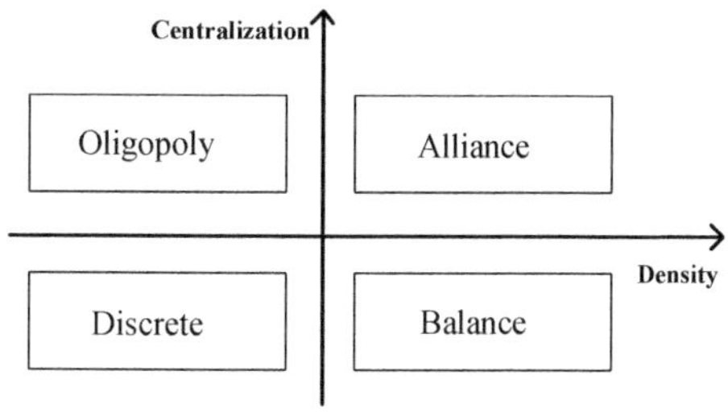

Fig. 10.1 A structural classification of project governance structure

structure; strong relationship-weak center which is the balance structure; weak relationship-strong center which is the oligopoly structure; and weak relationship-weak center which is the discrete structure.

Balanced project governance structure means that there is no node with extremely high network centrality in a closely connected PGN (shown in Fig. 10.2). The characteristics of the relationship and network status of various stakeholders in a balanced PPP project governance structure are as follows. Stakeholders have close relationships, and different stakeholders can fully interact and exchange resources and information. Different project governance roles can obtain the external support required to perform their role tasks through this structure. The role of each stakeholder in the project governance is relatively equal, which leads to effective containment and restraint between all parties. However, there are also problems with this type of project governance structure. Governance subjects cannot fully exert the influence of their power and status, and it is easy to produce a stalemate in the decision-making of the project.

Alliance project governance structure refers to the PGN where there exists two or more core stakeholders, and marginal stakeholders tend to form alliances with core stakeholders in order to alleviate their project pressure (shown in Figs. 10.3 and 10.4). It has obvious characteristics: the network centralization of the project governance structure is high, and the network density is high. It means that stakeholders in the project are closely connected, and there are two or several nodes occupying a high degree of centrality in the network. In PPP projects, stakeholders who play the role of project governance planning and operation often occupy a central position. The stakeholders in the core position of the network often have most of the resources or the power of resource exchange in the project. Other stakeholders in the network tend to establish alliances with the stakeholders occupying the core position in order to

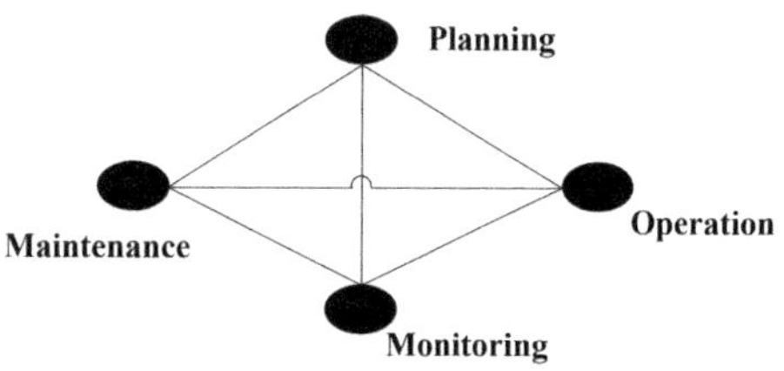

Fig. 10.2 Diagram of balanced project governance structure

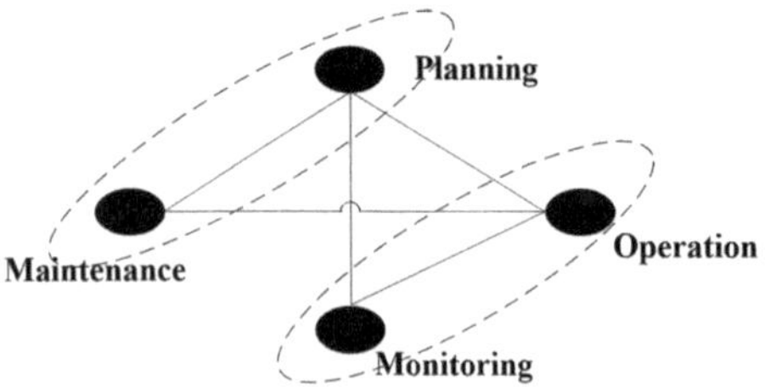

Fig. 10.3 Diagram of one alliance project governance structure

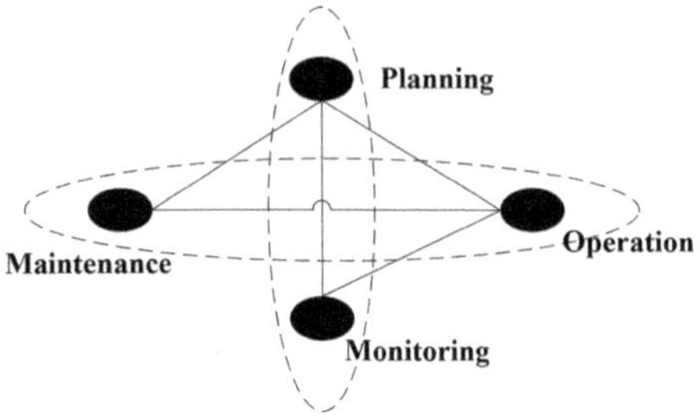

Fig. 10.4 Diagram of another alliance project governance structure. *Note* The dotted circle indicates possible alliances

effectively respond to the pressure of the network structure, to effectively grasp the project's network resources. In order to effectively exercise their privileges, the core parties in the network also tend to establish alliances with other parties in the network to cope with the pressure of other network core nodes. Therefore, the supervision nodes and maintenance nodes in the network often establish alliance relationships with the planning and operation nodes, respectively, thereby forming a situation of multi-center containment on the network, which promote the balance of the network structure. However, project risks are easily generated under this alliance structure. For example, the alliance between supervision and operation nodes will hamper the ability of the planning nodes to effectively control the behavior of operation nodes, which is not conducive to the overall control of the project.

Oligopoly project governance structure is defined as a project governance structure where there is only one core node in the project network, and the interaction between other nodes needs to go through the core node (shown in Figs. 10.5 and 10.6). In this type of network structure, the network density is small and the network is gathered to the core nodes. The project stakeholders of the core nodes have taken the lead of the entire project. The interaction and communication of resources and information

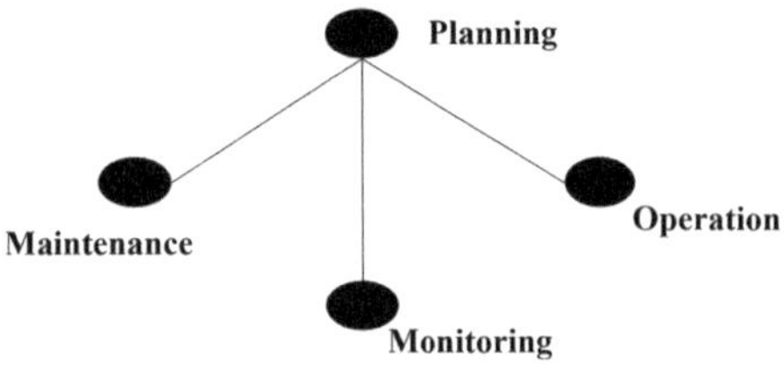

Fig. 10.5 Diagram of one oligopoly project governance structure

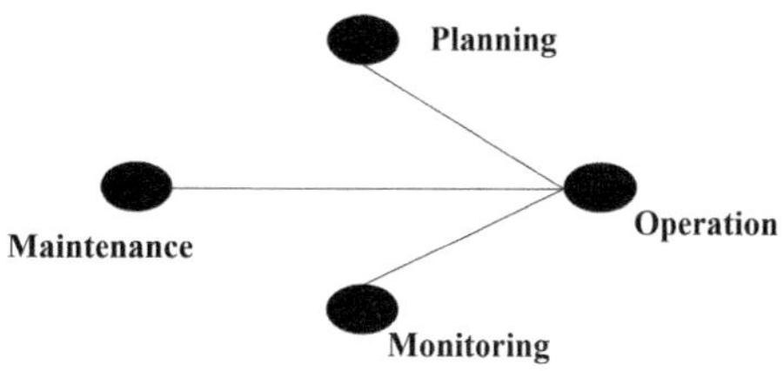

Fig. 10.6 Diagram of another oligopoly project governance structure

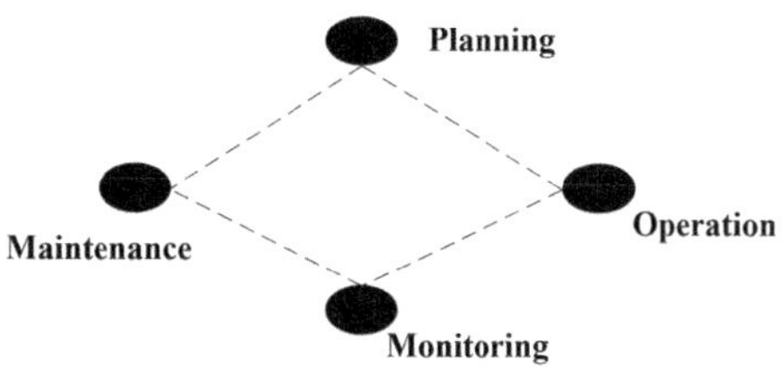

Fig. 10.7 Diagram of discrete project governance structure. *Note* Dotted line means there are weak link between the node

between other stakeholders need to go through the transition of core stakeholders, and the core project stakeholders have a network core position to control and influence all project stakeholders. The project governance role that often serves as the core node in a project, is a planning role and an operation role. This type of project governance structure can facilitate the main governance subject control of the project and ensure that project results do not deviate from project expectations. However, this type of project structure also has project governance risks. In a PPP project, the government department that assumes the planning role is too powerful to occupy the core nodes of the network, which will not be conducive to the protection of the rights and interests of private investors. If the private investor who assumes the operation role occupies the core position of project governance, the enterprise with the purpose of obtaining profit will not be conducive to the public welfare.

Discrete project governance structure refers to the project network that only has a potential cooperation intention between stakeholders, and the stable governance relationship has not been formed (Fig. 10.7). A discrete project governance structure often appears in the early stages of project planning. The density of the PGN is low and the network centralization is low, indicating that there is a cooperation intention between the parties, but a stable project governance organizational structure has not been formed yet. The role positioning and relationship need to be further clarified. With the development of the project, the project governance structure will actively evolve into the other three types of project governance structure. Without sufficient resources and support to promote this evolution, the project will eventually fail.

10.4 Case Study

A sewage treatment plant project will be selected as a case to establish its project governance structure at each stage of the project and identify the risks of each stage according to different project governance structures. The selected project fits for the establishment of the project governance network due to its long duration, clear declared phases, and clear definition of the relationship between the various parties.

The DY project, which located in a province in central China, is a sewage treatment facility construction project carried out by the government in order to solve the sewage treatment problem of enterprises and residents in industrial parks. The project is carried out using the PPP model. The objective of the project is to build a sewage treatment facility with a daily treatment capacity of 40,000 tons, as well as other infrastructures such as office buildings and auxiliary workshops. The total investment in the project is about $28.3 million.

The government selected the BT Environmental Company as the private investor through public bidding. BT Environment Company and DY City Wastewater Treatment Company jointly funded the BR Water Company (the project company). The government and the project company signed a PPP project contract; the project company is responsible for the construction, operation, and maintenance of the project; and the government exercises supervisory power during the project construction and operational processes. After the cooperation period expires, the project company will transfer the project assets and related rights and interests to the government or its designated agency for free.

Case studies typically combined data collection methods such as interviews, questionnaires, and observations. Three interviews set from January to March 2020, due to the COVID-19 pandemic, the interviews conducted remotely. We interviewed five of the project key contributors which have deep understanding of the project: the project manager, the project engineer, and three of the project's consultants. The project manager was involved in the project construction and operation and the project's consultants conducted a mid-term evaluation of the project after the end of the project construction period.

We analyzed the interviews outcome by using a table to sort the key issues mentioned in the interviews. The interviewees reviewed the results and confirmed the forms. After two rounds of interaction, the identification results of the key stakeholders of this project were finally obtained (Table 10.1).

The total duration of the project is 27 years, and 29 key stakeholders were identified through our interviews participating in the project in three stages, including the planning stage, the construction stage, and the operational stage (Table 10.2). Since the project planning begun in January 2015, it has gone through the first two phases and entered the formal operational stage in July 2018.

(1) Planning stage

The PGN at the planning stage is shown in Fig. 10.8. The network was set based on the interview result and was modified by the academic experts, who are familiar with this project.

Table 10.1 Stakeholders and their code

Stakeholders	Code	Stakeholders	Code	Stakeholders	Code
DY Government	S1	RH Consulting company	S11	Material Supplier I	S21
Project Leading Group of DY government	S2	Project Manager of RH Consulting company	S12	Material Supplier J	S22
DY Housing Construction Bureau	S3	BT Environmental Corporation	S13	YJ Pharmaceutical Company	S23
DY Finance Bureau	S4	DY City Sewage Treatment Company	S14	WH Testing Company	S24
DY Finance Bureau PPP Center	S5	BR Water Company (Project Company)	S15	SJ Certified Public Accountants	S25
DY Development and Reform Bureau	S6	BR Water Company Project Department	S16	BMW Consulting company	S26
DY Natural Resources Bureau	S7	GD Bank	S17	Enterprises and residents of DY Industrial Park	S27
DY Environmental Protection Bureau	S8	HY Construction Company	S18	Production and Operations Department of BR Water Company	S28
DY Audit Bureau	S9	Project Manager of HY Construction Company	S19	DY Water Company	S29
ZN Municipal Design Institute	S10	DY Supervision Company	S20		

Table 10.2 Project stage division and identification of stakeholders in each stage

Stage	Stage task	Stakeholders
Planning stage (2015.01–2017.08)	The government initiates projects, selects social investors, and establishes project companies	S1, S2, S3, S4, S5, S6, S7, S8, S9, S10, S11, S12, S13, S14, S15
Construction stage (2017.09–2018.07)	Project construction according to project objectives	S3, S13, S14, S15, S16, S17, S18, S19, S20, S21, S22
Operational stage (2018.07–)	Maintain the operation of project results and obtain project benefits	S23, S24, S25, S26, S27, S14, S13, S15, S28, S8, S3, S4, S29

It can be calculated by UCINET6.0, which is a software commonly used to calculate network indicators, that $Den = 0.09$ and $C_B = 0.71$. The two nodes with the highest centrality in the network are S2 and S14, with $C_{BS2} = 69$ and $C_{BS14} = 24$. According to the quantitative results, the network density is relatively low and the network centrality is high, the S2 node occupies the single core position of the network, so the network has obvious characteristics of oligopoly project governance structure. The S2 node is the oligopoly node of the network.

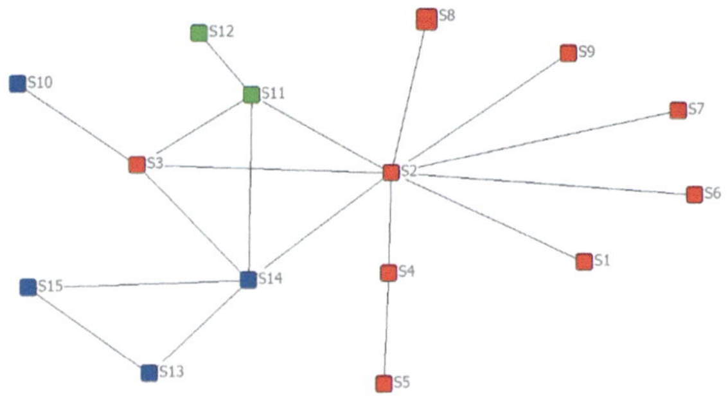

Fig. 10.8 PGN of DY project at the planning stage. *Note* Red nodes indicate planning roles; blue nodes indicate operation roles; green nodes indicate maintenance roles; and yellow nodes indicate monitoring roles

Through the identification and analysis of the project governance role and project governance structure, the PPP project has many governance risks at the planning stage. Firstly, from the perspective of the project governance role, there is a clear lack of roles in the project stage, that is, the lack of independent stakeholders who assume the monitoring role, which will eventually lead to uncontrolled project execution and greater process risks, especially regarding S2, which is at the core position of communication and resource conversion between various parties. The lack of supervision by the monitoring role will lead to unrestricted behavior of the node, which will lead to inefficient project execution.

Secondly, from the perspective of the project governance structure, S2 occupies the core position of the network and has the power of transmitting information and resources at the project stage. Although an important participant in the project, private investors are in a marginal position in the PGN. The status of both parties in the network shows that the interests of private investors may not be fully protected. In addition, the S2 node, which is at the core of the network, is a centralized node in relation to all the planning nodes in the network, and the question of whether it has the ability to assume corresponding responsibilities also poses a risk to the project.

(2) Construction stage

The PGN at the construction stage is shown in Fig. 10.9. The quantitative results of the PGN during the construction phase of the PPP project shows that $Den = 0.17$

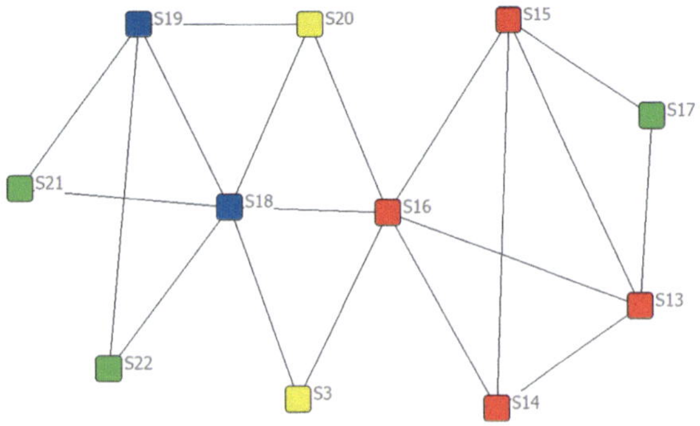

Fig. 10.9 PGN of DY project at the construction stage. *Note* Red nodes indicate planning roles; blue nodes indicate operation roles; green nodes indicate maintenance roles; and yellow nodes indicate monitoring roles

and $C_B = 0.50$. The betweenness centrality of key nodes indicates that $C_{BS16} = 24.5$ and $C_{BS18} = 17.5$. According to the calculation results, all stakeholders in this network are connected more closely, and the exchange of information and resources between stakeholders is more effective. At the same time, in this network, S16 and S18 nodes have a high degree of betweenness centrality, forming a dual-center structure in the network, which is in line with the characteristics of the alliance project governance structure.

At this stage, there is no indication of missing project governance role, but there are problems regarding the project governance role transition. Among these, the role of nodes S13, S14, and S15 has changed from operation to planning role, and the role of S3 has changed from planning role to monitoring role. When roles change at the project governance, the tasks and responsibility undertaken by stakeholders also change, which might pose a risk to the project. This is manifested by the fact that S3 nodes intervene in the project execution during the project process. The planning objectives of S13, S14, and S15 at this stage are inconsistent with the objectives of the operational tasks undertaken in the first stage.

From the perspective of governance structure, due to the existence of a dual-center structure, other network nodes tend to form an alliance with one of the nodes, such as node S20. When S20 forms an alliance with the operation node S18, the alliance will lead to insufficient guarantee of construction quality and damage the interests of planning nodes. When S20 and S16 nodes form an alliance relationship, the alliance of supervision nodes and planning nodes will exert pressure on the behavior of operating nodes. The current PGN lacks the structure that inhibits the S20 node and the two nodes from forming an alliance, so there is a greater risk to the project governance.

Operational stage

The PGN of the PPP project during the operational stage is shown in Fig. 10.10, according to which the $Den = 0.10$ and $C_B = 0.66$; the betweenness centrality of key nodes are as follows: $C_{BS15} = 47$ and $C_{BS28} = 24$, the network density is low and the network center potential is high. Accordingly, S15 has an obvious core position in the network, and the network at this stage has obvious characteristics of oligopoly project governance structure.

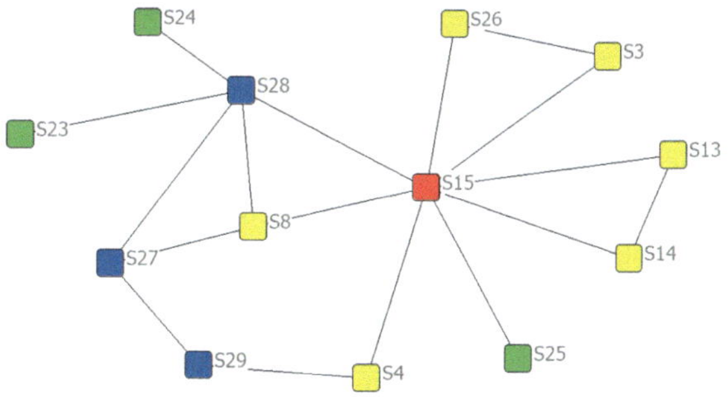

Fig. 10.10 PGN of DY project at the operational stage. *Note* Red nodes indicate planning roles; blue nodes indicate operation roles; green nodes indicate maintenance roles; yellow nodes indicate monitoring roles

From the perspective of the project governance role, the tasks and responsibilities of each role in the project governance structure at this stage are clear, and there is no obvious problem of missing roles. The roles of S13 and S14 nodes have changed from planning roles to monitoring roles. At this stage, the role of the private investor S13 has changed from planning the project construction to supervising the project execution; it benefits from the project, and there is no risk of affecting the operation of the project.

From the perspective of the project governance structure, the behavior of S15, which occupies the core position of the network, is fully supervised by many monitoring nodes, so its behavior will not lack constraints due to the particularity of its status, thus avoiding the problem of oligopoly project governance structure. Therefore, the project is operating normally during the operation phase. During the operation of the project, there was a problem of stealing sewage from operation node S27, but due to the existence of supervision node S3, the risks was controlled on time.

From the overall perspective of the project, the project governance risks mainly appear in the project planning stage and the execution stage. Since the relationship between the stakeholders has been relatively stable during the project operation phase, the risk of project governance is relatively low. From the perspective of project governance role, the main governance risks in the project planning stage are due to the project planning role. They are in the critical network position of the project, but there is a potential problem of mismatching capabilities and responsibilities, resulting in project governance risks. The governance risks in the project execution stage are mainly due to the stakeholders' role changes. As the roles change, the form of project tasks undertaken by them also changes, and the relationship between each other's powers and responsibilities also changes. Project governance risks are produced when changes have been ignored. The decision of project governance risk management can be made based on these key issues.

10.5 Discussion

There are a number of theoretical findings from the results of this research. First, in contrast to the previous studies, this paper provides a new perspective to identify project risks by investigating the governance role and relationship structure of project stakeholders. Second, different from the previous one-off risk identification, this study considers the project life cycle and identifies the governance risk in the project by stages. Therefore, the identification results from the proposed model are more accurate.

This study generates many important managerial implications as well, for example, the mismatching of stakeholder role and structure will cause project risk, and the lack of project governance role will also affect project implementation. These results will help the researchers and practitioners have a better understanding of the governance risk in the project.

The governance risks are different in different life-cycle stage. This is also justifies why we should consider the dynamic characteristics of the project. In the PPP projects in China, for example, in the planning stage, the government department will take on a very important role in project governance and have a great influence on the project. They are often at the core node of the whole governance network, controlling the progress of the project approval, so they need to invest a lot of energy in the project investigation. However, in the actual situation, the government departments often do not have enough energy to complete these tasks. Also, there are lacks of effective constraints and supervision on whether the government has completed the tasks. These comprehensive factors will have a negative impact on the project. In order to solve this problem, it is necessary to introduce government internal supervision mechanism for risk management when government departments are rarely be supervised by external supervision.

In the construction stage, lack of awareness of governance role transformation and lack of structure to avoid interest alliance are the main reasons for project risk. In this stage, the project company which plays the planning role has replaced the government department becoming the main body of project governance. If the government can't change its role effectively, it will interfere with the implementation of project company's governance measures. In this respect, the government should avoid to give a lot of instructions to the project, to avoid conflicts with the project company. For the project company, the key to reduce the project risk is to put forward effective measures to avoid the alliance between the supervision company and the construction company.

In the operational stage, as the project task is easier than the planning and operational stage, and the relationship between stakeholders is relatively clear, so the possibility of project governance risk is small, and the risk in this stage is often due to the problems of some project stakeholders themselves. Advice for the project company in this aspect is that they should pay attention to the possible changes of each stakeholder and keep the project governance structure stable.

10.6 Conclusions

Stakeholders and their relationships are important parts of the concept of project governance, and they are also the basis for analyzing project governance risks. Due to the particularity of each project, different projects have different project governance risks. Therefore, in order to improve the accuracy of project governance risk analysis, it is necessary to establish a unified process of risk analysis, identify the basic risk models, and their different combinations. The contribution of the present research derives from the fact that it begins from the perspective of project stakeholders, unify different types of stakeholders into four basic governance roles, and then divide the project governance structure into four basic types according to the different combinations of governance roles in the project governance structure. The project risk was analyzed according to the role and structural characteristics. Finally, the DY

PPP project was used to verify that the method can effectively identify the project governance risks. In this case, project governance risks were identified, and the risk identification results were verified in our subsequent interviews with project related person. In order to improve the project's governance risk identification method, we are currently conducting further case studies to validate the results. Future research will include more case studies on different type and scale of PPP projects to generalize the results of this study. Furthermore, quantitative analysis through SNA could be done to complete the qualitative aspect of the governance structures that were shown in this study.

Ethical Statement

Conflict of Interest: The authors declare that they have no conflict of interest.

Waiver: The authors declare that they complied with all guidelines given by the Ethics Committee at Project Management Institution of Shandong University. Informed consent was obtained from all participants and all data were anonymized. The research does not require ethics approval, as it mentioned in the waiver issued by the Project Management Institution of Shandong University Ethics Committee under the registered number 20200501–01, May 1, 2020.

Informed consent: An informed consent was obtained from all individual participants included in the research and the data used in this study are completely anonymized.

References

1. Amadi C, Carrillo P, Tuuli M (2020) PPP projects: improvements in stakeholder management. Eng Construct Architect Manage 27(2):544–560
2. Chan APC, Yeung JFY, Yu CCP, Wang SQ, Ke Y (2011) Empirical study of risk assessment and allocation of public-private partnership projects in China. J Manage Eng 27(3):136–148
3. Cheng Z, Ke Y, Lin J, Yang Z, Cai J (2016) Spatio-temporal dynamics of public private partnership projects in China. Int J Project Manage 34(7):1242–1251
4. Cui C, Liu Y, Hope A, Wang J (2018) Review of studies on the public-private partnerships (PPP) for infrastructure projects. Int J Project Manage 36(5):773–794
5. Dewulf G, Garvin MJ (2020) Responsive governance in PPP projects to manage uncertainty. Construct Manage Econ 38(4):383–397
6. Ding R, Wang L, Gao S, Sun T (2018) A new risk management framework for collaborative innovation projects based on network dynamic analysis. In: IAMOT 2018 conference. Birmingham, United Kingdom
7. Effah Ameyaw E, Chan APC (2013) Identifying public-private partnership (PPP) risks in managing water supply projects in Ghana. J Facilities Manage 11(2):152–182
8. Gao S, Song X, ASCE M, Ding R (2020) Promoting information transfer in collaborative projects through network structure adjustment. J Construct Eng Manage 146(2):04019108
9. Hoefer RL, Green SEJ (2016) A rhetorical model of institutional decision making: the role of rhetoric in the formation and change of legitimacy judgments. Acad Manag Rev 41(1):130–150
10. Ke Y, Wang S, Chan APC (2010) Risk allocation in public-private partnership infrastructure projects: comparative study. J Infrastruct Syst 16(4):343–351
11. Kerkhove LP, Vanhoucke M (2016) Incentive contract design for projects: the owner's perspective. Omega 62:93–114

12. Liu Y, Sun C, Xia B, Liu S, Skitmore M (2018) Identification of risk factors affecting PPP waste-to-energy incineration projects in China: a multiple case study. Adv Civil Eng 1–16
13. Muller R (2012) Project governance. Gower Publishing, Ltd.
14. Musawir AU, Abd-Karim SB, Mohd-Danuri MS (2020) Project governance and its role in enabling organizational strategy implementation: a systematic literature review. Int J Project Manage 38(1):1–16
15. Owolabi HA, Bilal M, Oyedele LO, Alaka HA, Ajayi SO, Akinade OO (2018) Predicting completion risk in PPP projects using big data analytics. IEEE Trans Eng Manage 1–24
16. Pollack J, Matous P (2019) Testing the impact of targeted team building on project team communication using social network analysis. Int J Project Manage 37(3):473–484
17. Pu W, Xu F, Chen R, Marques RC (2020) PPP project procurement model selection in china: does it matter? Construct Manage Econ 38(2):126–139
18. Rowley TJ (1997) Moving beyond dyadic ties: a network theory of stakeholder influences. Acad Manage 22(4):887–910
19. Ruan X, Ochieng EG, Price ADF, Egbu CO (2012) Knowledge integration process in construction projects: a social network analysis approach to compare competitive and collaborative working. Construct Manage Econ 30(1):5–19
20. Sanderson J (2012) Risk, uncertainty and governance in megaprojects: a critical discussion of alternative explanations. Int J Project Manage 30(4):432–443
21. Sarvari H, Valipuor A, Yahaya N, Noor NM (2014) Risk ranking of malaysian public private partnership projects. Appl Mech Mater
22. Shrestha A, Tamošaitienė J, Martek I, Hosseini MR, Edwards DJ (2019) A Principal-agent theory perspective on PPP risk allocation. Sustainability 11(22):6455
23. Xu Y, Lu Y, Chan APC, Skibniewski MJ, Yeung JFY (2012) A computerized risk evaluation model for public-private partnership (PPP) projects and its application. Int J Strategic Property Manage 16(3):277–297
24. Yang RJ, Zou PXW (2014) Stakeholder-associated risks and their interactions in complex green building projects: a social network model. Build Environ 73:208–222
25. Yu T, Shen GQ, Shi Q, Lai X, Li CZ, Xu K (2017) Managing social risks at the housing demolition stage of urban redevelopment projects: a stakeholder-oriented study using social network analysis. Int J Project Manage 35(6):925–941
26. Yuan J, Chen K, Li W, Ji C, Wang Z, Skibniewski MJ (2018) Social network analysis for social risks of construction projects in high-density urban areas in China. J Clean Prod 198:940–961
27. Zhang X, Wang L (2018) Causal relationships of risk factors in PPP WTE incineration projects. Int J Architect Eng Construct 7(3):56–65
28. Zheng X, Le Y, Chan APC, Hu Y, Li Y (2016) Review of the application of social network analysis (SNA) in construction project management research. Int J Project Manage 34(7):1214–1225

Chapter 11
Revisiting Shenhar and Dvir's Diamond Model: Do We Need an Upgrade?

Anne-Sofie Hansen, Per Svejvig, and Lars K. Hansen

Abstract In 2017, Shenhar and Dvir released their Diamond Model as a typology for project categorization with the following dimensions: novelty, technology, complexity, and pace. The Diamond Model is useful for uncovering the project type at hand with a view to selecting a suitable management style. The objective of the model is to be universal and context-free to capture a broad spectrum of projects. However, the model was built on military and commercial market product projects primarily in the United States and Israel, calling into question the validity of the model in other settings. This study addresses this problem and seeks to evaluate the Diamond Model in different settings. The study uses a mixed-methods approach and evaluates data from 62 projects in 16 project-based organizations. The study points to several ways to upgrade the model, such as splitting the pace dimension into two dimensions: pace (time) and impact. The study contributes to a broader discussion of the categorization of projects.

Keywords Agile practices · Diamond model · Mixed-methods · Project categorization typology

11.1 Introduction

The understanding that different types of projects requires different management styles has become mainstream in project management theory and practice [10]. This is expressed nicely in the aphorism that "one size does not fit all" [15]. Therefore, Shenhar and Dvir developed the Diamond Model, referred to as one of the most

A.-S. Hansen (✉) · P. Svejvig · L. K. Hansen
Aarhus Business and Social Sciences, Fuglesangs Allé 4, 8210 Aarhus V, Denmark
e-mail: ash@mgmt.au.dk

P. Svejvig
e-mail: psv@mgmt.au.dk

L. K. Hansen
e-mail: lkh@mgmt.au.dk

R. Ding et al. (eds.), *Research on Project, Programme and Portfolio Management*,
Lecture Notes in Management and Industrial Engineering,
https://doi.org/10.1007/978-3-030-86248-0_11

eloquent typologies for project categorization [9]. Accordingly, projects can be categorized into the four dimensions of novelty, technology, complexity, and pace. Each dimension is split into three or four levels, which together illustrate the project type in a diamond-shaped figure. Recognizing the project type allows recommending a suitable management style, and this may result in increased probability of project success [14]. The objective of the Diamond Model was "to build a context-free framework that would not depend on the industry, technology, or specific organization and would be universal enough to capture the wide spectrum of projects" [14]. Nonetheless, the project sample for testing the Diamond Model included only military and commercial market projects [8, 14] and they were primarily located in the United States and Israel [14]. Therefore, the validity of the Diamond Model in other industries and geographical areas can be questioned. The following research questions thus guide the study:

(1) Which challenges may the Diamond Model face when used in another setting?
(2) To what degree is the Diamond Model a context-free typology?

To address these questions, we used a mixed-methods approach [18], collecting data from 62 projects in 16 project-based organizations, most of them with a global presence. For each project, we collected data about the Diamond Model, duration, cost, project management practices, and contextual information. We analyzed the data and compared them with earlier studies on the Diamond Model. The findings contribute to the discussion about the categorization of projects and the development of project management theories [4, 5, 8].

The paper is structured as follows. Section 2 introduces the categorization of projects, followed by a review of Shenhar and Dvir's Diamond Model. Section 3 describes the mixed-methods research design used in this study. Section 4 outlines the qualitative results, while Sect. 5 describes the quantitative results. Finally, Sect. 6 discusses the findings, conclusions, and limitations of the study.

11.2 Theoretical Background

One of the myths about projects is that all projects are the same, and you can use similar tools for all project activities. This is called the "project is a project is a project" syndrome, and this understanding may lead to project failure because projects differ on various aspects [5, 12]. To achieve project success, it is vital to use a suitable management style for the project type in question [2, 6]. In other words, it is important to categorize which type of project you want to manage. Various project categorization typologies have been put forward over the years; however, the project management literature is still in need of more clear-cut distinctions [13, 15].

11.2.1 Shenhar and Dvir's Diamond Model

One prominent model addressing this call is the Diamond Model [14], which seeks to undercover the underlying dimensions differentiating projects [14]. The categorization suggested is based on four dimensions: novelty, technological uncertainty, complexity, and pace (thus, it is also called the NTCP model).

Technological uncertainty is built on the level of technological uncertainty at the time of project initiation [13]. It depends on the availability of new or mature technology for use in the project [11, 14].

Complexity represents the nested nature of systems and subsystems. It is suggested that complexity be addressed on three levels: assembly projects, system projects, and array projects. An assembly project is a collection of single elements and modules combined into a single unit; a system project is characterized by a pool of subsystems that perform a wide range of functions; and an array project is made up of a number of systems that inter-operate to achieve a common purpose [14, 16].

The *pace* dimension considers the criticality of meeting the time goals of the project—in other words, the urgency of the project's delivery [14]. Projects are distinguished on four levels of pace: regular, fast/competitive, time critical, and blitz projects, blitz projects being the most time critical ones in terms of product launch [15].

The *novelty* dimension is considered in terms of the project's product output and reflects the newness of the product. Here, it is necessary to ascertain how new the product is to the market, customers, and potential users. This dimension considers external uncertainty in the market conditions and uncertainty in how well the project goal can be defined from the beginning. The level of novelty varies from derivative projects to platform projects, and further to breakthrough projects.

11.2.2 Challenges of the Diamond Model

Even though the Diamond Model is argued to be one of the most eloquent frameworks for project categorization [9], it has been challenged. Firstly, it can be questioned whether the Diamond Model includes enough dimensions and ideal types to imply the type of project and the appropriate management style [1]. Secondly, the Diamond Model is criticized because the project sample included only military and commercial market projects [8], and primarily projects located in the United States and Israel [14]. Recent research suggests that the Diamond Model should be reconsidered and tested in other project types arguing that it is conceivable that today's extensive use of agile methods affects the ideal types and dimensions in the Diamond Model [8].

11.3 Methodology

The challenges mentioned motivated us to engage in this study of revisiting the Diamond Model. The empirical study was approached by a mixed-methods research design [3, 18] drawing on data from 62 projects in 16 Danish project-based organizations, most of them with a global presence and typically four projects per organization. The data are distributed across a wide range of technological uncertainty, complexity, pace, and novelty.

Data collection was performed from 2015 to 2019, considering projects in a wide range of industries and involving different project types. The duration of the projects investigated ranged from 28 days to 2,337 days and in financial cost from EUR 0 to 40 m. The projects were chosen due to their availability and comparability to other projects investigated. The data collection was performed in organizations in a Western culture and, due to their global presence, they may in some respects be considered representative of a wide range of project management settings in Western cultures.

For each project, we collected quantitative and qualitative data about the Diamond Model in addition to duration, cost, project management practices, and contextual information.

11.4 Qualitative Results

The analysis of the 62 projects resulted in the identification of various contextual situations, practices, and patterns regarding project management, which in various aspects support and challenge the four dimensions of Shenhar and Dvir's (2007) model.

11.4.1 Technological Uncertainty

As we looked along the technological uncertainty dimension, we found three dominating patterns in the arguments for scoring the projects: typical knowledge about the technology, the extent of change the technology caused, and the degree of modification of existing technologies. Table 11.1 shows these patterns mapped according to the level of technological uncertainty, where the argumentation for scoring the technological uncertainty dimension differs along with the level of this type of uncertainty.

Low-tech projects. Most of the projects scoring low-tech were argued to be projects based on known and well-established technologies. The projects only caused changes in respect to the work with the specific technology in the organization, and primarily involved standard technologies with few modifications.

Table 11.1 Technological uncertainty dimension

Level/variable	Low-tech (12 projects)	Medium-tech (34 projects)	High-tech (16 projects)	Super-high-tech (0 projects)
Novelty of technology	Known and well-established technologies	Existing but not well-established technologies	Relatively new technologies—unknown to the organization	
Changes caused	The project only changes work with the technologies	No changes in organizational structure, but some tasks are done differently because of the technology		
Modification of technology	Standard technologies with few modifications	Customized technology	The technology was developed but used in a new way	

Medium-tech projects. The medium-tech projects were characterized by using technologies that were not very established, but existing. The technologies used were customized for the projects.

High-tech projects. Projects scoring high-tech were using relatively new technology that was unknown to the organization before the project was initiated.

Super-high-tech projects. Not surprisingly, there were no projects scoring super-high-tech. The definition of a super-high-tech project is that "[k]ey project technologies do not exist at the time of project initiation" [14], and the number of projects that fall under this definition was expected to be close to zero as they are hard to find.

11.4.2 Pace

Distinctions along the pace dimension typically related to the time criticality or meeting a specific deadline as well as the impact on the organization regarding its competitiveness or overall success if it did not meet the deadline.

Table 11.2 illustrates the various patterns and levels on the pace dimension found inductively based on the interviewees' arguments for the scorings.

Regular projects. The projects scored as regular were characterized by not being time critical because of the capacity to adjust if any time pressure was experienced. It was not important to the success of the organization if the project was delayed, and several projects were postponed or delayed for years. The respondents experienced some internal time pressure but stated no pressure from the external environment.

Fast/competitive projects. Fast/competitive projects were argued to be time critical for some parts of the organization. These projects were perceived as important for the organization's competitiveness and were to be executed as quickly as possible to

Table 11.2 Pace dimension

Level/variable	Regular (10 projects)	Fast/competitive (30 projects)	Time critical (22 projects)	Blitz (0 projects)
Time criticality	Not time critical	Time critical for one section: the faster the better	Time critical for the organization	
Deadline		No critical project deadline	Important to reach the initial deadline	
Competitiveness	Possible to adjust the capacity if needed to become more competitive	Pace important for competitiveness		
Success/consequences for the organization	Pace not critical for the success of the organization	The faster the project is completed, the more money is saved, but this is not critical for the organization's success	The pace is important for the success of the organization	
Pressure	Internal pressure to complete the project fast but no pressure from external stakeholders	Pressure from customers (external stakeholders) to execute the project fast	High degree of time pressure to meet the deadline	
Reaction to time pressure	The project was postponed or the project was delayed for years	Did not meet the initial deadline	Did not meet the initial deadline but could not be done faster  ambitious time plan	

meet the market demands before the competitors, although it was not critical if the organizations did not meet the specific deadline. They not only experienced internal pressure to meet the deadline, but also some pressure from customers and other external stakeholders to execute the project quickly, even though the projects were characterized by being postponed or on hold for some time.

Time-critical projects. In the time-critical projects, it was important for the organization to reach the initial deadline for a project to be successful. This was also the reason why a higher degree of time pressure from stakeholders was experienced compared to the regular and fast/competitive projects. However, several of the projects investigated diverged from the initial deadline.

Blitz projects. There were no blitz projects, which was not surprising as they are defined as "crisis projects" of the "utmost urgency; project should be completed as soon as possible" [14]; blitz projects are considered to be very rare.

According to Shenhar and Dvir, the pace dimension considers the criticality of meeting the time goals of the project and the urgency of delivering the project [14]. This element is also found in this analysis where time criticality and deadlines were evident. Shenhar and Dvir further state that the pace of a project requires a specific project management style, with a higher degree of involvement from the management as the pace level increases [14]. Our analysis showed a distinction in pressure from both internal and external stakeholders as the level of pace increased. This might be another way of framing that the level of management involvement changes with the level of pace, as stated by Shenhar and Dvir [14]. This might indicate that the scoring may be company or skill dependent [12], and the arguments for the scoring can vary, too. Another reason might be the inconsistencies and multiple interpretations of the pace dimension [5], indicating that the pace dimension is not sufficient and might lead to distorted or unreliable scorings.

11.4.3 Novelty

Distinctions along the novelty dimension typically related to the project purpose, who it was new to, and the effect on the organization's work procedures. These patterns are shown in Table 11.3 where they are mapped against the level of novelty ideal types as stated by Shenhar and Dvir [14].

Derivative projects. The purpose of the projects that scored relatively low on the novelty dimension was to optimize or improve existing products, services, or

Table 11.3 Novelty dimension

Level/variable	Derivative (19 projects)	Platform (27 projects)	Breakthrough (16 projects)
Purpose of the project	Optimization of existing project or improvement	Optimization of existing project with some new elements	Improved existing processes but did it in a new way
Degree of novelty/who is it new to	Not new to the world or the organization	Not new to the world, but new to one department/location of the organization	Not new to the world, but brand new to the organization
Work procedures		The project resulted in new ways of working	New architecture—extensive transformation of the organization
Technology	No new technology used in the project	Used existing technologies in the project	

processes. The product of the project was neither new to the world nor to the organization.

Platform projects. The platform projects were also characterized as optimization projects, but they were argued to involve some new elements, which meant that the product of the project was experienced as new to some parts of the organization.

Breakthrough projects. The breakthrough projects had the purpose of improving processes within an existing portfolio, but in a new way. The project product was still characterized as not being new to the world, but it was considered groundbreaking internally in the organization.

The Diamond Model ranks novelty according to how new the product of the project is to the market, customers, and potential users [14], which was reflected in our study. Our data further showed new ways of working, which is consistent with the literature stating that the novelty depends on the extent of product and process change [1, 19]. The technology pattern found in this study may be due to confusion about the difference between the novelty dimension and the technological uncertainty dimension as the two were dealt with in a single dimension of uncertainty in the previous version of the Diamond Model [14]. Crawford et al. argue that technology and technological uncertainty are two separate attributes for categorizing projects [5]. The arguments for a breakthrough project consider "Improve existing processes in a new way", "Not new to the world, but brand new to the organization", and "New architecture—extensive transformation of the organization". The definition of a breakthrough project in Shenhar and Dvir [14], on the other hand, is "Introducing a new-to-the-world product or concept, a new idea, or a new use of a product that customers have never seen before". This indicates a mismatch between the two definitions of the breakthrough project, and the data in this study indicate a need for another novelty type, meaning it is recommended that the novelty dimension be expanded to a four-level scale.

11.4.4 Complexity

The complexity dimension was scored according to the complexity schema from the Danish Project Management Association [7] to increase the level of detail of the dimension than in Shenhar and Dvir's categorization.

As Table 11.4 shows, the patterns found in the arguments for the scoring of the complexity dimension vary across the levels of complexity.

Low complexity projects. The projects scoring low on the complexity dimension were characterized as having a project environment with a straightforward process and no changes during execution. The related tasks were known and predictable, and no subprojects were connected to the projects. The resources used for the projects were few, and the project teams were considered uniform and in one geographical location.

Medium complexity projects. Medium complexity projects were typically experienced as being of higher importance compared to the low complexity projects

Table 11.4 Complexity dimension

Assessment of degree of variable/	Level 1 (4 projects)	Level 2 (36 projects)	Level 3 (21 projects)	Level 4 (1 project)
Project environment	Not critical. Straightforward decision process. Predictable parties. No changes caused	High prioritization but not strategically important. High degree of agreement and uncomplicated process, but with several stakeholders. Few and well-known parties or many and well-known parties. Minor changes caused	Important for the organization. More complicated decision-making process. Unpredictable parties. Many changes caused, mainly internally	Vital part of the organizational strategy. Did not succeed because of political conditions. Many, partly unpredictable parties. Extensive changes caused both internally and externally
Project tasks and processes	Well-known and predictable outcome. None or only one subproject	Known outcome, but with unpredictable elements. Few subprojects	Known problem, but unknown or complex outcome. Many subprojects	Outcome unknown. Many subprojects that involve many departments/sections
Project resources and organization	Low investment and few people involved, with uniform competencies. Few and stable organizational units. One location	Limited investment, but several people with different competencies involved, though few different cultures. Several stable units involved. More locations, but primarily located in Denmark	Relatively large investment, but have done bigger projects. Different competencies and cultures of participants. Many units involved both externally and internally. Located in different countries in one or two time zones	Big investment with many resources. Diverse competencies, but supplementing one another. Many units involved, including other countries and subcompanies. Great distance between participants within several time zones

and involved some subprojects. The outcomes were known from the beginning but involved a few unpredictable elements. The project teams were primarily located in Denmark and involved several stable units and people from different companies.

High complexity projects. These involved unpredictable parties and caused many changes during the project. The projects had the purpose of solving a known problem, but the outcome was unknown at project initiation. The project teams were located in different countries and involved many organizational units and various competencies.

Super high complexity projects. These were experienced as being vital for the organization and resulted in extensive changes during project execution. Such projects were characterized by involving many subprojects and using many resources. The

Table 11.5 Duration and cost

Technology		A	B	C	D	Correlation
Cost	Mean	1,75	3,36	2,67		
	SD	0,89	1,75	2		0,199
Duration	Mean	1,9	3,07	3,44		
	SD	1,1	1,21	0,88		0,486**
Complexity		**1**	**2**	**3**	**4**	**Correlation**
Cost	Mean	1	2,47	3,75		
	SD	0	1,5	1,91		0,478**
Duration	Mean	1,67	2,76	3,33		
	SD	0,58	1,22	1,22		0,350*
Pace		**Regular**	**ist/competiti**	**Time critical**	**Blitz**	**Correlation**
Cost	Mean	2	2,56	3,11		
	SD	1,73	1,71	1,83		0,202
Duration	Mean	3,33	3,06	2,38		
	SD	0,58	1,25	1,26		-0,285
Novelty		**Derivative**	**Platform**	**Breakthrough**	**Correlation**	
Cost	Mean	1,56	2,91	3,63		
	SD	1,01	1,64	1,92	0,480**	
Duration	Mean	1,82	3,36	3,25		
	SD	1,08	1,01	1,04	0,478**	

*p < 0.05; **p < 0.01; ***p < 0.001

Table 11.6 Agile practices

	Technology			Complexity			Pace		
	ANOVA		Correlation	ANOVA		Correlation	ANOVA		Correlation
	df	F		df	F		df	F	
Co-location	3, 52	7,643**	-0,349**	3, 52	1,381	-0,092	3, 52	3,703	0,272*
Visual tools and planning	3, 54	0,026	-0,001	3, 54	1,719	0,23	3, 54	10,234**	0,423**
Reducing complexity	3, 54	0,318	-0,042	3, 54	2,753	0,198	3, 54	2,09	0,214
Focus on a fixed rhytm	3, 29	0,032	0,033	3, 29	2,103	0,321*	3, 29	8,577**	0,509**
Leadership									
Involvement of project board	3, 52	1,523	0,042	3, 52	13,003**	0,432**	3, 52	0,009	0,072

*p < 0.05; **p < 0.01; ***p < 0.001; Yellow indicates relationships which are significant

project teams were located not just in different countries, but also across different time zones. The complexity dimension is considered in different ways in the project management literature. Shenhar and Dvir [13, 14] score complexity according to three levels of system complexity. Abell [1] argues for the benefit of looking at complexity from a system and an organizational perspective [1]. Crawford et al. state that complexity should be scored based on five attributes [6]. Based on 12 questions, we scored complexity on three overall dimensions—project environment, project tasks and processes, and project resources and organization—with 12 attributes [7]. This allowed for a deeper understanding of a project's complexity. For example, such an analysis indicates a differentiation of the number of changes made during the project. This point of view would not be captured in the complexity dimension in the present Diamond Model. Furthermore, variables like geographical location, resources, and strategic importance are not explicitly reflected in the Diamond Model but are attributes used to categorize projects [5]. One potential solution could be to implement these factors in the complexity dimension, as shown in Table 11.4.

11.5 Quantitative Results

This part of the analysis had the objective of investigating two areas: (1) the relationship between duration and cost in the Diamond Model to replicate the study by Shenhar and Dvir [14] and (2) the use of selected agile practices with the Diamond Model.

11.5.1 Duration and Cost

Table 11.5 includes descriptive statistics on the resources that were used in the projects and Pearson's correlation coefficients between the four dimensions and the cost and duration of the projects investigated. The duration of the projects was recoded into 1: less than 100 days; 2: 100–200 days, 3: 201–500 days; 4: 501–1000 days; and 5: 1001 days or more. The project's cost was recoded into 1: less than DKK 500,000; 2: DKK 500,000–1 million; 3: DKK 1–5 million; 4: DKK 5–10 million; and 5: more than DKK 10 million. It should be noted that level 1 (less than DKK 500,000) primarily contains projects with zero costs. This may be because the cost variable includes internal and external costs but not labor costs, meaning a project can have zero costs when using this definition.

The analysis shows that when the level of technological uncertainty or the level of complexity increases, the duration and the cost of the project also increase. This confirms what might be common sense, and is consistent with the results in Shenhar and Dvir [13, 14], as they, too, find an association between complexity and size. Both this analysis and that of Shenhar and Dvir [13, 14] show that the project costs increase relatively more with the level of complexity than with the project duration. The increase in project cost and duration and the level of the technological uncertainty is also consistent with the findings in Shenhar and Dvir [13, 14]. The consistency of the results in Shenhar and Dvir [13, 14] and this analysis indicates that the technological uncertainty dimension and complexity dimensions in the Diamond Model are also useful for differentiating projects in terms of cost and duration in other settings.

Furthermore, the novelty of the projects investigated is significantly positively correlated with both the cost of the projects ($R = 0.480$) and the duration ($R = 0.488$), indicating the higher degree of novelty in the product of the project, the higher cost, and the longer duration of the project execution.

Now we have looked at the impact on a project's resources along the four dimensions. The next section will look into how the use of practices may vary with the dimensions.

11.5.2 Agile practices

This part of the analysis focuses on agile project management practices, which is a new point of view in this analysis compared to the analysis of project management practices by Shenhar and Dvir [14], and relates to self-organizing in projects in this study with focus on agile practices. The variables were four-level scale measures representing the extent to which agile practices were applied in the projects. The analysis includes descriptive statistics and Pearson's correlations for the various levels of the four dimensions related to the agile practices performed in the projects. An ANOVA test was added to confirm the consistency of the results. Table 11.6 shows the practices that were significantly correlated with one of the dimensions and also significant in the ANOVA test ($p < 0.001$). The full matrix can be seen in Appendix A.

Table 11.6 shows that there was significant correlation between the agile practices and three of the four dimensions.

Technology. In the dataset, it was found that higher technology projects use less effort to co-locate the project team compared to lower technology projects. The co-location practice covers allocating a minimum of 50 percent of the core team's time to the specific project. The negative correlation between the two variables may appear surprising. One possible explanation could be that high-tech projects often require more specialist knowledge, which might be harder to co-locate. Shenhar and Dvir [13, 14] found that the percentage of employees with academic degrees in the projects increased with the level of technological uncertainty. This finding supports the argumentation of more expert knowledge in higher tech projects, which may challenge the agile practice of co-location.

Complexity. The complexity dimension was highly positively correlated with the agile practice of involving the project board (0.432), meaning the higher the project complexity, the greater the involvement of the project board as a guiding support. This indicates a need for the project board to be closer to the project if the project is complex, and the role of the project board should take a supportive rather than a controlling role when the project becomes more complex. Shenhar and Dvir (1996, 2007) do not explicitly consider the role of the project board; they focus more on the project manager's role in the project. The project board tends to be a vital part of a project and this should be reflected when distinguishing among projects. As the dataset shows that the role of the project board changes with the level of complexity, it might be relevant to implement this point of view in the Diamond Model.

Pace. Faster paced projects use more visual tools in project planning. To explain this trend, one must look at the nature of the regular, fast/competitive, and time critical projects analyzed in the previous section. The regular projects were not given much attention to complete the project faster because the deadline was not critical to the organization. One potential explanation could be that the speed of the regular projects was not valued, which Svejvig et al. [17] argue is vital for actually speeding up the project. Working with visual tools requires the project team to put continuous effort into planning, which might be less prioritized if the progress of the project is not considered important. On the other hand, the interviewees experienced the pace dimension as important for the success of the organization in time-critical projects. Therefore, one potential explanation for the use of visual tools increasing along the pace dimension could be the prioritization of project speed. The pace dimension was further significantly positively correlated with the agile practice of focusing on a fixed rhythm in the project. In other words, the more time critical the project was perceived to be, the more focus the project team had on keeping a fixed rhythm in events like status meetings and agile sprints. This finding may not seem surprising. The prioritization of higher speed is very likely also associated with reaching milestones from the sprint plan, which is part of keeping a fixed rhythm in key events. According to Shenhar and Dvir [14], regular projects are characterized by paying no specific attention to procedures and processes, while the more time-critical projects use more structured procedures to shorten development cycles and have tighter schedule control [14]. This aligns with the finding of keeping a fixed rhythm in key events.

11.6 Concluding Remarks

This study set out by introducing the Diamond Model and the challenges it faces. The Diamond Model has been criticized for being too simple and for its homogeneous data sample. This motivated our study of the suitability of the Diamond Model for use in a Danish project-based context. We did this by qualitatively mapping the arguments for scores along the four dimensions, and offering a quantitative comparison of distributions and correlations of time and cost variables with Shenhar and Dvir [13, 14]. Lastly, the study looked into trends of agile practices along the dimensions in question. The first research question regarded which challenges the Diamond Model may face when used in another setting. This is presented in the following four findings.

Firstly, the original pace dimension in Shenhar and Dvir's Diamond Model is too tight to capture both the time aspect and the impact of time criticality [15]. Our data showed a need to consider both time and impact. Both aspects reflect the urgency of the project. Therefore, an overall dimension of urgency with sub-dimensions of pace (time) and impact is suggested. This expansion of the pace dimension further meets the criteria according to the simplicity of the Diamond Model.

Secondly, the value of speed shows a high impact on the practices used for project execution. Time-critical projects tend to prioritize agile practices that contribute to the flow of the project, like visual planning and keeping a fixed rhythm in key events, which potentially speeds up the project.

Thirdly, the evaluation of novelty shows that this dimension is insufficient as the novelty dimension solely reflects the product aspect. Our analysis showed a need for a process perspective in the novelty dimension to account for the changes in work procedures derived from the product novelty. With advantage, the product/process elements could be implemented in the technological uncertainty dimension. Furthermore, it is suggested that the novelty dimension be expanded to four types, as the three levels in the present Diamond Model are shown to be insufficient to account for the newness of the product.

The last major finding relates to the complexity dimension, which, based on the empirical data and the related literature, is too simple. Therefore, it is suggested that the complexity dimension be expanded to capture not only system complexity but also complexity in respect to the organization, the project task, and its environment.

The second research question concerned the statement of the Diamond Model as a context-free typology. This study finds that the Diamond Model presented by Shenhar and Dvir captures some important elements such as the novelty of the product of the project. On the other hand, some challenges when using the Diamond Model in a Danish project-based context were found. Furthermore, significant relations between agile practices and the dimensions of technology, complexity, and pace showed up. This could indicate that the Diamond Model is not entirely context-free as an upgrade embracing agile practices is needed.

One important practical implication of our research is that a lack of reflection on the key dimensions in projects might have consequences for the project, as there is a risk of following a less optimal management style decided from the Diamond Model at project initiation.

This study has some limitations that give rise to future research opportunities, as it concentrated on showing how suitable the Diamond Model is to differentiate projects when using it in a different context than those presented by Shenhar and Dvir. Future research could investigate the fit between the Diamond Model and other settings, as this research was focused on a Danish project-based context only.

11.7 Ethical Statement

We declare that we ensured the objectivity and transparency in our research and that accepted principles of ethical and professional conduct have been followed. Prior informed consent was obtained from individual participants included in the study before the research. No sensitive personal data was accessed. Anonymity of individual participant data is maintained.

The research does not require ethics approval, as it is mentioned in the waiver issued by the **Aarhus University's Research Ethics Committee** under the number 2019–616-000,009, issued on February 8, 2021.

Appendix A: Agile practices related to the four dimensions in the Diamond Model

Technology

	A Mean (S.D.)	B Mean (S.D.)	C Mean (S.D.)	D Mean (S.D.)	ANOVA df	ANOVA F	Correlation
Imapct							
Business case	2,1	2	2,23		3, 39	1,357	
	1,52	1,45	1,36				0,039
Deliver impact as soon as possible	2,2	1,65	2,08		3, 39	0,155	
	1,75	1,63	1,44				-0,015
Pulse check	1,6	1	0,85		3, 39	2,64	
	1,35	1,45	1,28				-0,194
Focus on customer value	3,25	2,67	3,13		3, 56	2,384	
	0,87	1,05	0,83				-0,018
Flow							
Co-location	3,3	2,19	1,8		3, 52	7,643	
	0,95	1,3	1,42			**	-0,349**
Fast learning	2,91	1,79	2,67		3, 56	0,167	
	1,04	1,2	0,9				-0,019
Fewer people with more time	2,4	2,09	2,4		3, 55	1,182	
	1,17	1,22	0,91				0,021
Visual tools and planning	2,55	2,5	2,53		3, 54	0,026	
	1,29	1,27	1,55				-0,001
Reducing complexity	2,55	2,63	2,47		3, 54	0,318	
	0,82	0,79	0,83				-0,042
Focus on a fixed rhytm	2,6	2,69	2,7		3, 29	0,032	
	1,26	0,95	1,49				0,033
Leadership							
People before models	2,55	2,36	2,87		3, 55	0,034	
	1,13	1,32	1,19				0,101
Freedom with responsibility	3,27	2,81	3,2		3, 55	0,208	
	0,65	1,07	0,86				0
Lead inwards	2,92	2,52	3		3, 56	0,043	
	0,29	1,06	0,93				0,049
Work with project leadership	2,9	2,35	2,85		3, 39	0,547	
	0,74	1,24	1,14				0,004
Engagement of project owner	2,7	2,05	2,25		3, 38	0,011	
	1,57	1,64	1,71				-0,093
Involvement of project board	2,17	1,76	2,27		3, 52	1,523	
	1,11	1,33	1,28				0,042
customization to the unique projec	2,5	2,07	2,3		3, 31	0,015	
	0,97	1,28	1,34				-0,064

Complexity

	1 Mean (S.D.)	2 Mean (S.D.)	3 Mean (S.D.)	4 Mean (S.D.)	ANOVA df	ANOVA F	Correlation
Imapct							
Business case	1	2,12	2,29		3, 39	1,31	
	1,73	1,42	1,33				0,289*
Deliver impact as soon as possible	1	2	1,93		3, 39	0,089	
	1,73	1,7	1,38				0,129
Pulse check	1	0,96	1,36		3, 39	1,606	
	1,73	1,37	1,39				0,042
Focus on customer value	2,75	2,72	3,21	4	3, 56	3,136	
	1,26	1	0,85	-			0,264*
Flow							
Co-location	3,67	2,09	2,39	3	3, 52	1,381	
	0,58	1,38	1,33	-			-0,092
Fast learning	2,25	2,08	2,47	2	3, 56	0,634	
	1,26	1,25	1,12	-			0,1
Fewer people with more time	1,75	2,09	2,5	4	3, 55	2,274	
	1,26	1,13	1,04	-			0,261*
Visual tools and planning	2,5	2,26	2,94	4	3, 54	1,719	
	0,58	1,4	1,21	-			0,23
Reducing complexity	2,67	2,39	2,94	2	3, 54	2,753	
	0,58	0,84	0,64	-			0,198
Focus on a fixed rhytm	2	2,5	3,2		3, 29	2,103	
	1	1,19	1,14				0,321*
Leadership							
People before models	2,25	2,25	3,11	3	3, 55	3,524	
	1,5	1,36	0,76	-			0,284*
Freedom with responsibility	3,25	2,83	3,28	3	3, 55	0,908	
	0,96	1,03	0,83	-			0,117
Lead inwards	3	2,64	2,84	2	3, 56	0,183	
	0	0,99	0,96	-			-0,005
Work with project leadership	2,75	2,48	2,92		3, 39	0,335	
	0,5	1,22	0,79				0,113
Engagement of project owner	2,75	2,12	2,42		3, 38	0,531	
	1,89	1,61	1,68				-0,002
Involvement of project board	2	1,48	2,74	4	3, 52	13	
	1,73	1,12	1,05	-		**	0,432**
customization to the unique projec	3	2,18	2,2		3, 31	0,011	
	0	1,3	1,24				-0,118

Pace

	Regular Mean (S.D.)	t/competi Mean (S.D.)	ime-critic Mean (S.D.)	Blitz Mean (S.D.)	ANOVA df	ANOVA F	Correlation
Imapct							
Business case	3	1,68	2,41		3, 39	0,408	
	0	1,36	1,5				0,047
Deliver impact as soon as possible	2,5	1,36	2,47		3, 39	0,449	
	1,29	1,36	1,74				0,17
Pulse check	0,75	0,59	1,82		3, 39	2,861	
	1,5	0,91	1,59				0,374**
Focus on customer value	2,8	2,79	3,09		3, 56	0,116	
	1,03	0,96	1,02				0,126
Flow							
Co-location	1,86	2	2,77		3, 52	3,703	
	1,57	1,36	1,19				0,272*
Fast learning	1,89	2,28	2,27		3, 56	0,217	
	0,78	1,28	1,24				0,086
Fewer people with more time	1,89	1,89	2,77		3, 55	3,611	
	0,78	1,2	0,97				0,331**
Visual tools and planning	1,57	2,24	3,18		3, 54	10,23	
	1,27	1,35	1,01			**	0,423**
Reducing complexity	2,29	2,48	2,77		3, 54	2,09	
	0,95	0,83	0,69				0,214
Focus on a fixed rhytm	1,33	2,41	3,31		3, 29	8,577	
	1,15	1,18	0,85			**	0,509**
Leadership							
People before models	2,13	2,55	2,64		3, 55	0,092	
	1,64	1,21	1,18				0,115
Freedom with responsibility	3,13	2,9	3,09		3, 55	0,01	
	0,64	1,08	0,092				0,026
Lead inwards	2,9	2,52	2,91		3, 56	0,193	
	0,6	1,12	0,75				0,07
Work with project leadership	2,67	2,5	2,8		3, 39	0,002	
	1,03	1,22	0,86				0,078
Engagement of project owner	2,2	2,5	2,29		3, 38	0,196	
	1,79	1,52	1,79				0,019
Involvement of project board	1,67	2	2,05		3, 52	0,009	
	1,03	1,31	1,33				0,072
customization to the unique projec	2,6	2,35	2		3, 31	0,793	
	1,67	1,27	0,91				-0,180

Novelty

	Derivative Mean (S.D.)	Platform Mean (S.D.)	eakthro Mean (S.D.)	ANOVA df	ANOVA F	Correlation
Imapct						
Business case	1,43	2,4	2,43	2, 40	3,618	
	1,5	1,18	1,4			0,289*
Deliver impact as soon as possible	1,64	1,93	2,14	2, 40	0,394	
	1,78	1,39	1,66			0,129
Pulse check	1,07	1	1,21	2, 40	0,019	
	1,38	1,36	1,48			0,042
Focus on customer value	2,53	3	3,2	2, 57	3,642	
	1,22	0,94	0,56			0,264*
Flow						
Co-location	2,53	2,16	2,21	2, 53	0,096	
	1,55	1,21	1,42			-0,092
Fast learning	2,26	2,15	2,27	2, 57	0,033	
	1,15	1,19	1,33			-0,003
Fewer people with more time	1,9	2,19	2,67	2, 56	2,071	
	1,41	0,98	0,9			0,257*
Visual tools and planning	2,63	2,36	2,64	2, 55	1,413	
	1,16	1,38	1,5			-0,007
Reducing complexity	2,83	2,36	2,6	2, 55	1,784	
	0,71	0,81	0,83			-0,124
Focus on a fixed rhytm	2,91	2,31	2,9	2, 30	0,279	
	0,94	1,37	1,17			-0,022
Leadership						
People before models	2,58	2,28	2,87	2, 56	0,005	
	1,3	1,34	0,99			0,074
Freedom with responsibility	2,95	3,04	3	2, 56	0,004	
	1,08	0,89	1			0,023
Lead inwards	2,53	2,81	2,8	2, 57	0,573	
	1,07	0,75	1,08			0,116
Work with project leadership	2,33	2,69	2,92	2, 40	1,879	
	1,23	1,01	0,9			0,22
Engagement of project owner	3,08	2,18	1,62	2, 39	6,122	
	1,31	1,67	1,61		(*)	-0,351*
Involvement of project board	1,72	2,16	2	2, 53	0,002	
	1,18	1,34	1,29			0,094
customization to the unique projec	2,73	2,07	2	2, 32	1,579	
	0,79	1,44	1,15			-0,242

References

1. Abell AF (2016) Predicting the future and learning from the past: developing a framework for evaluating success in IT projects. PhD thesis, Department of Business Communication, Aarhus School of Business and Social Science
2. Burgan SC, Burgan DS (2014) One size does not fit all: choosing the right project approach. In: PMI Global Congress 2014 North America Project Management Institute Phoenix
3. Cameron R, Sankaran S, Scales J (2015) Mixed methods use in project management research. Proj Manag J 46:90–104
4. Crawford L, Hobbs JB, Turner JR (2005) Project categorization systems: aligning capability with strategy for better results. Project Management Institute
5. Crawford L, Hobbs B, Turner JR (2006) Aligning capability with strategy: categorizing projects to do the right projects and to do them right. Proj Manag J 37:38–50
6. Dvir D, Lipovetsky S, Shenhar A, Tishler A (1998) In search of project classification: a non-universal approach to project success factors. Res Policy 27:915–935
7. Fangel M (2005) Competences in project management, Hillerød, Denmark, Danish Project Management Association
8. Niknazar P, Bourgault M (2017) Theories for classification vs. classification as theory: implications of classification and typology for the development of project management theories. Int J Project Manage 35:191–203
9. Orhof O, Shenhar A, Dori D (2013) A model-based approach to unifying disparate project management tools for project classification and customized management. In: INCOSE international symposium, 2013. Wiley Online Library, pp 960–972
10. Orhof O, Shenhar A, Dori D (2014) The role of subproject task-specific attributes in managing enterprise-wide projects. In: CESUN 2014, 4th international engineering systems symposium. Stevens Institute of Technology
11. Sauser BJ, Reilly RR, Shenhar AJ (2009) Why projects fail? How contingency theory can provide new insights–a comparative analysis of NASA's Mars Climate Orbiter loss. Int J Project Manage 27:665–679
12. Shenhar AJ (1993) From low-to high-tech project management. R&D Management 23:199–214
13. Shenhar AJ, Dvir D (1996) Toward a typological theory of project management. Res Policy 25:607–632
14. Shenhar AJ, Dvir D (2007) Reinventing project management: the diamond approach to successful growth and innovation. Harvard Business Review Press
15. Shenhar AJ, Dvir D, Lechler T, Poli M (2002) One size does not fit all: true for projects, true for frameworks. In: Proceedings of PMI research conference, 2002 Seattle, Washington. Newtown Square. Project Management Institute, pp 99–106
16. Shenhar AJ, Dvir D, Shulman Y (1995) A two-dimensional taxonomy of products and innovations. J Eng Tech Manage 12:175–200
17. Svejvig P, Geraldi J, Grex S (2019) Accelerating time to impact: deconstructing practices to achieve project value. Int J Project Manage 37:784–801
18. Tashakkori A, Teddlie C (1998) Mixed methodology: combining qualitative and quantitative approaches. Sage Publications Inc., Thousand Oaks
19. Wheelwright SC, Clark KB (1992) Revolutionizing product development: quantum leaps in speed, efficiency, and quality. The Free Press, New York

Part IV
Self-Organizing in Different Types of Projects

Chapter 12
A New Model of a Project, Program, and Portfolio Recovery to Tackle COVID-19 in Construction Projects

Lukas Beladi Sihombing and Jiwat Ram

Abstract The outbreak of the coronavirus disease (COVID-19)—which started to spread in Indonesia at the beginning of March 2020, has affected Indonesia's economy and business sectors. The construction industry is no exception. The purpose of this paper, therefore, is to develop a new disaster recovery model that can be used in situations such as COVID-19 pandemic in the construction industry. The data collected through survey was analysed using frequency analysis. The results of the first survey revealed that the impact during and post-COVID-19 have greatly influenced the construction phase. Although a risk mitigation plan has been implemented, less than 50% of the respondent companies were found to use it. In the second survey, we observed that the presence of COVID-19 has not resulted in any force majeure claims to project owners. The results show that there are high expectations that projects can still be carried out with the established health protocols, and that the government will provide stimulus assistance. Also, there is a rather significant desire for a recovery disaster plan to be enacted, especially for the prioritized business functions. We also found that there is a great need for artificial intelligence (AI) applications to be applied in construction projects. The results of this study will help project owners, contractors, and the government to devise policies.

Keywords COVID-19 · Construction industry · Recovery

12.1 Introduction

The coronavirus pandemic has affected every industrial sector, and construction industry is no exception. The outbreak has significant implications for construction industry and could lead to project completion delays, difficulties for workers to enter project locations, and delays in the procurement of goods and services [1]. To prevent

L. B. Sihombing (✉)
University of Pelita Harapan, Tangerang, Banten 15811, Indonesia

J. Ram
Excelia Group, La Rochelle, France

R. Ding et al. (eds.), *Research on Project, Programme and Portfolio Management*, Lecture Notes in Management and Industrial Engineering,
https://doi.org/10.1007/978-3-030-86248-0_12

the spread of COVID-19 at project locations, the government has issued "Instructions of the Minister of Public Works and Public Housing (PUPR)" No. 02/IN/M/2020, which involve health protocols for: projects, temporary stoppages due to the high risk of project locations being at the epicenter of the outbreak, workers who have been found to be positive and/or patients under observation, or leadership from ministries/institutions/region heads that have enacted regulations to temporarily halt activities due to a force majeure condition.

According to the Indonesia National Construction Implementation Alliance, out of 30,763 Construction Service Business Agencies, 82% of them are micro-, small, and medium enterprises (MSMEs) such as the materials, workers, equipment, transportation, time, and mobility. These enterprises have experienced the most significant impact due to onset of COVID-19 crisis [2].

Numerous construction projects have been affected by the pandemic due to either a total shutdown or delays and disruptions. According to Russo et al. [3], the project team should consider the following actions to address and mitigate the project impacts, such as identify and assess relevant local and state restrictions on construction activities; identify and assess relevant contractual provisions; identify, assess, and mitigate project impacts; consider contract notice requirements; consider project suspension and termination options; and consider claims for time extensions and delay damages.

In addition, based on Junkin [4], there are five best practices to protect construction workers during a pandemic, such as ensure adherence to reliable prevention recommendations, understand and comply with OSHA standards, leverage standardized training, consider social distancing measures, and build a pandemic preparedness plan.

Given the scant knowledge on the subject and the importance of construction industry to economic and social fabric of the country, a number of questions need to be addressed to build knowledge on disaster recovery planning. These include (RQ1) What extent the impact of COVID-19 on the construction industry?; (RQ2) How many lawsuits have been filed against project owners due to COVID-19 in the construction industry?; (RQ3) What aspects of disaster recovery plan have been used during and post-COVID-19 in the construction industry?; and (RQ4) What is the disaster recovery plan model to deal with COVID-19 in construction projects?

We aims to identify impact of COVID-19 on the construction industry; to identify lawsuits have been filed against project owners; to identify aspects of disaster recovery plan; and to develop a disaster recovery plan model in construction projects in Indonesia.

12.2 Literature Review

Despite the difficult circumstances during COVID-19 pandemic, the construction industry has played an important role in responding to the crisis facilitating construction of hospitals in short duration and even donating lifesaving equipment. According to McKinsey [5], the construction industry represents 13% of the global GDP. But, due to the crisis, construction work was halted severely affecting the economic growth. Also, most sites that were open faced disrupted supply chains and operational restrictions.

In Indonesia, according to BPS-Statistics Indonesia [6], construction has a significant role in the Indonesian economy, with 11.26% contributing to Indonesia's gross domestic product (GDP) in the fourth quarter of 2019. Meanwhile, the growth rate and GDP growth sources in the construction field in the first quarter of 2020 from the fourth quarter of 2019 experienced a reduction of 6.38% [7].

Furthermore, according to Bank Indonesia [8], which conducted a survey on business global activities, the weighted net balance (WNB) of business activities in the construction sector in the first quarter of 2020 indicates a contraction growth with a weighted net balance of −0.08% compared to the previous quarter. This delay in business activities was due to the weak market for domestic construction/infrastructure project requests, as well as due to the impact of the COVID-19 outbreak.

Therefore, to minimize the unexpected impact due to risks and calamities, a disaster recovery plan [9–11] and a disaster risk management plan are needed. These should also include a social recovery process which can be made part of disaster governance and policy implementation [12] with the disaster cycle: mitigation, planning, response, and recovery [13].

In addition, the construction sector plays a significant role in the post-disaster by improving the resilience of the construction, for instance, by encouraging the identification of business risks [14, 15], the planning and establishment of a temporary reverse logistics system [16], artificial intelligence (AI) and big data [17], public health readiness [18, 19], management accountants and finance professionals [20], a holistic approach of crisis management[21], and a renegotiation of contracts [22].

12.3 Methodology

The study to develop a disaster recovery model for project, program, and portfolio management for construction projects was conducted in four steps. First, we examined secondary data about the issues related to COVID-19 and impacts on construction industry. Second, we conducted literature review to identify factors for model development. Third, we conducted survey 1 to answer RQ 1, survey 2 to answer RQs 2 and 3, and the fourth, build model to answer RQ 4, as shown at Fig. 12.1.

The first survey was conducted from 11 to 15 May 2020, to discover what extent the impact was during and post-COVID-19 on the planning, construction, and operational phases. This survey utilized a Google Form application. The study received 86 responses, and majority of the respondents were contractors with over 15 years of experience in the road and bridge field. The details of sample profile is given in Table 12.1. A description of the survey question in Bahasa Indonesia but for this paper it was translated in English. The survey can be seen at Appendix 12.1.

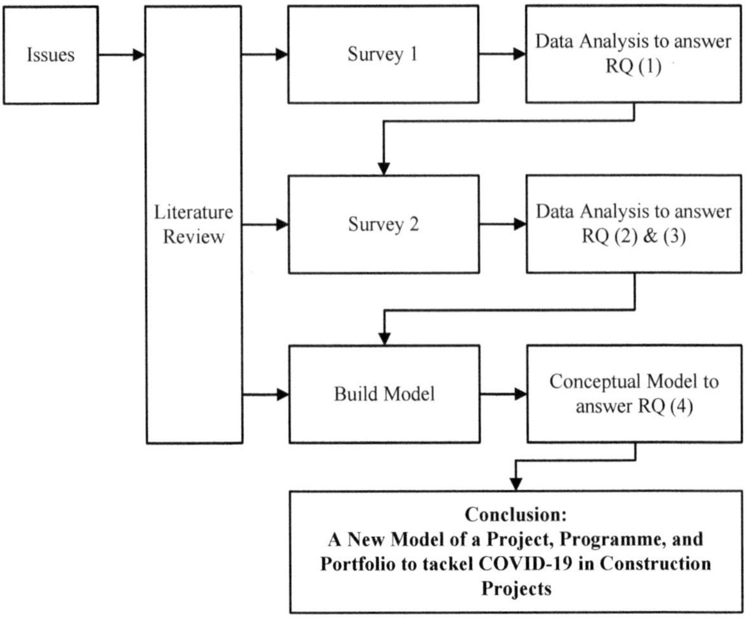

Fig. 12.1 Research process

Table 12.1 Profile of respondents in survey 1, N = 86

Kind of construction industry		Respondent area		Experience	
Road and bridge	29%	Contractor	64%	<5 years	10%
Commercial	20%	Designer	10%	5–10 years	23%
General contractor	17%	Supervisor	8%	11–15 years	20%
Industrial	9%	Owner	13%	16–20 years	15%
Home	8%	Government	3%	>20 years	31%
Institution	6%	Academician	1%		
Oil and gas piping	5%				
Power plant	3%				
Others	2%				

The second survey was done from 20 to 25 May 2020. The purpose was to obtain information about handling of the situation in construction sector during and post-COVID-19. This survey conducted using Google Form application. We received 53 respondents. Majority of the respondents were contractors and had over 15 years of work experience in the road and bridge field. Then it was followed by those in the commercial field. The details can be seen in Table 12.2. A description of the survey question in Bahasa Indonesia but for this paper it was translated in English. The survey can be seen at Appendix 12.2.

Table 12.2 Profile of respondents in survey 2, N = 53

Kind of construction industry		Respondent area		Experience	
Road and bridge	13%	Contractor	36%	<5 years	11%
Commercial	15%	Designer	11%	5–10 years	21%
General contractor	9%	Supervisor	11%	11–15 years	23%
Industrial	11%	Owner	11%	16–20 years	11%
Home/building	19%	Government	19%	>20 years	34%
Institution	26%	Academician	11%		
Oil and gas piping	2%				
Power plant	2%				
Others	2%				

12.4 Results and Discussion

12.4.1 Results and Discussion for Survey 1

The results of survey 1 reveal that the COVID-19 greatly impacted planning, construction, and operational phases of construction industry projects as depicted in Figs. 12.2 and 12.3. Figure 12.2 displays the impact during COVID-19 in the construction industry in Indonesia.

In the above figure, it shows that the biggest effect among the three phases is on the construction phase. Fifty-five percent of the respondents mentioned that COVID-19

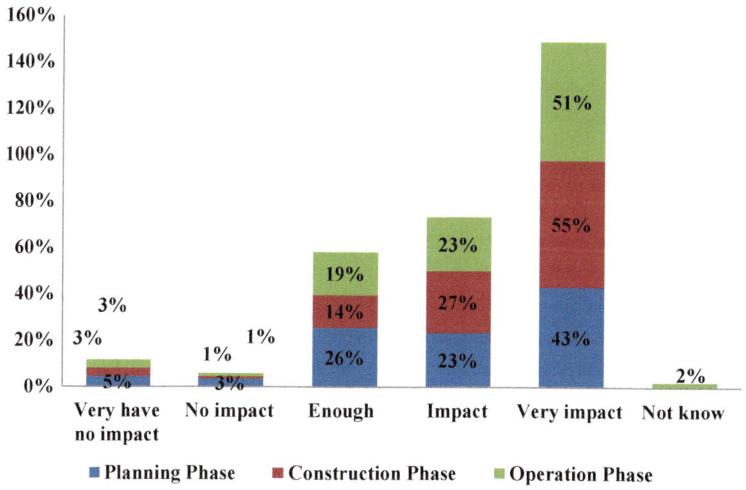

Fig. 12.2 Results of survey 1 on the impact during COVID-19 on the construction industry in Indonesia

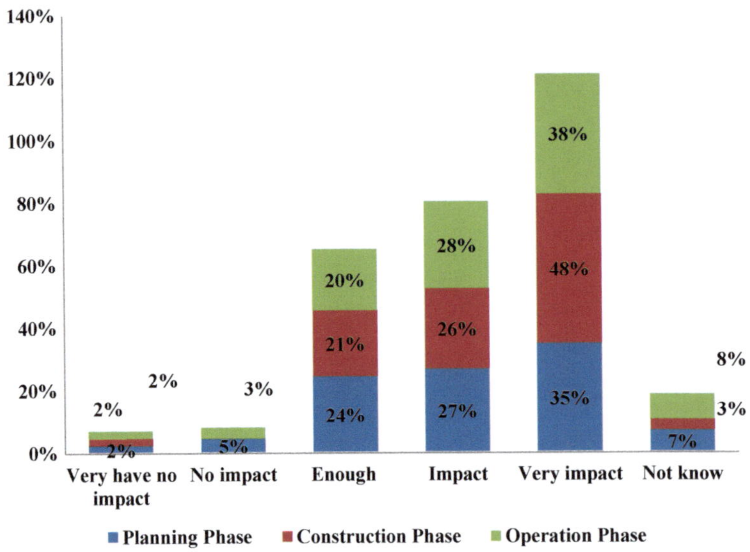

Fig. 12.3 Results of survey 1 on the post-COVID-19 on the construction industry in Indonesia

crisis has a large impact on construction phase, whereas 51% said the largest impact was on operation phase and 43% thought the same for planning phase.

This findings align with the fact that GDP growth in the construction field in quarter I of 2020 experienced a reduction of 6.38% [7], and business activities in the construction sector in quarter I of 2020 indicate a contracted growth with WNB of −0.08% from the previous quarter [8]; as well as from 30,763 Construction Service Business Agencies, 82% of them are in the MSME scale and experience the most significant impact [2].

Therefore, it is pertinent to know how the respondents consider the impact in post-COVID-19 in Indonesia's construction industry. Their viewpoints can be observed in Fig. 12.3.

Figure 12.3 displays the differences between during and post-COVID-19 pandemic, because at this time it is ongoing and the government has not stated that the pandemic is over. Therefore, there are respondents who do not know how COVID-19 will affect the construction industry. The survey results mention that the biggest post-COVID-19 effects will be in the construction phase (48%), then the operational phase (38%), and the planning phase (35%). In prediction by GlobalData [23], the global construction industry will be revised down to 0.5%.

According to Biörck et al. [24], in order that a project can manage a current crisis and can actively improve the project's success post-COVID, several things need to be done: accelerate the rollout and adoption of digitization; invest in the culture and skills needed to operate in the next normal; build a control tower across the portfolio; bolster supply-chain resilience; redeploy capital and resources; identify opportunities to shift work off-site; and get closer to customers.

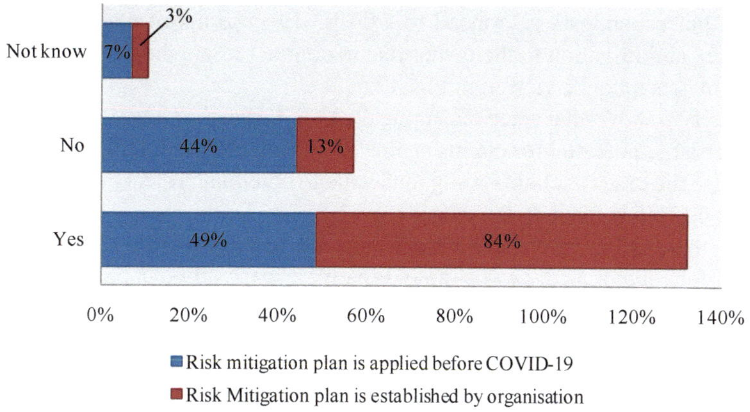

Fig. 12.4 Results of the risk mitigation plan

In addition, the respondents also were asked about the risk mitigation plan. The results can be seen in Fig. 12.4.

Figure 12.4 reveals that when asked if a risk mitigation plan was done before COVID-19, 49% of the respondents said Yes, 44% said No, and 7% said Do Not Know. This means that a large number of respondents still do not execute a risk mitigation plan, even though many of the respondents (84%) stated that it has been established by their organization.

According to ICB 4.0 [25], "risk and opportunity management helps decision makers to make informed choices, prioritize actions, and distinguish among alternative courses of action. Risk and opportunity management is an ongoing process taking place throughout the life cycle of the project".

Based on Young [26], "the risks that happen become the issues that you must promptly resolve to maintain the integrity of the project schedule. It is good practice to prepare risk mitigation plans for known major risks, taking early action to avoid the risk from occurring".

After analyzing the survey 1, in relation to RQ1, we can conclude that COVID-19 impacted construction industry both during and post-crisis and particularly construction phase was affected due to the crisis.

12.4.2 Results and Discussion of Survey 2

The results of survey 2 show respondents' opinion about handling of issues during and post-COVID-19 in the construction industry. There were 28 questions which were divided into three sections. The first section dealt with the expectations about the project owner. The second part was about the expectations in running the project. The third section focused on the disaster recovery plan.

How the respondents responded to COVID-19 in terms of claims, mediation, arbitration, and litigation to the owner due to contract rescheduling, project delays, or force majeure can be viewed in Fig. 12.5.

Figure 12.5 shows that when managing projects during the COVID-19 crisis, there is a rather large potential for claims against the owner due to contract rescheduling. This is also the case for claims being made due to potential project delays. However, there are no claims made to the owner regarding force majeure.

Then, there is the potential for mediation due to project delays, but there is no potential for force majeure to be done. Finally, there is no potential for arbitration and litigation to be done to the owner for project delays as force majeure.

According to Bleasby [27], parties may wish to address this concern by giving more thought to force majeure definitions. It is a typical condition of any force majeure entitlement that the event in question was unforeseen. Meanwhile, the delays and impacts on the project should be set out clearly in the delay notice such as the contract notification requirements of each contract [28].

The second part of this survey shows the expectations of operating a project during COVID-19 in the construction industry. The results are displayed in Fig. 12.6.

The figure shows that there is a strong desire to run projects during COVID-19 with the stipulation that there are health protocols and government permission, including when the government announces that the pandemic is over. However, the majority of the respondents want to obtain a stimulus from the government, and a large number of them want to adapt and operate a project in the new normal condition.

According to Katseff et al. [29], to provide an immediate economic boost to infrastructure sector, the federal government prepares to invest a portion of stimulus funding based on the principles: be strategic about state-of-good-repair investments, prioritize investment that reduce the cost of existing operations, accelerate

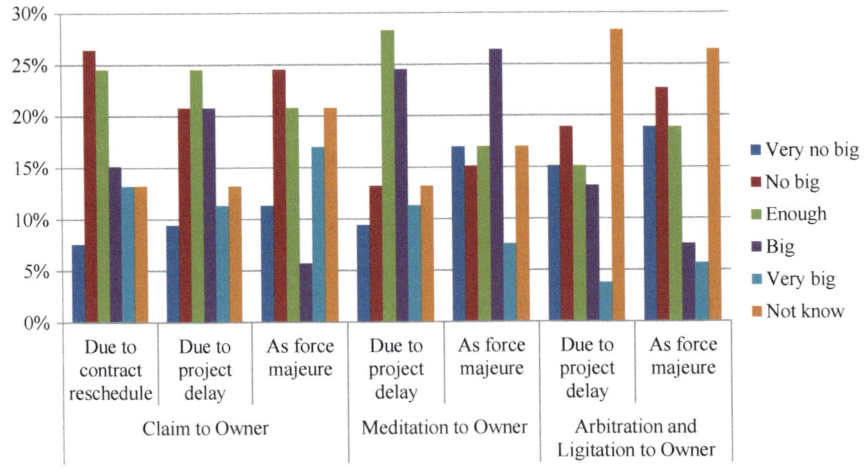

Fig. 12.5 Claims, mediation, arbitration, and litigation to the owner due to rescheduled contracts, project delays, or force majeure

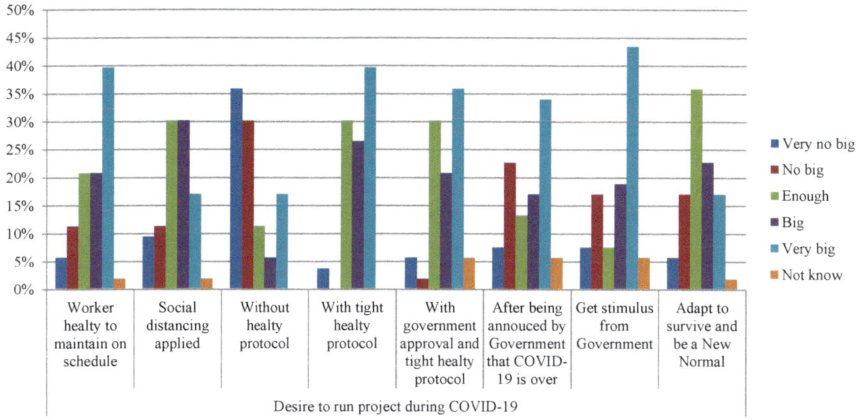

Fig. 12.6 Desire to run a project during COVID-19

transformational investments, capitalize technology investment, and incorporate decarbonisation.

The third part of this survey is about how the disaster recovery plan is implemented during and post-COVID19. The results can be viewed in Fig. 12.7.

From the survey results, it reveals that recovery disaster planning is done during and post-COVID-19. These plans include nine factors ranked in order of importance: prioritization of business functions; communication plans, such as for employees, customers, vendors, the public, and the media; prevention and mitigation strategies; an emergency response checklist; potential threats, vulnerabilities, and risks; periodic

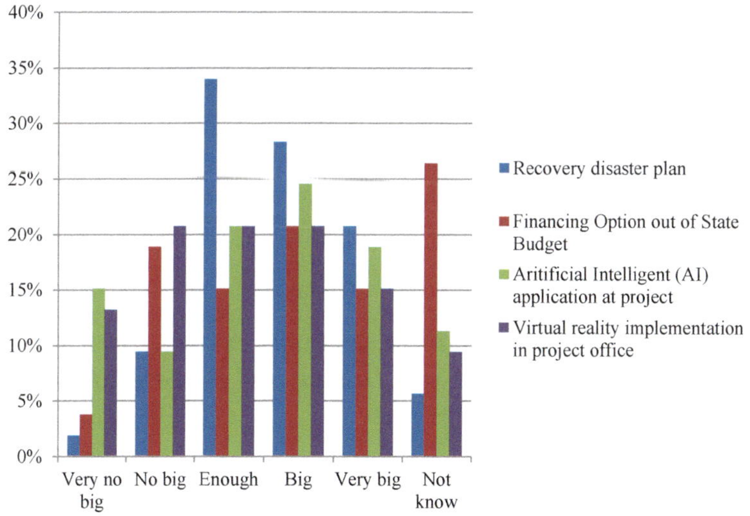

Fig. 12.7 Tackling options during and post-COVID-19 in the construction industry

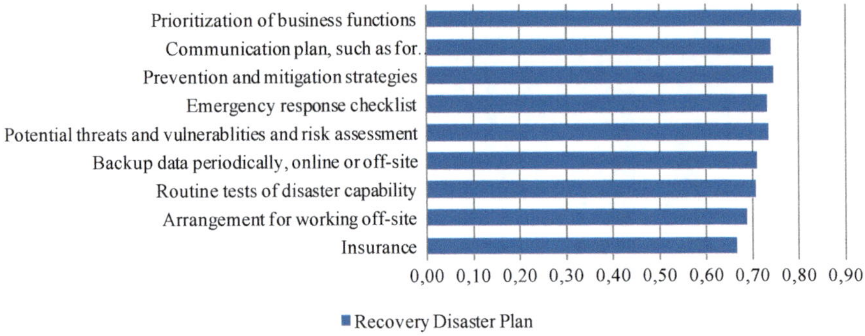

Fig. 12.8 Recovery disaster plan during and post-COVID-19 in the construction industry

backup data, online or off-site; routine tests of disaster capability; arrangements for working off-site; and insurance. The details are provided in Fig. 12.8.

Then to handle COVID-19, the majority of the respondents stated that they Do Not Know (26%) for the financing option out of the state budget. However, there is great potential for it to be done (21%). Besides that, there is great potential for an artificial intelligence (AI) application to be applied at a project (25%). In spite of this, there is not much potential for virtual reality to be implemented in the project office. The details can be observed in Fig. 12.7.

Based on the result of survey 2, as an answer to research question 2, it can be concluded that COVID-19 is not a force majeure but at the special case the contractors need a mediation to claim a force majeure to owner. Then, to run the project, construction industry needs a stimulus from government to overcome the cash flow, for instance lower taxes, lower interest, although these stimulus has to be deeply researches.

In addition, to answer research question 3, based on data analysis, it is concluded that construction industry needs a recovery disaster plan to tackle COVID-19 both during and after crisis. As stated above that the three highest rank of recovery disaster plan such as: prioritization of business function, communication plan, and prevention and mitigation strategies. Therefore, to proof that construction industry need a recovery disaster plan, at the next section will discuss the model of recovery disaster for project, program, and portfolio.

12.4.3 *Recovery Model of a Construction Project, Program, and Portfolio Life Cycle During and Post-COVID-19 in a Construction Project*

Based on the survey results, we argue that COVID-19 has significant impact on the construction industry, both during and post-COVID-19. Despite that, the risk mitigation plan was not completely used before COVID-19. In addition, other factors to tackle COVID-19 in construction industry such as make claims to the owner due to the contract rescheduling, implement the project with permission from the government and follow the health protocols, obtain a stimulus from the government, and devise a recovery disaster plan.

A preliminary causal loop diagram demonstrating recovery model of a construction project, program, and portfolio life cycle can be viewed in Fig. 12.9. The recovery model conceptual is divided into four cycles in one system. First sub-system is project life cycle; second sub-system is recovery life cycle; third sub-system is program life cycle, and fourth sub-system is portfolio life cycle.

The first sub-system is project life cycle, if there is a project distraction (COVID-19) at the construction stage 1. Based on the survey 1 that project at the construction phase is very impact during and post-COVID-19. In addition, the second survey said that implementation of recovery disaster plan is significant to tackle the disruption either during or post-COVID 19. Therefore, to recover project distraction as COVID-19, the model using the second life cycle is recovery life cycle.

The second sub-system is recovery life cycle. Recovery life cycle divided into six phase, such as: understand, audit, trade-off, negotiate, restart, and execute [30]. The first phase is understanding the project's history, targeted value as well as expected benefit. At this phase, expected value is to complete the project under COVID-19 outbreak, for example, claims to owner due to contract rescheduling and project delays, but a strong desire to adapt and operate a project in the new normal with health protocols and government permission.

The second phase is audit. This phase is to assess the actual performance to date, identify the flaws, and perform a root cause analysis.

The third phase is trade-off. At this point, we have the necessary information for decision making as well as the team's support for the recovery. At the case of COVID-19, there is contract reschedule but there is no potential for arbitration and litigation to be done to the owner for project delays as force majeure. Meanwhile, recovery disaster plan needs prioritization of business function, communication plans, and prevention and mitigation strategies.

The fourth phase is negotiation. At this stage, part of the stakeholder negotiations include identifying items important to the stakeholders (e.g., time, cost, and value); and negotiating for the needed sponsorship and stakeholder support. Negotiation during the pandemic is claim against the owner due to contract reschedule, and project delay; run a project with government approval and tight health protocols; and get stimulus from government.

The fifth phase is restart. This point includes making sure the team learns from past mistakes; restoring team confidence; and getting buy-ins for the new action plan for a rapid recovery. The sixth phase is execution. At this stage, he project manager must focus on certain back-to-work implementation factors.

The third sub-system is program life cycle. After the execution phase in project recovery life cycle, impact of COVID-19 either during or after at construction stage in project life cycle will reduce and finally the project can be close successfully and project outcomes and benefit will impact positively to strategic objection in portfolio life cycle.

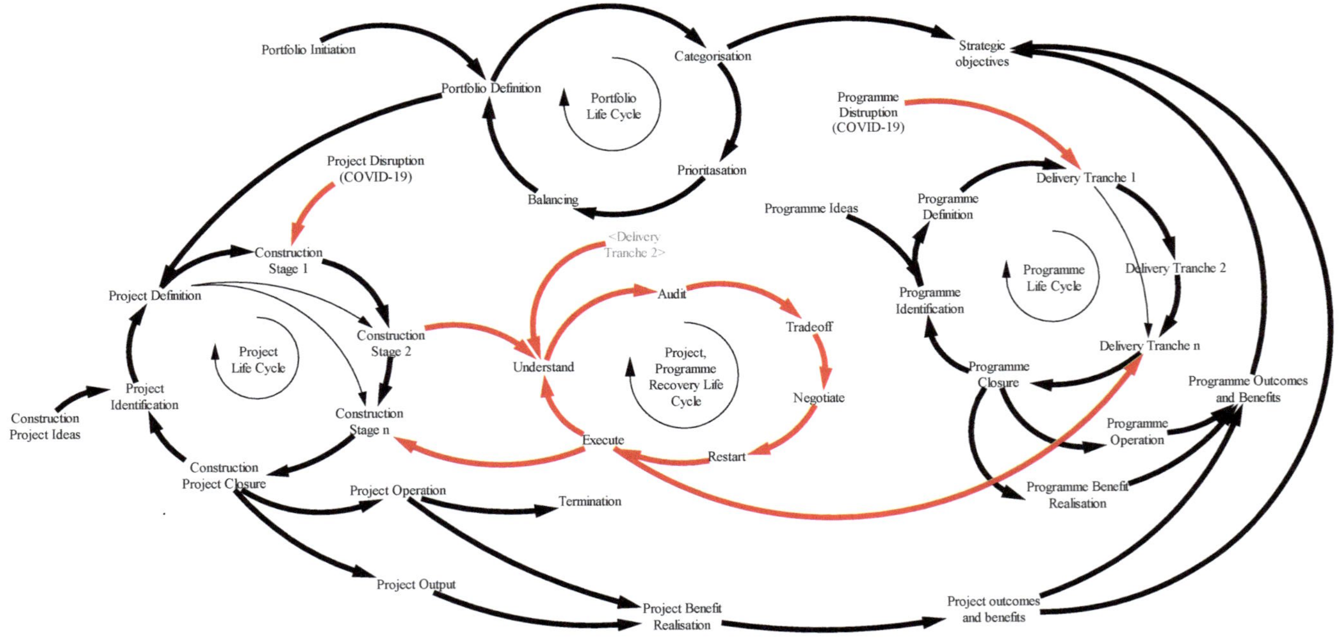

Fig. 12.9 A preliminary causal loop diagram demonstrating recovery model of a construction project, program, and portfolio life cycle

Relationship between the project life cycle and recovery project life cycle can also be applied to relationship between program life cycle and recovery program life cycle. The life cycle of program and recovery as same as the life cycle of project and recovery that has been described above.

Finally, program outcome and benefits will effect successfully to strategic objective in portfolio life cycle. This phase is segmented to reflect the change in management necessary to realize that benefits are not constant and will fluctuate in levels [31].

The fourth sub-system is portfolio life cycle. This sub-system will prioritise and balance project and program recovery life cycle. Portfolio life cycle begin in initiation, definition phase; categorization; prioritization, and balancing. At this phase, the portfolio sub-system must be balanced in terms of risks, resource usage, cash flow, and impacts across the business. And finally, in one system, project disruption at construction stage, for example COVID-19 outbreak, will be managed by recovery project and program to leverage business strategic.

12.5 Conclusion

The purpose of this study was to develop a disaster recovery model so that a project can run smoothly during a crisis and not be disadvantageous by not attending-to the following things: prioritizing business functions; having a communication plan (e.g. for employees, customers, vendors, the public, and the media); having prevention and mitigation strategies; having an emergency response checklist; determining the potential threats, vulnerabilities, and risks; backing up data periodically, both online and off-site; conducting routine tests of disaster possibilities; arranging for working off-site; and having insurance.

The lessons learned from this survey and model compared with IPMA ICB 4.0 can be viewed in Table 12.3.

Compliance with Ethical Standards (or Ethical Statement)

The authors declare that they complied with all guidelines given by the Center for Research & Community Development (CRCD) of Universitas Pelita Harapan. Informed consent was obtained from all participants and all data was anonymized. The research does not require ethics approval, as it mentioned in the waiver issued by the Center for Research & Community Development (CRCD) under the number 031/LPPM-UPH/II/2021, February 18, 2021.

Table 12.3 IPMA ICB 4.0 alignment of lessons learned

Lessons learned	IPMA ICB 4.0 project, program, and portfolio
All recovery processes should be aligned to the strategic business objectives	4.3.1.; 5.3.1.; 6.3.1
Recovery cannot be done in isolation. Involvement by the operational stakeholders is necessary	4.4.5; 5.4.5; 6.4.5; 4.5.12; 5.5.12; 6.5.12
Decisions made during the recovery process must still focus on the creation of business value	4.3.4; 5.3.4; 6.3.4; 4.4.10; 5.4.10; 6.4.10; 4.5.7; 5.5.7; 6.5.7
Enterprise environmental factors may have changes	4.3.5.; 5.3.5; 6.3.5; 4.4.4.; 5.4.4.; 6.4.4; 4.5.2.; 5.5.2.; 6.5.2
The definition of success may change during the recovery process	4.3.5.; 5.3.5.; 6.3.5; 4.4.7.; 5.4.7.; 6.4.7; 4.5.2.; 5.5.2.; 6.5.2
A well-structured scope change control process should be in place	4.5.3.; 5.5.3.; 6.5.3
A new work statement and possibly a new business case may need to be developed	4.3.2.; 5.3.2.; 6.3.2; 4.4.8.; 5.4.8.; 6.4.8; 4.5.8.; 5.5.8.; 6.5.8

Appendix 12.1

Survei: Dampak Pandemi COVID-19 pada Industri Konstruksi Indonesia (*Survey: Impact of COVID-19 pandemic at Construction Industry in Indonesia*)
https://forms.gle/C9sxCTKGsYYEssaLA

Pendahuluan

Pandemi coronavirus disease (COVID-19) sudah mulai melanda Indonesia pada awal Maret 2020, di mana di semua sektor terpengaruh termasuk di industri konstruksi.

Tujuan survei adalah untuk mengetahui dampak pandemi COVID-19 pada industri konstruksi di Indonesia sehingga dapat mengetahui kebijakan yang akan dilakukan di masa mendatang pasca pandemi COVID-19.

Atas partisipasinya kami ucapkan terima kasih.

Peneliti:

Dr. Ir. Lukas Beladi Sihombing, MT, MPU, A.M. ASCE

Introduction

Coronavirus diseases (COVID-19) pandemic has begun to hit Indonesia in early March 2020, where in all sectors affected including in the construction industry.

The aim of this survey is to know impact of COVID-19 pandemic at construction industy in Indonesia, so it can know the policy that will be conducted in the future at post COVID-19.

Researcher:

Dr. Ir. Lukas Beladi Sihombing, MT, MPU, A.M. ASCE
Nama (*Name*): …
Alamat E-mail (*E-mail address*): …

Pekerjaan (*Profession*): *

- Pegawai Negeri/Pegawai PUPR (*Government*)
- Pemilik (*Owner*)
- Kontraktor (*Contractor*)
- Konsultan Perencana (*Designer*)
- Konsultan Pengawas (*Supervision*)
- Akademisi (*Academics*)
- Lainnya (*Others*): …

Pendidikan (*Education*):

- D3 (*Diploma 3 years*)
- D4 (*Diploma 4 years*)
- Sarjana Strata −1 (S-1) (*Bachelor*)
- Sarjana Strata −2 (S-2)/Master (*Master*)
- Sarjana Strata −3 (S-3)/Doktoral (*Doctoral*)

Lama Kerja (*Expereience*):

- <5 tahun (*years*)
- 5–10 tahun (*years*)
- 10–15 tahun (*years*)
- 15–20 tahun (*years*)
- >20 tahun (*years*)

Nama Perusahaan (*Company Name*): …

Jenis Industri Konstruksi Anda (*Kind of Construction Industry*):

- Tempat Tinggal (*Home*)
- Industrial (*Industrial*)
- Komersial (*Commercial*)
- Kelembagaan (*Institution*)
- Pemipaan Oil & Gas (*Piping Oil & Gas*)
- Pemipaan Air Bersih (*Piping Water*)
- Jalan dan Jembatan (*Road and Bridge*)
- *Power Plant*
- *Plumbing*
- *Electrical Works*
- *Masonry, carpentry dan roofing*
- Lainnya (*Others*): …

1. Seberapa besar dampak pandemi COVID-19 pada fase perencanaan di industri konstruksi Anda? (What extent impact of COVID-19 pandemic at planning phase in your construction industry?)

1. Sangat tidak berdampak (*Very have no impact*)
2. Tidak berdampak (*No impact*)
3. Cukup berdampak (*Enough*)
4. Berdampak (*Impact*)
5. Sangat berdampak (*Very impact*)
6. Tidak tahu (*Do not know*)

2. Seberapa besar dampak pandemi COVID-19 pada fase pelaksanaan/konstruksi di industri konstruksi Anda? (What extent impact of COVID-19 pandemic at construction phase in your construction industry?)

   1. Sangat tidak berdampak (*Very have no impact*)
   2. Tidak berdampak (*No impact*)
   3. Cukup berdampak (*Enough*)
   4. Berdampak (*Impact*)
   5. Sangat berdampak (*Very impact*)
   6. Tidak tahu (*Do not know*)

3. Seberapa besar dampak pandemi COVID-19 pada fase pelaksanaan/konstruksi di industri konstruksi Anda? (What extent impact of COVID-19 pandemic at construction phase in your construction industry?)

   1. Sangat tidak berdampak (*Very have no impact*)
   2. Tidak berdampak (*No impact*)
   3. Cukup berdampak (*Enough*)
   4. Berdampak (*Impact*)
   5. Sangat berdampak (*Very impact*)
   6. Tidak tahu (*Do not know*)

4. Seberapa besar dampak pandemi COVID-19 pada fase operasional di industri konstruksi Anda? (What extent impact of COVID-19 pandemic at operational phase in your construction industry?)

   1. Sangat tidak berdampak (*Very have no impact*)
   2. Tidak berdampak (*No impact*)
   3. Cukup berdampak (*Enough*)
   4. Berdampak (*Impact*)
   5. Sangat berdampak (*Very impact*)
   6. Tidak tahu (*Do not know*)

5. Apakah rencana mitigasi risiko dilakukan sebelum pandemi COVID-19 terjadi di industri konstruksi Anda? (Was risk mitigation plan been conducted before COVID-19 pandemic in your construction industry?)

   1. Ya (*Yes*)
   2. Tidak (*No*)
   3. Tidak tahu (*Do not know*)

6. Apakah rencana mitigasi risiko ditetapkan pada instansi Bapak/Ibu? (*Was risk mitigation plan been determined at your company/institution*?)
   1. Ya (*Yes*)
   2. Tidak (*No*)
   3. Tidak tahu (*Do not know*)
7. Siapa yang bertanggung jawab rencana mitigasi risiko di instansi Bapak/Ibu? (*Who is responsibility of risk mitigation plan at your company/institution*?)
   1. Pemegang Saham (*Shareholders*)
   2. Para direksi (*Directors*)
   3. General Manager (*General Manager*)
   4. Setingkat Manajer/Kepala Divisi (*Manager/Division Head*)
   5. Manajer Risiko (*Risk Manager*)
   6. Setingkat Kepala Departemen (*Head of Department*)
8. Seberapa besar dampak pasca pandemi COVID-19 pada fase perencanaan di industri konstruksi Anda? (What extent impact of post-COVID-19 pandemic at planning phase in your construction industry?)
   1. Sangat tidak berdampak (*Very have no impact*)
   2. Tidak berdampak (*No impact*)
   3. Cukup berdampak (*Enough*)
   4. Berdampak (*Impact*)
   5. Sangat berdampak (*Very impact*)
   6. Tidak tahu (*Do not know*)
9. Seberapa besar dampak pasca pandemi COVID-19 pada fase konstruksi di industri konstruksi Anda? (What extent impact of post-COVID-19 pandemic at construction phase in your construction industry?)
   1. Sangat tidak berdampak (*Very have no impact*)
   2. Tidak berdampak (*No impact*)
   3. Cukup berdampak (*Enough*)
   4. Berdampak (*Impact*)
   5. Sangat berdampak (*Very impact*)
   6. Tidak tahu (*Do not know*)
10. Seberapa besar dampak pasca pandemi COVID-19 pada fase operasional di industri konstruksi Anda? (What extent impact of post-COVID-19 pandemic at operational phase in your construction industry?)
    1. Sangat tidak berdampak (*Very have no impact*)
    2. Tidak berdampak (*No impact*)
    3. Cukup berdampak (*Enough*)
    4. Berdampak (*Impact*)
    5. Sangat berdampak (*Very impact*)
    6. Tidak tahu (*Do not know*)

Terima kasih atas partisipasi Anda, silahkan memberi masukan bila ada.
(*Thank you for your participations, please give comments if any*).

Appendix 12.2

Survei: Penanganan Dampak Pandemi COVID-19 pada Industri Konstruksi Indonesia.
(*Survey: Tackling Impact of COVID-19 pandemic at Constructio Industry in Indonesia*).
https://forms.gle/9GuhWMzjqHFRpgfc6

Pendahuluan.

Pandemi coronavirus disease (COVID-19) sudah mulai melanda Indonesia pada awal Maret 2020, di mana di semua sektor terpengaruh termasuk di industri konstruksi.

Tujuan survei adalah untuk mendapatkan penanganan yang tepat dalam mengatasi dampak pandemi COVID-19 pada industri konstruksi di Indonesia sehingga dapat mengetahui kebijakan yang akan dilakukan di masa mendatang pasca pandemi COVID-19.

Atas partisipasinya kami ucapkan terima kasih.
Peneliti:
Dr. Ir. Lukas Beladi Sihombing, MT, MPU, A.M. ASCE.
Introduction.

Coronavirus diseases (COVID-19) pandemic has begun to hit Indonesia in early March 2020, where in all sectors affected including in the construction industry.

The aim of this survey is to tackle impact of COVID-19 pandemic at construction industy in Indonesia, so it can know the policy that will be conducted in the future at post COVID-19.

Thank you for your participations.
Researcher:
Dr. Ir. Lukas Beladi Sihombing, MT, MPU, A.M. ASCE.
Nama (*Name*):
Alamat E-mail (*E-mail address*):
No. HP/WA (*Mobile Phone/Whatsapp*):

Pekerjaan (*Profession*):

- Pegawai Negeri/Pegawai PUPR (*Government*)
- Pemilik (*Owner*)
- Kontraktor (*Contractor*)
- Konsultan Perencana (*Designer*)
- Konsultan Pengawas (*Supervision*)
- Akademisi (*Academics*)
- Lainnya…

Pendidikan (*Education*):

- D3 (*Diploma 3 years*)
- D4 (*Diploma 4 years*)
- Sarjana Strata -1 (S-1) (*Bachelor*)
- Sarjana Strata -2 (S-2)/Master (*Master*)
- Sarjana Strata -3 (S-3)/Doktoral (*Doctoral*)

Lama Kerja:*

- <5 tahun (*years*)
- 5–10 tahun (*years*)
- 10–15 tahun (*years*)
- 15–20 tahun (*years*)
- >20 tahun (*years*)

Nama Perusahaan (*Company Name*): …

Jenis Industri Konstruksi Anda (*Kind of Construction Industry*):

- Tempat Tinggal (*Home*)
- Industrial (*Industrial*)
- Komersial (*Commercial*)
- Kelembagaan (*Institution*)
- Pemipaan Oil & Gas (*Piping Oil & Gas*)
- Pemipaan Air Bersih (*Piping Water*)
- Jalan dan Jembatan (*Road and Bridge*)
- *Power Plant*
- *Plumbing*
- *Electrical Works*
- *Masonry, carpentry dan roofing*
- Lainnya (*Others*): …

1. Seberapa besar dampak pandemi COVID-19 pada kesehatan tenaga kerja untuk menjaga proyek tepat waktu? (*What extent impact of COVID-19 pandemic at Workers health to maintain project on time*?)

    1. Sangat tidak berdampak (*Very have no impact*)
    2. Tidak berdampak (*No impact*)
    3. Cukup berdampak (*Enough*)
    4. Berdampak (*Impact*)
    5. Sangat berdampak (*Very impact*)
    6. Tidak tahu (*Do not know*)

2. Seberapa besar klaim dilakukan kepada Pemilik Proyek karena ada penjadwalan kembali kontrak akibat pandemi COVID-19?(*What extent claim was conducted to Owner due to contract reschedule caused by COVID-19 pandemic*?)

1. Sangat tidak besar (*Very no big*)
2. Tidak besar (*No big*)
3. Cukup besar (*Enough*)
4. Besar (*Big*)
5. Sangat besar (*Very big*)
6. Tidak tahu (*Do not know*)

3. Seberapa besar klaim dilakukan kepada Pemilik Proyek karena keterlambatan proyek akibat pandemi COVID-19?(What extent claim was conducted to Owner due to project delay caused by COVID-19 pandemic?)

    1. Sangat tidak besar (*Very no big*)
    2. Tidak besar (*No big*)
    3. Cukup besar (*Enough*)
    4. Besar (*Big*)
    5. Sangat besar (*Very big*)
    6. Tidak tahu (*Do not know*)

4. Seberapa besar klaim dilakukan kepada Pemilik Proyek sebagai force majeure akibat pandemi COVID-19? (*What extent claim was conducted to Owner as force majeure due to COVID-19 pandemic*?)

    1. Sangat tidak besar (*Very no big*)
    2. Tidak besar (*No big*)
    3. Cukup besar (*Enough*)
    4. Besar (*Big*)
    5. Sangat besar (*Very big*)
    6. Tidak tahu (*Do not know*)

5. Seberapa besar mediasi dilakukan kepada Pemilik Proyek karena keterlambatan proyek akibat pandemi COVID-19? (What extent mediation was conducted to Owner due to project delay caused by COVID-19 pandemic?)

    1. Sangat tidak besar (*Very no big*)
    2. Tidak besar (*No big*)
    3. Cukup besar (*Enough*)
    4. Besar (*Big*)
    5. Sangat besar (*Very big*)
    6. Tidak tahu (*Do not know*)

6. Seberapa besar mediasi dilakukan kepada Pemilik Proyek sebagai force majeure akibat pandemi COVID-19?(*What extent mediation was conducted to Owner as force majeure causec by COVID-19 pandemic*?)

    1. Sangat tidak besar (*Very no big*)
    2. Tidak besar (*No big*)
    3. Cukup besar (*Enough*)
    4. Besar (*Big*)
    5. Sangat besar (*Very big*)

6. Tidak tahu (*Do not know*)

7. Seberapa besar arbitrase dan ligitasi dilakukan kepada Pemilik Proyek karena keterlambatan proyek akibat pandemi COVID-19? (*What extent arbitration and ligitation was conducted to Owner due to project delay caused by COVID-19 pandemic*?)

   1. Sangat tidak besar (*Very no big*)
   2. Tidak besar (*No big*)
   3. Cukup besar (*Enough*)
   4. Besar (*Big*)
   5. Sangat besar (*Very big*)
   6. Tidak tahu (*Do not know*)

8. Seberapa besar arbitrase dan ligitasi dilakukan kepada Pemilik Proyek sebagai force majeure akibat pandemi COVID-19? (*What extent arbitration and ligitation was conducted to Owner as force majeure caused by COVID-19 pandemic*?)

   1. Sangat tidak besar (*Very no big*)
   2. Tidak besar (*No big*)
   3. Cukup besar (*Enough*)
   4. Besar (*Big*)
   5. Sangat besar (*Very big*)
   6. Tidak tahu (*Do not know*)

9. Seberapa besar social distancing diterapkan di lapangan proyek? (What extent social distancing was implemented in project field?)

   1. Sangat tidak besar (*Very no big*)
   2. Tidak besar (*No big*)
   3. Cukup besar (*Enough*)
   4. Besar (*Big*)
   5. Sangat besar (*Very big*)
   6. Tidak tahu (*Do not know*)

10. Seberapa besar keinginan agar pekerjaan di proyek tetap dilaksanakan selama pandemi COVID-19 tanpa protokol kesehatan ? (*How much desire that work on the project will continue during the COVID-19 pandemic without health protocols*?)

   1. Sangat tidak besar (*Very no big*)
   2. Tidak besar (*No big*)
   3. Cukup besar (*Enough*)
   4. Besar (*Big*)
   5. Sangat besar (*Very big*)
   6. Tidak tahu (*Do not know*)

11. Seberapa besar keinginan agar pekerjaan di proyek tetap dilaksanakan selama pandemi COVID-19 tetapi dengan protokol kesehatan yang ketat? (*How much desire that work on the project will continue during the COVID-19 pandemic with tight health protocols*?)
    1. Sangat tidak besar (*Very no big*)
    2. Tidak besar (*No big*)
    3. Cukup besar (*Enough*)
    4. Besar (*Big*)
    5. Sangat besar (*Very big*)
    6. Tidak tahu (*Do not know*)

12. Seberapa besar keinginan agar pekerjaan di proyek tetap dilaksanakan selama pandemi COVID-19 dengan persetujuan pemerintah dan dengan protokol kesehatan yang ketat? (*How much desire that work on the project will continue during the COVID-19 pandemic with government approval and tight health protocols*?)
    1. Sangat tidak besar (*Very no big*)
    2. Tidak besar (*No big*)
    3. Cukup besar (*Enough*)
    4. Besar (*Big*)
    5. Sangat besar (*Very big*)
    6. Tidak tahu (*Do not know*)

13. Seberapa besar keinginan agar pekerjaan di proyek dilaksanakan setelah pandemi COVID-19 dinyatakan tidak ada oleh pemerintah ? (*How much desire that work on the project will continue after the COVID-19 pandemic is stated no pandemic by government*?)
    1. Sangat tidak besar (*Very no big*)
    2. Tidak besar (*No big*)
    3. Cukup besar (*Enough*)
    4. Besar (*Big*)
    5. Sangat besar (*Very big*)
    6. Tidak tahu (*Do not know*)

14. Seberapa besar keinginan agar mendapatkan bantuan stimulus dari pemerintah akibat COVID-19? (*How much desire to get stimulus assistance from the government due to COVID-19*?)
    1. Sangat tidak besar (*Very no big*)
    2. Tidak besar (*No big*)
    3. Cukup besar (*Enough*)
    4. Besar (*Big*)
    5. Sangat besar (*Very big*)
    6. Tidak tahu (*Do not know*)

15. Seberapa cepat adaptasi dilakukan selama COVID-19 untuk terus bertahan dan menjadi new normal? (*How quickly the adaptation was carried out during COVID-19 to continue to survive and become new normal*?)

    1. Sangat tidak besar (*Very no big*)
    2. Tidak besar (*No big*)
    3. Cukup besar (*Enough*)
    4. Besar (*Big*)
    5. Sangat besar (*Very big*)
    6. Tidak tahu (*Do not know*)

16. Seberapa besar rencana recovery disaster akan dilakukan selama dan pasca COVID-19? (*What extent disaster recovery plan will be conducted during and post-COVID-19*?)

    1. Sangat tidak besar (*Very no big*)
    2. Tidak besar (*No big*)
    3. Cukup besar (*Enough*)
    4. Besar (*Big*)
    5. Sangat besar (*Very big*)
    6. Tidak tahu (*Do not know*)

17. Seberapa besar rencana recovery disaster akan dilakukan selama dan pasca COVID-19 untuk backups data secara berkala baik online maupun off-site? (*What extent disaster recovery plan will be conducted during and post-COVID-19 for backup data periodically, online or off-site*?)

    1. Sangat tidak besar (*Very no big*)
    2. Tidak besar (*No big*)
    3. Cukup besar (*Enough*)
    4. Besar (*Big*)
    5. Sangat besar (*Very big*)
    6. Tidak tahu (*Do not know*)

18. Seberapa besar rencana recovery disaster akan dilakukan selama dan pasca COVID-19 untuk membuat check list respons emergensi? (*What extent disaster recovery plan will be conducted during and post-COVID-19 for emergency response checklist?*)

    1. Sangat tidak besar (*Very no big*)
    2. Tidak besar (*No big*)
    3. Cukup besar (*Enough*)
    4. Besar (*Big*)
    5. Sangat besar (*Very big*)
    6. Tidak tahu (*Do not know*)

19. Seberapa besar rencana recovery disaster akan dilakukan selama dan pasca COVID-19 untuk pengaturan working off-site? (*What extent disaster recovery*

plan will be conducted during and post-COVID-19 for arrangement for working off-site?)

1. Sangat tidak besar (*Very no big*)
2. Tidak besar (*No big*)
3. Cukup besar (*Enough*)
4. Besar (*Big*)
5. Sangat besar (*Very big*)
6. Tidak tahu (*Do not know*)

20. Seberapa besar rencana recovery disaster akan dilakukan selama dan pasca COVID-19 untuk rencana komunikasi disaster seperti untuk karyawan, pelanggan, vendor/subskontraktor, publik dan media? (*What extent disaster recovery plan will be conducted during and post-COVID-19 for communication plan, suchas for employee, customers, vendor/subcontractor, public and media?*)

    1. Sangat tidak besar (*Very no big*)
    2. Tidak besar (*No big*)
    3. Cukup besar (*Enough*)
    4. Besar (*Big*)
    5. Sangat besar (*Very big*)
    6. Tidak tahu (*Do not know*)

21. Seberapa besar rencana recovery disaster akan dilakukan selama dan pasca COVID-19 untuk penilaian risiko, kerentanan, dan ancaman yang potensial? (*What extent disaster recovery plan will be conducted during and post-COVID-19 for potential threats and vulnerabilities and risk assessment?*)

    1. Sangat tidak besar (*Very no big*)
    2. Tidak besar (*No big*)
    3. Cukup besar (*Enough*)
    4. Besar (*Big*)
    5. Sangat besar (*Very big*)
    6. Tidak tahu (*Do not know*)

22. Seberapa besar rencana recovery disaster akan dilakukan selama dan pasca COVID-19 untuk prioritas fungsi bisnis? (*What extent disaster recovery plan will be conducted during and post-COVID-19 for prioritization of business functions?*)

    1. Sangat tidak besar (*Very no big*)
    2. Tidak besar (*No big*)
    3. Cukup besar (*Enough*)
    4. Besar (*Big*)
    5. Sangat besar (*Very big*)
    6. Tidak tahu (*Do not know*)

23. Seberapa besar rencana recovery disaster akan dilakukan selama dan pasca COVID-19 untuk asuransi? (*What extent disaster recovery plan will be conducted during and post-COVID-19 for insurance?*)

    1. Sangat tidak besar (*Very no big*)
    2. Tidak besar (*No big*)
    3. Cukup besar (*Enough*)
    4. Besar (*Big*)
    5. Sangat besar (*Very big*)
    6. Tidak tahu (*Do not know*)

24. Seberapa besar rencana recovery disaster akan dilakukan selama dan pasca COVID-19 untuk strategi mitigasi dan pencegahan? (*What extent disaster recovery plan will be conducted during and post-COVID-19 for prevention and mitigation strategies?*)

    1. Sangat tidak besar (*Very no big*)
    2. Tidak besar (*No big*)
    3. Cukup besar (*Enough*)
    4. Besar (*Big*)
    5. Sangat besar (*Very big*)
    6. Tidak tahu (*Do not know*)

25. Seberapa besar rencana recovery disaster akan dilakukan selama dan pasca COVID-19 untuk uji kemampuan recovery disaster secara rutin? (*What extent disaster recovery plan will be conducted during and post-COVID-19 forroutine tests of disaster capability?*)

    1. Sangat tidak besar (*Very no big*)
    2. Tidak besar (*No big*)
    3. Cukup besar (*Enough*)
    4. Besar (*Big*)
    5. Sangat besar (*Very big*)
    6. Tidak tahu (*Do not know*)

26. Seberapa besar opsi pendanaan di luar APBN dalam menangani proyek selama dan pasca pandemi COVID-19? (What extent funding/financing option out of State Budget in tackling project during and post COVID-19 pandemic?)

    1. Sangat tidak besar (*Very no big*)
    2. Tidak besar (*No big*)
    3. Cukup besar (*Enough*)
    4. Besar (*Big*)
    5. Sangat besar (*Very big*)
    6. Tidak tahu (*Do not know*)

27. Seberapa besar aplikasi Artificial Intelligent (AI) digunakan di proyek selama dan pasca COVID-19? (What extent Artificial Intelligent (AI) application was conducted in project during and post-COVID-19?)

1. Sangat tidak besar (*Very no big*)
2. Tidak besar (*No big*)
3. Cukup besar (*Enough*)
4. Besar (*Big*)
5. Sangat besar (*Very big*)
6. Tidak tahu (*Do not know*)

28. Seberapa besar virtual reality digunakan di kantor proyek selama dan pasca COVID-19? (What extent virtual reality was conducted in project office during and post-COVID-19?

    1. Sangat tidak besar (*Very no big*)
    2. Tidak besar (*No big*)
    3. Cukup besar (*Enough*)
    4. Besar (*Big*)
    5. Sangat besar (*Very big*)
    6. Tidak tahu (*Do not know*)

Terima kasih atas partisipasi Anda, silahkan memberi masukan bila ada. (*Thank you for your participation, please give your comments if any*).

References

1. Kompas (2020) Dampak Corona, PengusahaKonstruksiKeluhkanKeterlambatanPengerjaan-Proyek.https://money.kompas.com/read/2020/04/03/203100226/dampak-corona-pengusaha-konstruksi-keluhkan-keterlambatan-pengerjaan-proyek. Accessed 22 May 2020
2. Liputan6 (2020) Gara-Gara Corona, 30.763 JasaKonstruksiLesu. https://www.liputan6.com/bisnis/read/4218420/gara-gara-corona-30763-jasa-konstruksi-lesu. Accessed 22 May 2020
3. Russo SC, Mascialino JL, Epstein RC, Knaub ZD, Jensen DC (2020) How COVID-19 is affecting construction projects—and what you can do about it. Constr Exec. https://constructionexec.com/article/how-covid-19-is-affecting-construction-projectsand-what-you-can-do-about-it. Accessed 24 May 2020
4. Junkin J (2020) Five best practices to protect construction workers during a pandemic. Constr Exec. https://constructionexec.com/article/five-best-practices-to-protect-construction-workers-during-a-pandemic. Accessed 24 May 2020
5. McKinsey & Company (2020) How construction can emerge stronger after coronavirus
6. BPS-Statistics Indonesia (2020) Construction indicator, 4th Quarter—2019
7. BPS-Statistics Indonesia (2020) PertumbuhanEkonomi Indonesia Triwulan I—2020
8. Bank Indonesia (2020) SurveiKegiatanDunia Usaha Triwulan I—2020
9. McLean D (2004) Avoiding disaster: planning and testing are key parts of the disaster recovery process. Network World Canada 14(17):1–5
10. American Bankers Association (1992) Disaster preparation beats disaster recovery. ABA Bank J 84(12):40–46
11. Krumwiede K (2017) The road to disaster recovery. Strategic Fin. https://sfmagazine.com/technotes/may-2017-the-road-to-disaster-recovery/. Accessed 23 May 2020
12. Serrao-Neumann S, Crick F, Choy DL (2018) Post-disaster social recovery: disaster governance lessons learnt from Tropical Cyclone Yasi. Nat Hazards 93(3):1163–1180. https://doi.org/10.1007/s11069-018-3345-5

13. Coccolini F, Sartelli M, Kluger Y, Pikoulis E, Karamagioli E, Moore EE, Biffl WL, Peitzman A, Hecker A, Chirica M, Damaskos D, Ordonez C, Vega F, Fraga GP, Chiarugi M, Di Saverio S, Kirkpatrick AW, Abu-Zidan F, Mefire AC, Leppaniemi A, Khokha V, Sakakushev B, Catena R, Coimbra R, Ansaloni L, Corbella D, Catena F (2020) COVID-19 the showdown for mass casualty preparedness and management: the Cassandra Syndrome. World J Emerg Surg 15:26. https://doi.org/10.1186/s13017-020-00304-5
14. Wilkinson S, Chang-Richards AY, Sapeciay Z, Costello SB (2016) Improving construction sector resilience. Int J Disaster Resil Built Environ 7(2):173–185. https://doi.org/10.1108/IJDRBE-04-2015-0020
15. Deprest J, Marc Van R, Lannoo L, Bredaki E, Ryan G (2020) SARS-CoV2 (COVID-19) infection: is fetal surgery in times of national disasters reasonable? Prenat Diagn 4(11). https://doi.org/10.1002/pd.5702
16. Yu H, Xu S, Wei DS, Zhao X (2020) Reverse logistics network design for effective management of medical waste in epidemic outbreaks: insights from the coronavirus disease 2019 (COVID-19) outbreak in Wuhan (China). Int J Environ Res Public Health 17(5):1770. https://doi.org/10.3390/ijerph17051770
17. Bragazzi NL, Dai H, Damiani G, Behzadifar M, Martini M (2020) How big data and artificial intelligence can help better manage the COVID-19 pandemic. Int J Environ Res Public Health 17(9):3176. https://doi.org/10.3390/ijerph17093176
18. Smith N, Fraser M (2020) Straining the system: novel coronavirus (COVID-19) and preparedness for concomitant disasters. Am J Public Health 110(5):648–649. https://doi.org/10.2105/AJPH.2020.305618
19. Di Gennaro F, Pizzol D, Marotta C, Antunes M, Racalbuto V (2020) Coronavirus diseases (COVID-19) current status and future perspectives: a narrative review. Int J Environ Res Public Health 17(8):2690. https://doi.org/10.3390/ijerph17082690
20. Butcher D (2020) CFOs respond to the covid-19 pandemic. Strategic Finance. 101(11):24–31
21. Khan SM (2020) President Xi Jinping's crisis management model. Def J 23(8):30
22. Engineering Review (2020) COVID-19 and engineering. Construction 45(8)
23. GlobalData (2020) Global construction outlook to 2024 (COVID-19 Impact). https://www.reportlinker.com/p05879544/Global-Construction-Outlook-to-COVID-19-Impact.html?utm_source=GNW#backAction=2. Accessed 24 May 2020
24. Biörck J, Sjödin E, Blanco JL, Mischke J, Strube G, Ribeirinho MJ, Rockhill D (2020) How construction can emerge stronger after coronavirus, May 8, 2020, McKinsey's Company. https://www.mckinsey.com/industries/capital-projects-and-infrastructure/our-insights/how-construction-can-emerge-stronger-after-coronavirus. Accessed 24 May 2020
25. IPMA (2017) Individual competence baseline for project, programme, and portfolio 4th version
26. Young TL (2007) The handbook of project management: a practical guide to effective policies. Techniques and processes. Kogan Page Ltd., London, pp 99–125
27. Bleasby J (2020) The future of construction contracts post-COVID-19. Daily Commercial News. Markham 93(70):1–2
28. Hoal R (2020) Covid-19 lockdown: Impact on construction contract claims. Bizcommunity.com. SyndiGate Media Inc., Cape Town Cape Town
29. Katseff J, Peloquin S, Rooney M, Wintner T (2020) Reimagining infrastructure in the United States: how to build better, 5 July 2020. McKinsey' Company. https://www.mckinsey.com/industries/capital-projects-and-infrastructure/our-insights/reimagining-infrastructure-in-the-united-states-how-to-build-better. Accessed 5 July 2020
30. Kerzner H (2014) Project recovery: case studies and techniques for overcoming project failure. Wiley
31. Association for Project Management (APM) (2014) Praxis Framework: an integrated guide to the management of projects, programmes and portfolios. Association for Project Management (APM), Buckinghamshire, UK

Chapter 13
Let Us Integrate Self-Organization and Stakeholders into the Development of Infrastructure Projects, Because We Need More Creativity and Satisfying Solutions

Pia Herrmann, Reiner Singer, Philipp Kaufmann, and Konrad Spang

Abstract Developing public infrastructure projects is a crucial but challenging process, as these projects face various, sometimes conflicting, stakeholder expectations. In order to find good solutions in this challenging environment, we argue, that on the one hand, stakeholders should be involved in the development of public infrastructure projects for increasing the likelihood of good solutions and stakeholder satisfaction. On the other hand, we need a lot of creativity for creating, shaping, and negotiating good solutions. By drawing upon our initial understanding of self-organization and upon the relevance of self-organization, e.g., regarding agile project management, we discussed whether it is worth examining the integration of self-organization in the development of public infrastructure projects with stakeholders. Based on factors that increase the likelihood of creative and good solutions as well as on definitions of self-organization, we concluded that self-organization can support the required creativity. Based on literature concerning self-organization and governance and coordination of self-organization, we discussed how to integrate self-organization into the development of public infrastructure projects, outlined two possible applications and shared suggestions for further research.

Keywords Creativity · Infrastructure project · Project development · Self-organization · Stakeholder integration

P. Herrmann (✉) · R. Singer · P. Kaufmann · K. Spang
University of Kassel, Heinrich-Plett-Str. 40, 34109 Kassel, Germany
e-mail: p.herrmann@uni-kassel.de

R. Singer
e-mail: singer@uni-kassel.de

P. Kaufmann
e-mail: philipp.kaufmann@onlinehome.de

K. Spang
e-mail: spang@uni-kassel.de

R. Ding et al. (eds.), *Research on Project, Programme and Portfolio Management*,
Lecture Notes in Management and Industrial Engineering,
https://doi.org/10.1007/978-3-030-86248-0_13

13.1 Introduction

Since public infrastructure projects impact their environment [14], there are a lot of "individuals, groups or organisations participating in, affecting, being affected by, or interested in the execution or the result of the project": the stakeholders [20, p. 145]. As consequence, managers of public infrastructure projects find themselves in the middle of various, sometimes conflicting, expectations, project objectives, and project constraints [14]. Finding good solutions and satisfying stakeholders among this mass of various demands is difficult and the development of public infrastructure projects becomes challenging [14].

By drawing upon conducted interviews, we identified the development of projects, the early phases of the project life cycle, as important for contributing to stakeholder satisfaction. Furthermore, based on our understanding of the relevance of stakeholders, we came to two conclusions: In order to increase the likelihood of solutions that increase stakeholder satisfaction (we call them "good solutions"), (1) stakeholders need to be integrated in the development of projects and (2) a lot of creativity and good ideas are required.

Concerning the implementation of creativity in the early phases of public infrastructure projects, we discussed, among other things, whether we could enhance the required creativity by integrating self-organization. This idea emerged because, e.g., Takeuchi and Nonaka identified self-organized project teams as one characteristic of a set "that will make a difference" [32, p. 138] and, e.g., the Manifesto for Agile Software Development emphasizes the value of self-organization [6].

Based on these considerations, we wondered if self-organization might support creativity and good ideas and whether this would be the case, if we can integrate self-organization with stakeholders in the development of public infrastructure projects. During our first discussion on integrating self-organization, which was based on a vague understanding of self-organization, we wondered if integrating self-organization into a project imposed so many conditions and constraints that the remaining leeway would hinder self-organization. Therefore, within this contribution, we aim for answering the question whether it is worth examining further the integration of self-organization in the development of public infrastructure projects. In order to answer this superordinate question, we divided our interest in the following two research questions:

- Can self-organization increase the likelihood of creative and good solutions?
- Is it possible—and if so, how—to integrate self-organization into the collective development of public infrastructure projects?

In order to discuss these research questions, we start by arguing for integrating stakeholders in the development of public infrastructure projects. Subsequently, we discuss factors increasing the likelihood of creative and good solutions within the context of collective development. Afterwards, we discuss, if self-organization might support the defined factors. By presenting the results of a limited literature review concerning governance and coordination of self-organization thereafter, we conceptualize a framework for discussing whether and how self-organization could be integrated in the collective development of public infrastructure projects. Finally, we outline two ideas of self-organization in public infrastructure projects and share suggestions for further research.

13.2 Integrating Stakeholders into the Development of Public Infrastructure Projects

13.2.1 Why Are We Focusing on the Development of Public Infrastructure Projects?

For several years, public infrastructure projects have been challenged by conflicts with stakeholders. In order to prevent conflicts several efforts and research were conducted in Germany [e.g., 14]. However, despite these efforts made, we still face conflicts and dissatisfied stakeholders. Therefore, we initiated a research project aiming at increasing stakeholder satisfaction.

Considering that several research was conducted in Germany, we started our research project by conducting interviews with experienced practitioners ‘in and around’ projects on the current situation of stakeholder management in public infrastructure projects. As we aimed for understanding the current situation in practice, we asked for examples, challenges, and experiences regarding stakeholder management.

When analyzing our transcribed interviews, we had the impression that project stakeholder satisfaction is especially influenced during the time the project ‘evolves’. It is the time, when the project idea evolves and is specified into a detailed project scope definition. Furthermore, we got the impression that it is the action and behavior of the project representatives that influence the satisfaction of stakeholders. We called it “the development of a project”.

13.2.2 Why Are We Focusing on the Development of Public Infrastructure Projects?

According to literature [12, p. 9, exemplary referring to other sources], there are four reasons for the importance of project stakeholders: (1) “the project needs contributions”, (2) “stakeholders often establish the criteria for assessing the success of the project”, (3) “stakeholders’ (potential) resistance may cause various risks and negatively affect the success of the project”, and (4) “the project may affect stakeholders”. Additionally, these reasons interact with each other, as shown based on an example regarding the second and the first reason [18, p. 53]: “If the stakeholders do not believe that the project outcomes or process have the potential to meet or exceed their needs, they may refuse to contribute to the project". Furthermore, in Germany, various legal requirements regarding the involvement of directly and indirectly affected stakeholders exist [14].

Based on these four reasons, we propose that integrating stakeholders in the development of public infrastructure projects might increase stakeholder satisfaction:

- (1) "The project needs contributions" [12, p. 9]: The better the relations between project representatives and stakeholders and the better the project is explained to the stakeholders, the more contributions the project might get. To give an example: By presenting and discussing route alternatives, information, and advice can be collected from stakeholders. This is of special interest in public infrastructure projects, as stakeholders might have important regional knowledge [14, and based on the conducted interviews].
- (2) "Stakeholders often establish the criteria for assessing the success of the project" [12, p. 9]: The more and the better project representatives identify stakeholders' expectations, the better they can be met. Of course, only because of knowing expectations does not mean that expectations are fulfilled. However, if you do not know them, you cannot fulfil them at all, unless by chance [14, and based on the conducted interviews].
- (3) "Stakeholders' (potential) resistance may cause various risks and negatively affect the success of the project" [12, p. 9]: The better the relations between project representatives and stakeholders, the better the project is managed, and the lower the risk that stakeholders might harm the project [14, and based on the conducted interviews].
- Furthermore: the better the relations, the more identified expectations and the more contributions, the better and more satisfying solutions can be found [based on the conducted interviews].

Given the proposed benefits of integrating stakeholders in the development, we argue for stakeholder integration into the development of public infrastructure projects and for collectively developed (parts) of the project.

Since we are interested in how to support creativity and good solutions in the development of public infrastructure projects with stakeholders, we continue by presenting factors that increase the likelihood of creative and good solutions within the context of collective development.

13.3 Overview of Selected Factors Increasing the Likelihood of Creative and Good Solutions in a Collective Development of Infrastructure Projects

13.3.1 Introduction and Definition of Creativity

In this section, we will first attempt to approach the understanding of creativity by means of an accepted definition in literature. Subsequently, we will present our approach for the identification of factors that increase the likelihood of high creativity and idea generation in the context of stakeholders. Related to the term creativity, there are reviews of the research development to the terminology, which also include different approaches of some researchers e.g., [5, 37]. For this reason, we will not go into this in detail. One of the most important and most cited definitions for creativity is that of Amabile [1, p. 1001] which we use for our comprehension: "A product or response is creative to the extent that appropriate observers independently agree it is creative. Appropriate observers are those familiar with the domain in which the product was created or the response articulated. Thus, creativity can be regarded as the quality of products or responses judged to be creative by appropriate observers, and it can also be regarded as the process by which something so judged is produced".

13.3.2 Approach to Identify Factors for Collective Development

The emergence of creativity has been and is subject of psychological research in several fields, e.g., arts and science [37]. For this contribution, we focus on findings from organizational psychology. The number of factors examined and publications is high. Anderson et al. [5] provide a review of factors, which we use as point of departure besides the works of Amabile and other cited publications. For the analysis, we additionally set a filter and limit the result to those factors that are relevant for the emergence of creativity in situations in which stakeholders of a project interact. To give an example: We have not taken into account the organizational culture and its influence on creativity in our considerations, because there is no common organization in which a certain culture could have developed over time. However, we have considered the characteristics of jobs, especially job autonomy, and intended to adapt them to our context. For this purpose, we first outline the characteristics of stakeholders in comparison to employees of organizations: Based on definition, stakeholders are not necessarily…

- …in a common organization
- …bound to a leader authorized to give instructions
- …a team that has potentially already worked together over a certain period of time.

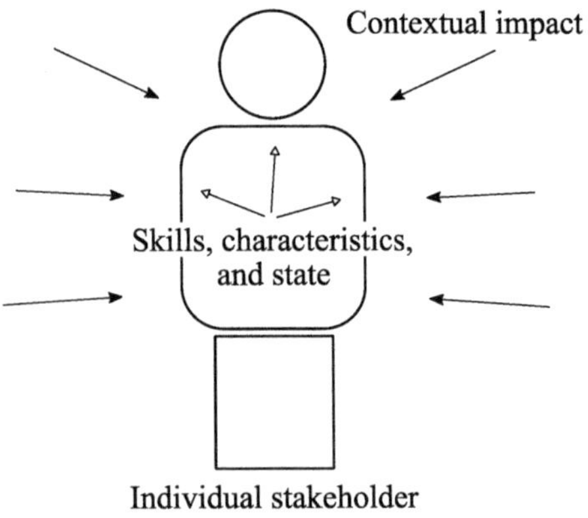

Fig. 13.1 Two perspectives on creativity (*source* authors)

Theoretical frameworks on creativity in organizational psychology are preferably structured into individual, group, and organizational levels [e.g., 35]. Since stakeholders are not in a common organization, we just adopt a simplified theoretical framework with an individual (skills, characteristics, and state of a person, e.g., personality, cognitive capabilities, or mood) and a contextual perspective including the work in a unit and the environment beyond, as shown in Fig. 13.1. An important aspect here is that both perspectives are not isolated, but influence each other in varying strong degrees [5]. In relation to the sheer volume of research results, it can be noted that the emergence of creativity is complex and has many dependencies and that examinations in this field do not always provide consistent results [e.g., 5, regarding team composition and structure].

13.3.3 Identification of Factors

The fact that there is no influence on the selection of stakeholders, as it might be with employees in organizations, implies that in the context of collectively developing infrastructure projects, influence on creativity is primarily possible by means of what we call the contextual perspective. However, according to research, the environment has a significant influence and any person with normal capacity can generate a reasonable or at least moderate result in creativity depending on the environment [3]. In theory, there are various models that structure the responsible components and their effects on creativity [e.g., 35]. One of the first and most cited theoretical frameworks in this field is the componential theory of creativity of Amabile [2]. Over the years, this approach has been further developed and was last revised substantially in 2016 towards the dynamic componential model [4]. In the model, the individual creativity is affected by three components [4]:

- "domain-relevant skills" [p. 164] ("one's expertise or factual knowledge about the domain, technical skills for doing work and advancing one's knowledge in the domain, and special domain-relevant talents" [p. 160])
- "creativity-relevant processes" [p. 164] ("cognitive styles, perceptual styles, and thinking skills that are conducive to taking new perspectives on problems, pivoting among different ideas, thinking broadly, and making unusual associations; personality processes, traits, and characteristics that lead the individual to take risks and eschew conformity; and persistent, energetic work styles" [p. 160])
- "intrinsic and synergistic extrinsic motivation" [p. 164] (processing a task for an interesting, involving, enjoying, personally challenging or satisfying reason; reward and recognition).

Amabile assumes that task motivation is the one of the three components that can be most strongly influenced from outside [3]. More recent research in respect of the former componential theory has shown that affective processes (discrete emotions, mood states) as well as further self-regulation processes (e.g., self-efficacy (beliefs), proactive personality) have a considerable impact on the creativity outcome, too [e.g., 5, 15, 29]. In summary, it can be said that there are numerous elements and components in the individual perspective that have an influence on the emergence of creativity.

From the contextual perspective, there are several factors that influence creativity and the individual perspective, too. For instance, research has shown that a certain degree of job-relevant diversity might be beneficial to creativity under certain conditions [34]. This is because individuals receive broader access to information and share different perspectives. The relevance of diversity for creativity in terms of different knowledge, information, and perspectives is also emphasized by contributions on communication inside and outside a team [5, 19]. Furthermore, Richter et al. [29] found in their research on team informational resources that shared "knowledge of who knows what' and the functional background diversity in a team have a determining influence on the creative self-efficacy and thus on the creativity of a person. However, a certain degree of job-relevant diversity appears to be only conductive to creativity under certain conditions, e.g., if there is a transformational leadership style. A transformational leader might be a moderator of educational specialization heterogeneity and creativity [30] and is characterized as motivating individuals, giving them a vision for the future, and inspiring them [25, cited from 9]. This is consistent with research on team climate: For creativity, it appears to be essential that there is an environment that supports creativity, where there are set goals and shared visions and where there is a high degree of task orientation, with individuals reviewing their performance, giving feedback [e.g., 36], and do not neglect or overreact to problems [4]. In addition, job autonomy (how to achieve goals) and job complexity (challenging tasks) might also have a positive impact on creativity under certain circumstances. These circumstances are determined by the fact that the task requirements and time pressure do not exceed a certain level [7].

Although the interaction of stakeholders on a particular issue is not similar to a team embedded in an organization, we suppose that these factors are transferable to

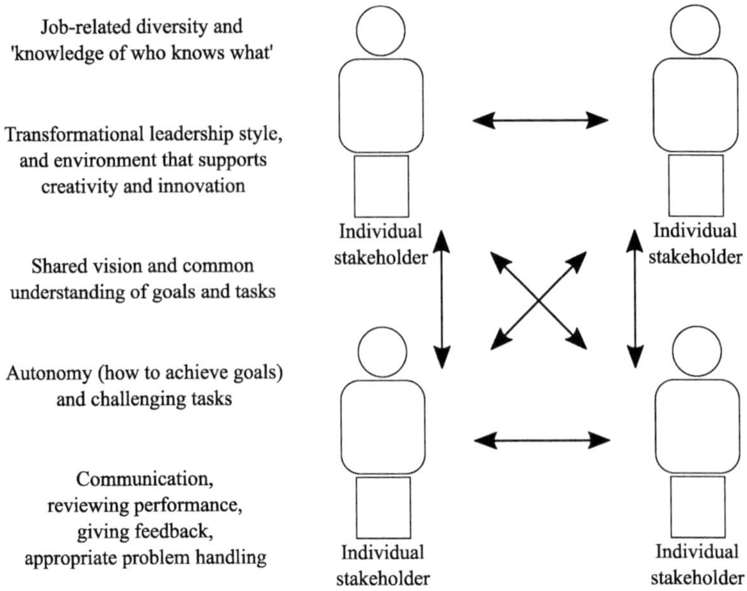

Fig. 13.2 Factors enabling creative and good solutions in collective developments (*source* authors)

the stakeholder context, i.e., when stakeholders interact in a unit in order to develop a project collectively. Based on the factors described above and presented in Fig. 13.2, we suggest that creativity in this context is likely to increase if stakeholders (viewed as individuals) have functional diversity, get a high degree of autonomy to achieve goals, share a common vision, and have a common understanding of the task, which should be positive challenging. Furthermore, there should be a fertilizing atmosphere that is reflective, supportive, and communicative.

13.4 Integrating Self-Organization and Stakeholders into the Development in Order to Enhance Creativity

13.4.1 What Does Self-Organization Mean?

Our question about integrating self-organization in projects in order to enable creativity originates mainly from two sources: Takeuchi and Nonaka [32] on product development and the discipline of agile project management [6].

Based on these as well as on further sources identified by an unstructured literature review, we first discuss the concept of self-organization, before we answer the question concerning its ability to enable creativity. A brief introduction of the papers is given in Table 13.1.

Table 13.1 Selected papers including self-organization

Authors, year	Of interest for us
Takeuchi and Nonaka (1986) [32]	The authors present an approach for product development, one of which characteristics are self-organizing project teams [p. 137], described as: "a group possesses a self-organizing capability when it exhibits three conditions: autonomy, self-transcendence, and cross-fertilization" [p. 139]. Autonomy is described as limited involvement of headquarter, "Proving guidance, money, and moral support" and teams, being "free to set its own direction" [p. 139]
Cockburn and Highsmith (2001) [10]	Concerning "the people factor" [p. 132], the authors explain, "agile teams are characterized by selforganization and intense collaboration" [p. 132] and "agility requires that teams have a common focus, mutual trust, and respect; (…)" [p. 132]
Jolivet and Navarre (1996) [21]	The authors present "a new approach to large-scale project management", which is "based on self-organization and meta-rules" [p. 265]. Concerning the requirement of "autonomous units" being "largely self-organized", the authors demand for "autonomy", "subsidiarity", and "cellular division" [p. 267]
Mahmud (2009) [22]	Mahmud is "transitioning and extending the concept of self-organization from the natural sciences to management" [p. 2] and lists the characteristics of "emergence of complex behavior through iteration, rules guiding behavior, and the presence of attractors" [p. 2]
Nachbagauer and Schirl-Boeck (2019) [27]	The authors examine, whether and if so, how management can "prepare for the unexpected in megaprojects" [p. 695] and argue for balancing "structure and self-organisation in various fields" [p. 695]
Volland (2019) [33]	Volland gives an example of self-organized rules: In the described case, project members criticized a "dominantly imposed" rule. This criticism was accepted and the rule was changed "in accordance with the team". Volland argues: "This is where the form of procedural control was modified through self-organization" [p. 477]

Based on the presented papers, we describe our first understanding of self-organization, structured in Definition and extent of self-organization as well as requirements for self-organization, structured in the context, the unit of self-organization, and individuals participating in self-organization. We conceptualized these three levels, because within the papers, interactions between "outside", "the unit", and members of the unit are described [e.g., 32, 33]. We assume that for self-organization, these three parts need to interact in certain constellations and need to meet certain requirements.

- **Definition**: For us, **self-organization** is an intense form of collaboration [e.g., 10, 32, 33], which is characterized by autonomous decisions in areas without external direction or domination [e.g., 21, 32, 33].
- **Extent of self-organization**: These autonomous decisions can relate to different extents, e.g., to rules [33] or how an ambitious purpose is to be achieved and thus contains objectives, tasks and procedures [21, 32].
- **Contextual requirements**: In order to enable self-organization, the context guides the collaboration by "signalling a broad goal or a general strategic direction" [32, p. 138], functioning as "attractors" [22, p. 2] as well as by "meta-rules" [21, p. 265] as boundaries. According to our understanding, how much autonomy these boundaries provide depends on the extent of self-organization.
- **Requirements for the unit of self-organization:** In order to enable self-organization, the unit of self-organization has to use the given autonomy and act collectively itself, meaning, that there is no domination within the unit [e.g., 33] as well as team members should respect and trust each other [e.g., 10]. Furthermore, the unit should consist of individuals with various functions and backgrounds [e.g., 21, 32] and searching "for 'the limit'" [32, p. 140]. Additionally, there should be respect and
- **Requirements concerning individuals participating in self-organization**: Individuals need to be honest and open [22], individually responsible [21] and "engage in a continual process of trial and error" [32, p. 141].

Based on this first understanding, we answer our first research question of "Can self-organization increase the likelihood of creative and good solutions?" by "Yes, it is possible". A brief summary of our conclusion is given in Table 13.2.

13.4.2 How to Integrate Self-Organization in the Development of Projects? Methodology

Although we have a first understanding about how to structure [27] self-organization, we are still uncertain about how to integrate self-organization. This uncertainty is particularly based on the fact that autonomy seems to be an important characteristic of self-organization. Since we are discussing the integration of self-organization in a given project, pursuing defined objectives, the potential autonomy seems to be

Table 13.2 Answer to our first research question: can self-organization increase the likelihood of creative and good solutions?

Creativity	Self-organization	Conformity
Functional diversity, shared 'knowledge of who knows what'	E.g., cross-fertilization, shared knowledge, transparency	Yes, because both include the need of different perspectives, and the knowledge about the knowledge of others
Transformational leadership style, and environment that supports creativity and innovation	E.g., coordinator	Yes, because both include the need of motivating and lack of threatening control
Communication, reviewing performance, giving feedback, appropriate problem handling	E.g., self-transcendence	Yes, because both include the need to put everything honestly on the table
Job autonomy (how to achieve goals) and job complexity (challenging tasks)	E.g., set goals	Yes, because both include the need of leeway and challenge
Vision and common understanding of the task	E.g., set goals	Yes, because both include the need of common mission

limited. Therefore, we summarized our interest in the specified research question of "How can we govern or coordinate self-organization with stakeholders in the development of public infrastructure projects?".

- We defined governance as search term, as governance is defined as creating "the formal context" [20, p. 28], consisting of "value systems, roles and responsibilities, processes and policies" [20, p. 45].
- We defined coordination as search term, as, for us, integrating self-organization and autonomy into existing projects is a matter of coordination.

In order to answer our research question, we conducted a limited literature review [e.g., 16] in EBSCO Host, Business Source Premier as follows:

- Search terms: (Project OR Organi*ation OR Network) AND (selforgani* OR self-Organi*) AND (Governance OR coordinat*)
- Searched within the search fields 'title' or 'abstract'.
- Searched for 'Academic Paper' in English.
- We limited the scope to papers published between 2009 and 2020.

We identified 121 papers. By reviewing the papers' abstracts, we conducted a first relevance check and refined the sample to 31 abstracts. To further refine our sample, we read the papers (if available) and examined, if these may help us answering our research question. Finally, we identified seven relevant papers and there was one that we cannot access.

As our review interest was wide and as we are still refining our first understanding, we selected papers with varying content, which, however, helped us to some degree answering our research question. A brief introduction of the relevant papers is given in Table 13.3.

13.4.3 How to Integrate Self-Organization in the Development of Projects? Results and Application

13.4.3.1 Results

By summarizing the results of our literature review and the sources, we read before, we conceptualize a framework for discussing whether and how self-organization could be integrated in the development of public infrastructure projects with stakeholders.

In order to answer our question of "How can we govern or coordinate self-organization with stakeholders in the development of public infrastructure projects?", we propose that first, three characteristics concerning the intended or emerging self-organization need to be discussed, since they influence the governance and coordination.

At first, the characteristic of the **extent of self-organization** needs to be defined. How much autonomy can and will be made available or will be claimed? On the one hand, we expect this extent to be influenced by the actual development stage of public infrastructure projects. On the other hand, if this stage provides leeway for various autonomy, one could aim for more or less autonomy. Volland describes an example, where in the beginning "only the teams were self-organized" [33, pp. 474–475] and in which this self-organization was extended to self-organizing the rules on which the collaboration was based on. Or, to give another example: Two of the problems defined by Martela [23], the problems of task division and task allocation, represent different areas for which autonomy could be provided. For us, the main challenge of governing and coordinating self-organization lies in balancing on the one hand the required autonomy (of varying extent) and, on the other hand, setting boundaries and providing direction. Therefore, we assume the following correlation: The wider the extent of autonomy, the wider the boundaries are defined and the more the given direction resembles a vision (instead of specific goals).

Although our research was initiated by the idea of planned self-organization, integrated top-down in order to support creative and good solutions, the literature review indicated further options. Two papers [11, 28] describe how to facilitate self-organized initiatives. For us, such initiatives are conceivable both inside and outside the project and are characterized by bottom-up impulses [11]. Nachbagauer and Schirl-Boeck [27] examined self-organization as reaction to unexpected events and Edelenbos, van Meerkerk, and Schenk explain community self-organization as

Table 13.3 Selected search results

Authors, year	Introduction and relevance
Buijs (2010) [8]	Buijs creates "a self-organization theory framework to analyze connective capacity in complex governance processes" [p. 29] and considers conservative and dissipative self-organisation "as two ends on a continuum" [p. 35]
Edelenbos, van Meerkerk and Schenk (2018) [11]	"How do community self-organization initiatives evolve vis-à-vis existing governmental institutions, and which factors influence their persistence or disappearance over time?" [p. 53]; definition of self-organisation "as bottom-up initiatives that are community-driven (…)" [p. 53]
Esposito and Evangelista (2014) [13]	"(…), the work provides a (…) analysis of the existing body of knowledge" [p. 146] in the field of virtual enterprises. The authors identify "the hierarchical and the holarchical" as extreme forms of virtual enterprises [p. 149]; the hybrid form shares "the relationships among peers with the holarchical model and the presence of a coordinating firm with the hierarchical model" [p. 155]
Martela (2019) [23]	Martela examines self-organizations as "an ideal type of organizational form" and builds on six problems, organizations need to solve: task division, task allocation, rewarding desired behaviour, eliminating freeriding, providing direction, and ensuring coordination" [p. 1]; Martela describes self-organized organizations as "radically decentralized model of authority" [p. 1]
Müller, Pemsel and Shao (2014) [26]	"RQ1: How can Organizational Enablers be conceptualized in project related studies? RQ2: What are the Organizational Enablers for governance in the realm of projects in project-based organizations?" [p. 1311]; of interest for us: "Mindfulness of the people" and "self-responsibility of the people" [p. 1316]
Nederhand, Klijn, van der Steen and van Twist (2019) [28]	Research question: "How do policy officials and key members of community-based collectives perceive the (ideal) governance relationship between government and collectives?" [p. 234], of interest for us, e.g., "removing barriers for collectives to function, supporting them by providing fast access to public decision making" [p. 237]

(continued)

Table 13.3 (continued)

Authors, year	Introduction and relevance
Snow, Fjeldstad and Langer (2017) [31]	The authors conceptualize a "framework for the design of effective digital organizations" and propose to focus on "actor-oriented" principles [p. 1]; actors "who have the capabilities and values to self-organize" [p. 6]

"often born out of discontent with the status quo or external threats" [11, p. 61]. As we can think of situations, in which self-organization born out of discontent might even positively influence the project development (e.g., initiatives demanding solutions that go beyond what is legally possible and whose activity enables new solutions), we include these different impulses in our framework. Therefore, we propose to define the characteristics of **impulse direction** and **location of self-organization** before discussing how to govern and coordinate self-organization, since we expect each of them to influence the required governance.

Once these proposed characteristics concerning the extent, impulse direction and location of self-organization, have been defined, we propose to discuss how to govern and coordinate self-organization based on four purposes, as given in Table 13.4, and three levels.

As the superordinate question of this article was initiated by the idea of planned and top-down integrated self-organization, the question of how to integrate self-governance was motivated by the question if integrating self-organization in a given project would prevent the required autonomy. This aspect is especially integrated in purpose one and two. However, based on the identified papers, we concluded that governance and coordination of self-organization should also support self-organization; therefore, we integrated the third purpose. Furthermore, the assignment of some examples to the four purposes, as diverse teams [e.g., 21, 32] or values [e.g., 10, 21, 22, 31] was difficult, as we intended to assign these examples to each purpose. We suppose there are many dependencies, as in the research on creativity.

Besides the four purposes, we differentiate the levels (1) context, (2) unit of self-organization as well as (3) individuals participating in self-organization. Concerning these levels, we suppose them to be "objects" that need to be governed and coordinated as well as sources of governance and coordination, and therefore playing a dual role. To give an example: On the one hand, self-organization within the unit can be facilitated and motivated by values of individuals who participate in self-organization. On the other hand, self-organization can be facilitated by the context, e.g., by providing the required resources or by communicating values that facilitate self-organization.

Furthermore, Buijs [8] argues that there are two forms of self-organization—dissipative and conservative. Buijs explains that these two forms should be balanced and conceived as "two ends on a continuum" [8, p. 35]. Therefore, we propose to consider both forms in discussing the governance and coordination of self-organization.

Consequently, we conceptualize the framework as illustrated in Fig. 13.3.

Table 13.4 Four purposes of governing and coordinating self-organization

Purpose	Examples (maybe overlapping each other, and usually assigned by the authors)
1. Setting narrower or wider boundaries to the provided autonomy [e.g., 10, 22, 23]	**Setting boundaries by, e.g.:** Creating an open work environment and open communication, mechanisms to resolve conflicts and negotiation [22, 23, 32, 33]; Elimination of freeriding [23, 32]; Checkpoints, audits and macromanagement [10, 21, 32]; Meta-rules and values [10, 21, 22, 31]; Mindfulness and self-responsibility [26]
2. Providing direction and guidance in varying degrees [e.g., 10, 21–23, 31, 32]	**Providing direction and guidance by, e.g.:** Setting goals, a global (multifunctional) mission, shared goals, attractors [10, 21–23, 32]; Ambition [32]; Transparency [23, 31]; Diverse teams, cellular division, setting common focus [10, 21, 32]
3. Facilitating and motivating collaboration within the unit of self-organization [e.g., 11, 13, 21, 23, 28, 31, 32]	**Facilitating and motivating by, e.g.:** Reward system [23]; Spanning boundaries and removing barriers [11, 28]; Enabling connections and interactions and access [11, 28]; Creating favourable conditions [28]; Resources [21, 32]; Trust [10, 21, 22]; Coordination, e.g., by a coordination unit [13, 23]; IT support, tools [23, 31]; Values [10, 21, 22, 31]
4. Ensuring and motivating coordination of interdependencies between the unit of self-organization and its context [13, 23, 26]	**Ensuring and motivating coordination of interdependencies by, e.g.:** Open communication, constant communication [22, 23]; Coordination by a coordination unit, communication, and IT support and tools [23, 31]

Therefore, we summarize the answer to our research question as follows: In order to integrate self-organization in the development of public infrastructure projects, the characteristics of extent, impulse direction, and location of self-organization have to be discussed, before the purposes of governance and coordination, divided into the levels, can be discussed.

13.4.3.2 Application

In order to verify, whether we can think of integrating self-organization and stakeholders into the development of public infrastructure projects, we present two ideas. Therefore, we formulate two situations and ask how self-organization and

1.) Extent of Self-Organisation				
Task Allocation	Task Division	Rules	...	Initiative

2.) Impulse Direction	
Bottom-up	Top-down
Within the Project	Outside the Project

3.) Location of the Unit of Self-Organisation	
Within the Project	Outside the Project

4.) Purposes and Levels			
	Level: Context	Level: Unit of Self-Organisation	Level: Individuum
1. Purpose: Setting Boundaries	Conservative? Dissipative?	...	
2. Purpose: Providing Direction and Guidance	...		
3. Purpose: Facilitating and Motivating Collaboration			
4. Purpose: Ensuring and Motivating Coordination			

Fig. 13.3 Framework for discussing how to integrate self-organization (*source* authors)

stakeholders could be integrated in these situations. While the formulated situations originate from projects, the applications are fictitious, simply created in our discussion.

Situation 1: Public infrastructure project, construction of a federal railway in Germany, the situation: First drafts for solutions, in our case, six alternatives for the location of the railway, are defined and are now to be optimized based on spatial conflict areas. The alternatives encompass small sections.

Idea of application 1: Guided by the idea of "cellular divisions" and "holons" [13, 21], we think of teams, composed of project representatives and various stakeholders. Each of these teams is composed by two local residents, one environmentalist, one farmer, one representative of the local businesses, one local politician and project representatives, representing the topics of noise, landscape and transport and operation. Concerning the composition of the teams, it is important to integrate conflicting interests into the team. The goal for each of these teams is to optimize one's own alternative in such a way that as few interests as possible are seriously harmed. Furthermore, a small competition between these optimizations could be established, e.g., "the team presenting a lot of optimisation-ideas, wins something". Furthermore, ideas of additional value are welcome. As this situation is located in the development process of a given public infrastructure project, being structured by legal proceedings, boundaries are set by the location within this development process and by the defined task. How each team optimizes their alternative, how it organizes its work, what additional values it is thinking about, is open and can be organized by itself. In order to support self-organization within each team, the teams are supported by a coordinator, who identifies different knowledge levels and different 'languages'. Furthermore, the coordinator motivates them to actively participate in the collaboration without dominating each other. However, in this example, one question is how collaboration can be established rapidly and how everyone can be motivated rapidly for the task.

Situation 2: Public infrastructure project, construction of a federal road in Germany, the situation: The existing road runs through the center of the city. The new road, build by the project, will pass alongside the city and thus reduce traffic in the city and especially in the center of the city. Due to the project, a design opportunity emerges, but designing the center is not the responsibility of the project.

Idea of application 2: In this case, we argue for exploiting the given design opportunity as we suppose that a creative and participative design of the center might have positive impact on the perception of the project. As project representatives should aim to exploit this opportunity, the opportunity should be actively promoted, for example, by communicating this opportunity in meetings with politicians or the public. Afterwards, project representatives should listen if they notice first activities concerning the design of this center. If first signs are identified, project representatives should facilitate the emergence of an initiative, e.g., by supporting the initiative through access to urban planning experts, or through providing resources for meetings (e.g., locations, technology, moderators).

We discussed these possible applications based on the conceptualized framework—at least, based on the characteristics and purposes—as presented in Table 13.5.

13.5 Conclusion

After we explained the relevance of project development and argued for integrating stakeholders in the development of public infrastructure projects, we identified factors that increase the likelihood of creative and good solutions. Based on our first understanding of self-organization, we concluded that with the help of self-organization, the realization of the described factors will be supported. Thereafter, we conceptualized a framework, on which the integration of self-organization can be based on. Finally, we applied this framework to two situations.

As we were wondering, whether it is worth investigating further the integration of self-organization into the development, we clearly affirm the research question. In our opinion, there are possibilities to integrate self-organization and stakeholders

Table 13.5 Exemplary framework application

	Application 1	Application 2
Characteristic extent of self-organization	Task division, rules of collaboration, and topics of added values	"Maximum"; the question whether or not the center of the city should be designed; how and by whom it should be designed, …
Characteristic impulse direction	Top-down	Bottom-up
Characteristic location	Inside the project	Outside the project
1. Setting boundaries	Bounded by the "location" within the development process, since, according to the location, questions as topics are given	–
2. Providing direction and guidance	The goal, in our situation: the competition for the best alternative; Bounded by the "location" within the development process of public infrastructure projects, as, according to the location, questions as topics are given	Guidance by the idea or vision of a more silent, more green and child-friendly center of the city (if individuals adopt this vision)
3. Facilitating and motivating collaboration	E.g., facilitating the collaboration by a good atmosphere and location	Promoting the idea, facilitating the emergence by, e.g., providing access and resources
4. Ensuring and motivating coordination	We suppose that there is no additional coordination needed, as the task, optimization of alternatives, is always included in the development of projects	Project representatives could use meetings with the public or politicians to ask for an update, but as there are no interdependencies to the project, only few coordination is required

in the development of public infrastructure projects or to support self-organization "next to" the project (e.g., application 2).

Before we summarize ideas for further investigating the integration of self-organization and stakeholders, we will discuss some limitations of this article. At first, the superordinate question of this paper was wide and initiated by a more practical discussion. Therefore, insights into this paper are derived from a first literature review, leaving several questions unanswered. Furthermore, we only gave an overview of factors increasing the likelihood of creativity, and finally our framework requires more discussion.

In order to share ideas for further investigating the integration of self-organization, we use the conceptualized framework to locate our ideas (Table 13.6). Since we need more examples in order to discuss possible applications, we integrated the column "general" into the framework.

Table 13.6 Suggestions for further research

	Context	Unit	Individual
General	Examples of self-organization: To which questions is self-organization applied and to what extent?	Examples of integrating "outsiders" (in our case stakeholders) in the unit of self-organization	Examples of factors increasing the likelihood of self-responsible working
1. Purpose: setting boundaries	What examples of failures exist, where boundaries of participation were clearly set but participation still did not work?	How can we support open communication with stakeholders and between stakeholders (as trust is lacking)?	How to set boundaries for independent stakeholders (in contrast to employees)?
2. Purpose: providing direction and guidance	Which examples of ambitious aims and shared goals exist?	How can we define shared goals if individuals participate for different reasons and are more or less autonomous?	How can we support the acceptance of the defined shared goals by stakeholders—also on the longer term?
3. Purpose: facilitating and motivating collaboration	Which examples of opportunities for self-organization "next to" projects exist?	How can we support participative and non-dominant behavior within the unit, if we think about units consisting of "project experts" and stakeholders, who, for example, have no experience with projects?	How can stakeholders be motivated to participate in self-organization?
4. Purpose: ensuring and motivating coordination	Which approaches exist concerning the coordination of interdependencies?	–	–

In order to start investigating these ideas, we suppose the existing literature on agile project management, motivation and literature as co-creation of value can be of great help.

One interesting source to base further research on is the paper of Moe, Dingsøyr and Dybå. The authors investigate barriers "with introducing self-organizing teams in agile software development" [24, p. 76], structure autonomy in the levels of external, internal, and individual autonomy [24, p. 78], and discuss the topic of redundancy. Furthermore, we are wondering whether we can identify the roles that are played by members of agile teams according to Hoda et al. [17].

13.6 Compliance with Ethics Standards

The authors declare that they complied with all guidelines given by the Ethics Committee at the University of Kassel. Informed consent was obtained from all participants and all data was anonymized. The research does not require ethics approval, as it mentioned in the waiver issued by the Office of the central ethics committee at the University of Kassel, dated: January 21, 2021.

Acknowledgements We wish to state that ongoing PhD studies take part in the mentioned research project. Furthermore, the work described in this paper is supported by the Karl-Vossloh-Foundation (reference number: S0047/10047/2018).

References

1. Amabile TM (1982) Social psychology of creativity: a consensual assessment technique. J Pers Soc Psychol 43(5):997–1013
2. Amabile TM (1988) A model of creativity and innovation in organizations. Res Organ Behav 10:123–167
3. Amabile TM (1997) Motivating creativity in organizations: on doing what you love and loving what you do. Calif Manag Rev 40(1):39–58
4. Amabile TM, Pratt MG (2016) The dynamic componential model of creativity and innovation in organizations: making progress, making meaning. Res Organ Behav 36:157–183
5. Anderson N, Potočnik K, Bledow R, Hülsheger UR, Rosing K (2018) Innovation and creativity in organizations. In: Ones DS et al (eds) The SAGE handbook of industrial, work and organizational psychology. Sage reference, Los Angeles, pp 161–186
6. Beck K, Beedle M, van Bennekum A, Cockburn A, Cunningham W, Fowler M, Grenning J, Highsmith J, Hunt A, Jeffries R, Kern J, Marick B, Martin RC, Mellor S, Schwaber K, Sutherland J, Thomas D (2001) Principles behind the Agile Manifesto. https://agilemanifesto.org/iso/en/principles.html. Accessed 13 Apr 2020
7. Binnewies C, Wörnlein SC (2011) What makes a creative day? A diary study on the interplay between affect, job stressors, and job control. J Organ Behav 32(4):589–607
8. Buijs J-M (2010) Understanding connective capacity of program management from a self-organization perspective. Emerg Complex Organ 12(1): 29–38
9. Burns JM (1978) Leadership. Harper & Row, New York
10. Cockburn A, Highsmith J (2001) Agile software development, the people factor. Computer 34(11):131–133
11. Edelenbos J, van Meerkerk I, Schenk T (2018) The evolution of community self-organization in interaction with government institutions: cross-case insights from three countries. Am Rev Public Admin 48(1):52–66
12. Eskerod P, Huemann M, Savage G (2015) Project stakeholder management—past and present. Proj Manag J 46(6):6–14
13. Esposito E, Evangelista P (2014) Investigating virtual enterprise models: literature review and empirical findings. Int J Prod Econ 148:145–157
14. Federal Ministry of Transport and Digital Infrastructure (eds) (2014) Manual for good public participation. Planning of major projects in the transport sector
15. George JM, Zhou J (2007) Dual tuning in a supportive context: joint contributions of positive mood, negative mood, and supervisory behaviors to employee creativity. Acad Manag J 50(3):605–622

16. Geraldi J, Maylor H, Williams T (2011) Now, let's make it really complex (complicated). Int J Oper Prod Manag 31(9):966–990
17. Hoda R, Noble J, Marshall S (2010) Organizing self-organizing teams. In: Kramer J et al (eds) Proceedings of the 32nd ACM/IEEE international conference on software engineering—ICSE'10. ACM Press, New York, NY, USA, pp 285–294
18. Huemann M, Eskerod P, Ringhofer C (2016) Rethink! Project stakeholder management. Project Management Institute, Inc., Newtown Square, PA
19. Hülsheger UR, Anderson N, Salgado JF (2009) Team-level predictors of innovation at work: a comprehensive meta-analysis spanning three decades of research. J Appl Psychol 94(5):1128–1145
20. International Project Management Association (IPMA) (eds) (2015) Individual competence baseline for project, programme & portfolio management
21. Jolivet F, Navarre C (1996) Large-scale projects, self-organizing and meta-rules: towards new forms of management. Int J Proj Manag 14(5):265–271
22. Mahmud S (2009) Framework for the role of self-organization in the handling of adaptive challenges. Complex Organ
23. Martela F (2019) What makes self-managing organizations novel? Comparing how Weberian bureaucracy, Mintzberg's adhocracy, and self-organizing solve six fundamental problems of organizing. J Organ Des 8(1):1–23
24. Moe NB, Dingsøyr T, Dybå T (2008) Understanding self-organizing teams in agile software development. In: 19th Australian conference on software engineering. IEEE Computer Society, Washington, pp 76–85
25. Mönninghoff M (2008) Effects of and influences on transformational leadership development
26. Müller R, Pemsel S, Shao J (2014) Organizational enablers for governance and governmentality of projects: a literature review. Int J Proj Manag 32(8):1309–1320
27. Nachbagauer AGM, Schirl-Boeck I (2019) Managing the unexpected in megaprojects: riding the waves of resilience. IJMPB 12(3):694–715
28. Nederhand J, Klijn E-H, van der Steen M, van Twist M (2019) The governance of self-organization: which governance strategy do policy officials and citizens prefer? Policy Sci 52(2):233–253
29. Richter AW, Hirst G, van Knippenberg D, Baer M (2012) Creative self-efficacy and individual creativity in team contexts: cross-level interactions with team informational resources. J Appl Psychol 97(6):1282–1290
30. Shin SJ, Zhou J (2007) When is educational specialization heterogeneity related to creativity in research and development teams? Transformational leadership as a moderator. J Appl Psychol 92(6):1709–1721
31. Snow C, Fjeldstad Ø, Langer A (2017) Designing the digital organization. J Organ Des 6(1):1–13
32. Takeuchi H, Nonaka I (1986) The new product development game. Harvard Business Review, pp 137–146
33. Volland MF (2019) How to intentionally forget rules in newly introduced agile projects. TLO 26(5):470–484
34. Webber SS, Donahue LM (2001) Impact of highly and less job-related diversity on work group cohesion and performance: a meta-analysis. J Manag 27(2):141–162
35. Woodman RW, Sawyer JE, Griffin RW (1993) Toward a theory of organizational creativity. Acad Manag Rev 18(2):293–321
36. Zhou J (2003) When the presence of creative coworkers is related to creativity: role of supervisor close monitoring, developmental feedback, and creative personality. J Appl Psychol 88(3):413–422
37. Zhou J, Shalley CE (2008) Organizational creativity research: a historical overview. In: Zhou J et al (eds) Handbook of organizational creativity. Taylor & Francis Group, LLC, New York, London, pp 3–32

Chapter 14
Self-Organization, Dynamic Meta-governance, and Value Creation in Megaprojects

Y. Li and Y. Han

Abstract The universal low performance of megaprojects forces us to think about its inherent complexity and the incompatibility of traditional project management methodology that emphasizes "control". Megaprojects are strongly influenced by goals or strategies; face unprecedented challenges in the long-term implementation process; lack experience to refer; are dominated by different key organizations at different stages of the project-life cycle; require flexible and efficient cooperation and innovation to cope with unanticipated and complex challenges that continue to emerge; and therefore, a complex dynamic behavior system. We need to re-examine this complex system and propose new governance strategies. Based on this perspective, in this chapter, we analyze the organizational behavior and self-organization phenomenon in megaprojects under the context of complexity, develop a megaproject governance portfolio strategy for self-organization and heterorganizations, propose a new path and direction towards more sustainable value creation in megaprojects to address the current and future megaproject challenges.

Keywords Megaprojects · Self-organization · Governance · Value

The low-performance problems that are commonly exhibited in megaprojects, including cost and schedule overruns, force us to reflect on the inherent complexity and challenges that are different from regular projects, and how to better manage and govern them. Due to the multidimensional complexity of megaprojects such as environment, technology, organization, and culture, we are unable to achieve the expected goals through pre-made plans in the dynamic and uncertain environment. Further, we sometimes could not even accurately define goals, which forces us to rethink some fundamental issues in the governance of megaprojects.

Y. Li (✉) · Y. Han
Tongji University, Tongji Building A, 1500 Siping Road, Shanghai 200092, China
e-mail: lyk@tongji.edu.cn

Y. Han
e-mail: yilong.han@tongji.edu.cn

R. Ding et al. (eds.), *Research on Project, Programme and Portfolio Management*,
Lecture Notes in Management and Industrial Engineering,
https://doi.org/10.1007/978-3-030-86248-0_14

In the era of VUCA (Volatility, Uncertainty, Complexity and Ambiguity), such challenging projects are becoming more and more prominent. The traditional project management methodology of emphasizing "control" is no longer suitable for megaprojects, and this is also the root cause of their "out of control". In practice, it is increasingly proven that in a complex environment, we need more agile organizations that provide flexible, adaptable, and rapid temporary solutions, short-term plans, and alternate strategies to deal with the complexity of megaprojects. Self-organization shows strong advantages in this situation and has been widely advocated in innovative enterprises and agile organizations.

However, the organizational environment of megaprojects is different from that of enterprises or public organizations. Generally speaking, megaprojects are strongly influenced by goals or strategies, dominated by key stakeholders (such as owners, general contractors, etc.), and are subject to institutional environment and contractual constraints. As a mixture of self-organization and heterorganizations, megaprojects frequently interact with the external environment, and the organizational tasks and goals have different challenges and are constantly changing at different stages of the project life cycle to realize project vision and strategy. Therefore, in the environment of megaprojects, organizations have complex behaviors and require more sophisticated governance strategies. However, there is limited research in this field, and megaprojects are to be investigated as key scenarios for the research of self-organization and new paradigms of governance.

In this chapter, we consider megaprojects as a complex system of organizational behavior. By integrating the observation and analysis of typical megaprojects, the organizational behavior and self-organization phenomena in megaprojects are examined under the background of complexity. Subsequently, we discussed the case of megaproject organizations as a hybrid system of organizational design, organizational control and self-organization, and developed a collective strategy of governance, meta-governance, and dynamic governance for megaprojects of self-organization and heterorganizations. Finally, the direction of self-organization of megaprojects is proposed, that is, towards a new paradigm of more sustainable value emergence [1].

14.1 Organizational Behavior, Self-Organization, and Leadership in Megaprojects

14.1.1 Context and Organizational Behavior in Megaproject

According to Griffin [2], organizational context is the setting of the environmental in which a phenomenon (such as an event, process, or entity) is located, and contexts can explain some significant aspects of the phenomenon. Therefore, as a type of collective organizational behavior system that has an important impact on the society, economy, and natural environment, we cannot ignore the special organizational context of megaprojects. In summary, organizational contexts of megaprojects include different aspects at three levels: macro, meso, and micro [3]. At the macro level, there are mainly external political, institutional, cultural, and legal contexts, which often have a fundamental impact on the governance model of megaprojects, and determine the basic attributes and overall framework of organizational governance. The meso-level context of megaprojects includes factors such as financing models, project strategies, related organizations and key stakeholders, project types, project delivery models, and project target pressures, which directly determine the specific model of governance mechanisms. The micro-level context is mainly the internal environment, including the internal scale and structure of participating organizations, organization culture, organizational capabilities, leadership, as well as management methods and tools. It is the detailed setting of organizational governance under macro- and meso-contexts. These contextual characteristics determine that the governance of megaprojects is different from government public governance, community governance, corporate governance, or the governance of regular project.

Under such contextual influences, megaprojects include different levels of organizational behavior types (systems) from individuals to organizational fields, and thus form a mixture of top-down heterorganizations and bottom-up self-organization [3]. However, the micro-, meso-, and macro-behaviors of megaprojects do not have clear boundaries, and each level is not isolated. The mutual interaction and influence of cross-level behaviors, the bottom-up behavior emergence (self-organizational behaviors, such as innovative behaviors), and the top-down behavior control (heterorganizational behaviors, such as public project procurement behaviors required by law) coexist in megaproject, forming a complex organizational behavior network. In addition, the organizational behavior of megaprojects is constantly evolving, which may evolve into positive citizenship behavior or negative behavior, such as anti-production behavior or corrupt behavior. Therefore, the organizational behavior system of megaprojects is diverse, networked and dynamic, that is, extremely complex.

14.1.2 Self-Organization in Megaprojects: Dynamics, Synergy and Evolution

In the past, we often paid too much attention to the design of the governance structure of megaprojects while neglecting their organizational behavior, especially the issues of self-organization. Hence, there is still a lack of in-depth understanding of self-organization, for example, in what ways do different levels of self-organization exist in megaprojects? What are the dynamics and influencing mechanisms of the emergence and evolution of self-organization? How does collective behavior emerge and evolve, and what impact does it have on the performance of megaprojects? It generally involves three important themes, namely dynamics, synergy and evolution.

Compared with enterprise organizations, there are more uncertainties in the environment and tasks of megaprojects, the urgency of crisis response is higher, and the importance of iterative plans is more prominent. Therefore, in some cases, it is necessary to formulate parallel trials to avoid failure of single plans and loss of key opportunities. For example, the Manhattan Project, the starting point of "modern" project management, used an implosion scheme as a backup plan for bomb design [4]. In some cases, the plan must be successful at the first time, such as the irreversibility of underground construction and offshore operations, which poses a major challenge to megaproject organization. The traditional hierarchical, functional and top-down inefficient and rigid control methods would not only provide limited help, but also further aggravate the risk of project failure. Therefore, it is necessary to find a more efficient, flexible and resilient organizational mechanism.

When encountering foreseeable and unpredictable challenges, organizations must quickly adjust or build a new task organization to form an "project organization in the project organization" to deal with these "projects in the project", which has typical self-organizing characteristics. In other words, the high challenge of the tasks, the variability of the environment, the lack of experience, and the high cost of failure have prompted the emergence of megaproject self-organization, and it is a result of efficient collaboration among participating organizations that has a positive effect in dealing with project complexity. Similarly, this type of self-organization must maintain efficient coordination and self-adjustment capabilities in the project implementation process to meet the challenges of specific complex tasks. The continuous emergence of formal or informal self-organization makes megaproject organizations present distributed network features, forming a meta-organization. As the project progresses, the continuous changes in project phases and tasks, as well as the adaptive adjustment of the organization, have promoted the simultaneous evolution of megaproject self-organization, making dynamics, synergy and evolution the three inseparable themes of megaproject self-organization.

14.1.3 Leadership in Megaprojects

Obviously, self-organization in megaprojects does not operate in a vacuum, as it would be disturbed by internal and external environments, nor is it appeared out of thin air and operates completely freely. Self-organization in megaprojects must have a "specific" mission and must be able to efficiently solve the complexity of megaprojects. Therefore, self-organization in megaprojects is an efficient organizational form that must complete specific tasks and achieve specific goals in a specific environment. To realize this capability or function, self-organizing leadership must be improved.

Let us revisit the Manhattan Project, widely known in the project management community owing to its prominent historical status. It even makes us begin to reflect on the rationality of the currently widely adopted project management body of knowledge (PMBOK) and project control methodology [4]. What should we do when "the problems we face have no ready-made solutions?" General Leslie Groves, the project manager of the Manhattan Project, not only provided us with practical experience in facing such projects, but also showed us how to win the "war" of this megaproject and the importance of leadership.

Compared with general projects, the leadership of megaprojects has significantly different requirements. As Flyvbjerg [5] analogizes, "you would not want someone with only a driver's license to fly a jumbo". Obviously, the environment faced in megaprojects is more complex and challenging, and the requirements for leadership are higher. In fact, leadership itself is also part of the solution to the complexity of megaprojects. Leadership plays an important role in megaprojects in terms of dealing with complexity, adapting to changes, and making up for the shortcomings of traditional project management methods. Therefore, Megaproject leader is a leader of leaders, self-organization in megaprojects needs leaders rather than managers.

14.2 Complexity and Organizational Adaptation in Megaprojects

14.2.1 Dealing with Complexity: The Root Cause of Self-Organization

Although the current research on complexity of megaprojects involves complexity sources, dimensions, measurement methods, and complexity management and governance, it is still in the exploratory stage, especially the "simple principles behind complexity" have not been found. However, the existing research outcomes still provide us important references to understand and manage the complexity of megaprojects. Taking a typical complexity dimension classification as an example, the complexity of megaprojects can be analyzed from the dimensions of composition, schedule, technology, institution, organization, and stakeholder. Even if the same type of megaprojects (e.g., water conservancy projects) are reflected in different dimensions of complexity, as factors like the spatial characteristics of the project, the crossing of administrative boundaries, and the involvement of immigration would all have an impact on the complexity of different fields in megaprojects.

As a typical type of complex adaptive system, or system of systems (SoS), megaprojects have the basic commonality of complex systems, that is, they have complex collective behaviors, signaling and information processing, and adaptation, and have nontrivial emergence and self-organizing behaviors [6]. According to PMI [7], human behaviors, system behaviors and ambiguity have been considered the three major causes of project complexity. From this perspective, if a megaproject is regarded as a complex system, or its complexity is examined, system behavior is then the key object to be investigated. Therefore, the complexity of project composition, large in-scale, and technological diversity are only the "physical elements" of megaproject complexity, and institutional complexity and uncertainty are only the "environmental elements", and the essential source of complexity is the collective adaptive social behavior of organizations to achieve the strategic goals of megaprojects. In this process, the typical characteristics of complex systems such as dynamics, evolution, association, network, self-organization, and emergence are fully reflected, and adaptation builds the complexity of megaprojects.

14.2.2 Adaptation: Self-Organizing Characteristics in Megaprojects

According to the theory of complex adaptive system (CAS), CAS can be regarded as a system composed of interacting agents described in terms of rules. With the accumulation of experience, these agents adapt by constantly changing their rules. The adaptation effort of any agent is to adapt to other adaptive agents, and this is the main source of the complex temporal patterns generated by CAS [8]. For megaprojects, if we consider individuals, teams, objects, tasks, resources, systems, environments, and among others as adaptive "agents", then these agents would continue to learn, interact, and adapt to each other under certain rules, which constitute the complexity of the megaproject system. CAS theory provides theoretical support for the study of megaproject complexity.

Take the adaptation of the decision-making organization for the construction of the Hong Kong–Zhuhai–Macau Bridge in China as an example, due to the continuous adjustments in decision making, the requirements for decision-making bodies, decision-making capabilities and decision-making mechanisms are also constantly changing. The Chinese Central Government and relevant departments of the State Council, as well as the governments of Guangdong, Hong Kong and Macau, continue to renegotiate to form different combinations of decision-making bodies, and constantly adjust decision-making powers to build an efficient decision-making mechanism to deal with the complexity of decision-making in the early planning and construction stages of the Hong Kong–Zhuhai–Macau Bridge [9] . Our long-term case study of the Shanghai World Expo park also proved this point. In order to cope with the complex challenges at different stages, the organizational structure of the Shanghai World Expo Bureau has undergone 19 adjustments, from the initial 7 departments to as many as 53 departments, and after the initial establishment of the Expo Bureau, before and after the start of the construction of the venues, and before the opening are the three stages that adjusted most [10]. Similar adjustments also occurred in the Manhattan Project. For example, in 1942, all research and production management were transferred to the Army, General Leslie Groves was appointed as the new project manager, and the Los Alamos Laboratory was expanded [11]. Thus, such kind of adaptation is an important mechanism for megaprojects to maintain the possibility of achieving the "set" goals, and it is also a manifestation of the high efficiency and resilience of megaproject organizations.

However, how is the organizational Adaptation of megaprojects achieved, active or passive? Don't megaprojects require external control? Is a megaproject organization a distributed network organization without a sole hub? Why are megaprojects generally out of control? These are all the key issues that need to be addressed in understanding megaproject self-organization.

14.3 Organizational Design, Organizational Control and Self-Organization

14.3.1 Where do Megaproject Organizations Come from, and Why are They Different?

Generally speaking, a megaproject has a long-term preliminary process, and is even a proposal for quite a long time. At this stage, the scale of the organization is very small, and even no fixed organization is formed. For example, the Three Gorges Project in China was proposed by Sun Yat-sen in 1919, the feasibility study was officially launched in 1958, and it was not until 1992 that the resolution was passed. The following year, the State Council Three Gorges Project Construction Committee and China Three Gorges Project Development Corporation were established, and the project entered the mass construction stage more than 70 years later from its initiation [9]. As megaprojects have huge investment, widespread impact, and important strategic significance and goals, once these goals are formally determined and launched, there must be a specific organization to achieve them. This is the birth of a megaproject organization.

However, the challenges encountered by megaprojects and the differences in context determine that the formation process, organizational structure, functions and mechanisms of the organization are also diverse and different. Even in the same country, the initial organizations of megaprojects show great differences. For example, in China, the initial organizations of megaprojects in different historical periods, with different project types, and at different project locations vary. Figure 14.1 illustrates the differences in the initial organizations of four typical megaprojects in China. It is demonstrated that some megaprojects are coordinated by the top-level government and implemented by public institutions (such as the Hong Kong–Zhuhai–Macau Bridge and Shanghai World Expo), while others are coordinated by the top-level government and implemented by state-owned enterprises (such as the Beijing-Shanghai High-speed Rail and the Three Gorges Project). There are certainly some projects that are mainly led by state-owned enterprises, for example, the Shanghai Tower. Strategic significance and goals, public attributes, investment and financing models, the difficulty of cross-regional coordination, and technical complexity would all affect the design of initial organizations in megaprojects. The path dependence formed by the setting of initial organizations affects the

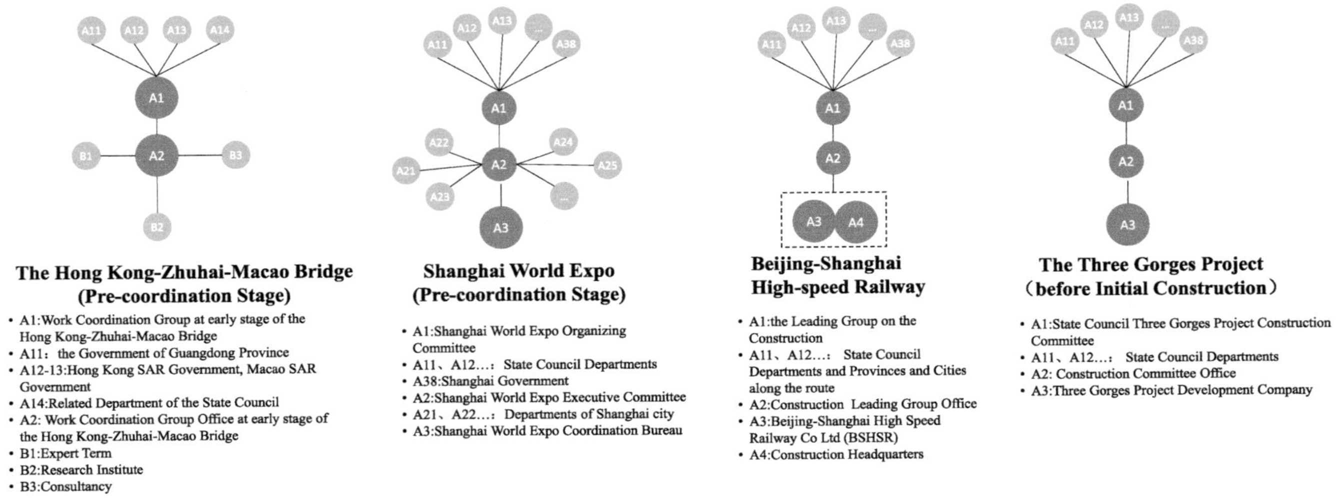

Fig. 14.1. Structural differences of initial organizations of typical megaprojects in China

evolution path of megaprojects, which makes the nonlinear evolution of megaproject organizations have the initial sensitivity.

There is no doubt that the initial formation of megaproject organizations was not spontaneous. It is an organizational design carried out by the higher-level institutions (such as the central government) under a specific institutional background in order to achieve megaprojects' strategic objectives. Organizations at this stage are often administrative and bureaucratic, and the mission is to discuss and make decisions on critical issues. Therefore, efficiency is not a key feature set by the initial organization of a megaproject, and even, professionalism is not the core issue considered by the organization at this stage. Once the project is decided to be implemented, the organization at this stage would transform into the top-level coordination or decision-making organization for project implementation, or gradually withdraw from the project organization.

14.3.2 Would Megaproject Organization Lose Control?

Obviously, although the decision-making at the initial stage of a megaproject is of great significance, there are few tasks involved, the scale of the organization is small, and the degree of complexity is not high. Therefore, the traditional organization mechanism has no obvious disadvantages at this stage, as it is more embodied in heterorganizations under a specific political, institutional and cultural environment, and the characteristics of self-organization are not obvious. However, as megaprojects progress, the multi-dimensional project complexity would increase significantly and rapidly. As a result, simply relying on a single top-level organization for traditional bureaucratic governance mechanisms could not cope with the unexpected challenges and complexity of megaprojects that continue to emerge. In order to achieve the goals of investment, schedule, quality, safety, environment in megaprojects, project management methods based on breakdown and control concepts have been applied. In the project implementation stage, megaprojects are broken down into several sub-projects and implemented by different organizations. The entire project organization gradually develops into a large-scale, complex organization system with diverse composition and relationships. During the operation of the organization, organizational tools are continuously applied to achieve the goal of "controllable" projects. For example, in the construction of the Shanghai World Expo, formal control methods of institutionalization, procedures, standardization, and informationization, as well as value-oriented and project-oriented informal control methods were integrated and applied, and a variety of control mechanisms were designed to ensure the smooth implementation of the World Expo construction and finally realize the grand opening on May 1, 2010 [12]. This mechanism based on the concept of "control" has certain reasons: (1) The implementation stage of megaprojects may still be affected by politics and the government, and bureaucratic mechanisms, especially top-level management mechanisms, will not disappear as the project enters the implementation stage; (2) Due to the rigid characteristics of the strategic objectives of some megaprojects, such as the opening of the Olympic Games, "control" is the classic idea of reducing the risk of loss of control. Project organizations would invest slack resources, including human resources; (3) Project organizers will adopt conservative strategies, that is, use past experience as much as possible and adapt to specific project situations through partial improvements or tailoring, instead of radical reforms. Therefore, even if it is inefficient, the vast majority of megaprojects would still adopt traditional organizational methods and invest as much resources as possible to ensure that the projects are "controllable." This is certainly another important reason for the cost overruns of megaprojects, but this method itself is obviously full of huge risks.

First is the issues of efficiency. Traditional organizational mechanisms place more emphasis on hierarchy, functions, rules, and processes, etc. This may be acceptable for a relatively stable administrative organization, as it may be more efficient after a certain degree of practice. However, for megaprojects that have a large number of uncertain events and require rapid response, this inefficiency may be catastrophic or fatal. Second is the issue of flexibility. Organizations can be designed in advance, but a large number of unexpected events may occur during the operation of the organization, so that the original organization can hardly adapt to changing needs. The bureaucratic mechanism is difficult to adjust quickly, which leads to the problem of insufficient organizational flexibility. This is the fundamental drawback of traditional organizations not adapting to complex environments. Third is the issues of organizational culture and behavior alienation. In addition to organizational structure and configuration, the prerequisite for efficient organization is also the core factor of organizational collective behavior. The ideal organization should have a good organizational culture and organizational citizenship behavior, which is the guarantee of the organization's ability to deal with uncertain environments, but traditional organizations may cause the alienation of organizational culture and organizational behavior, such as anti-productive behavior and even corruption. As a result, under the traditional organizational mechanism, megaprojects may run out of control when faced with a complex environment.

14.3.3 Continuously Evolving Hybrid Organizations

Since traditional top-down organizational design and control are difficult to deal with the complexity of megaprojects, a more complex organizational governance is needed. From a practical point of view, temporary task or goal-oriented self-organization shows great flexibility and high efficiency, and plays an important role in making up for the shortcomings of traditional organizations. For example, in the Hong Kong–Zhuhai–Macau Bridge, the Planning and Contract Management Department of the Administration had formed a learning organization, which constantly shares experience in similar domestic and international megaprojects through self-learning and seminars, and improves procurement and contract management capabilities to form collective wisdom and formulate innovative procurement strategies. The megaproject organization has never been a pure bureaucratic organization, nor has it completely self-organized, but a complex, constantly changing and evolving hybrid organization. However, in the course of evolution, the top-level strategic purpose of the organization is relatively stable, and adjustments mainly occur at the project implementation organization level to respond to varying environments, changing tasks, and adjusted phased goals.

In terms of form, megaproject organizations might be a mixture of hierarchical organizations, functional organizations, project organizations, and network organizations, or namely hybrid meta-organizations [13]. There is almost no single organizational form to deal with the practical cases of megaprojects. In project implementation, any level and any part of organizations may change, for example, forming a new temporary organization, and horizontal or vertical connections or integration may also occur to cope with the major challenges that are occurring and/or may arise. Figure 14.2 illustrates the adjustment process of key organizations at different stages in the construction of the Hong Kong–Zhuhai–Macau Bridge. This adjustment may come from external system and organizational design. To a certain extent, we can also regard this adjustment as the macro-level self-organization of megaprojects, reflecting the overall self-adjustment and adaptation of megaproject organizations. At the same time, as the project progressed, temporary tasks continued to emerge. Some special task groups with obvious self-organization properties began to appear, such as innovation teams. These organizations played a key role in effectively responding to environmental changes.

There is no doubt that self-organization is not a panacea for solving the complexity of megaprojects. Whether it is traditional bureaucratic organization, control, or adaptive self-organization, it is all about allocating resources more efficiently, responding to project challenges, and achieving project goals in the most "controllable" or innovative way to reflect the "legitimacy" of their own existence. From this perspective, a megaproject is a meta-organization that integrates multiple governance mechanisms. The purpose of megaproject governance is to form collective wisdom to deal with the high challenges of megaprojects, improve project performance or the possibility of success, and at least try to avoid project failures as much as possible. Therefore, the organization of megaprojects requires a goal-oriented innovative governance model, rather than "free" self-organization, and the organizational context is also very different from enterprise or social self-organization.

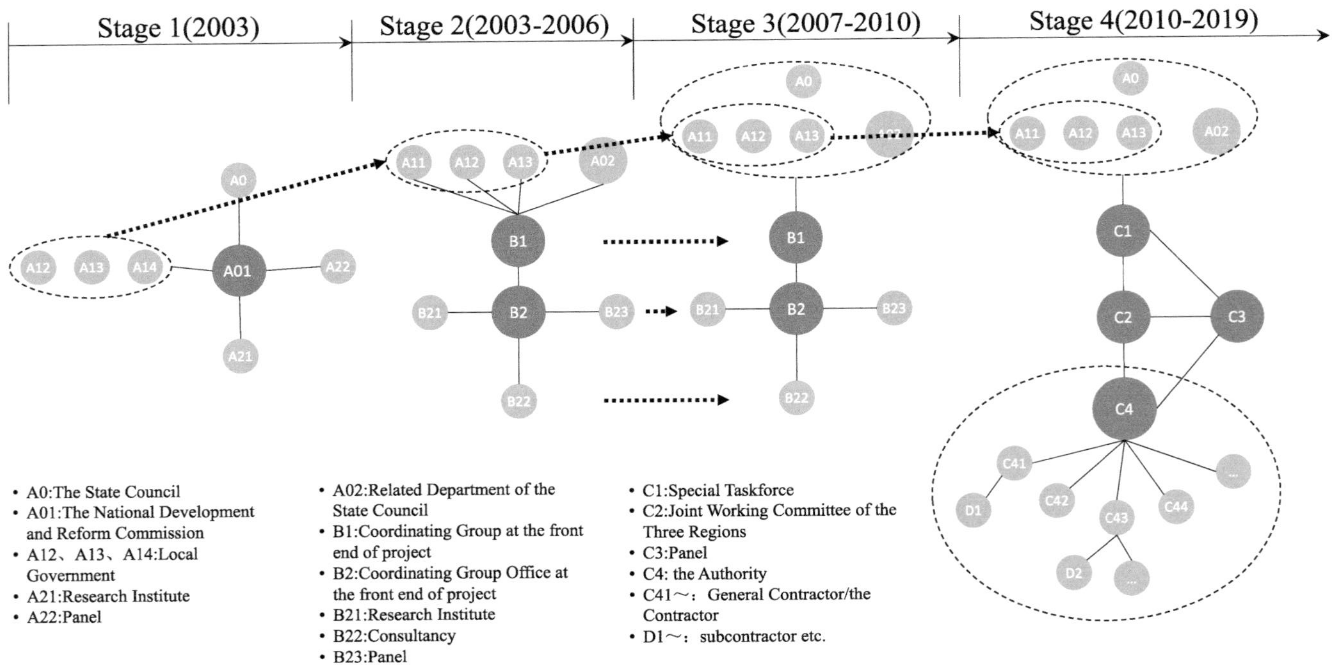

Fig. 14.2. Formation and adjustment process of the project organization during the construction of the Hong Kong–Zhuhai–Macau bridge

14.4 Governance, Meta-governance and Dynamic Governance of Megaprojects

14.4.1 Multi-layer and Network Governance: Governance System in Hybrid Organizations

The openness of megaproject systems, the influence of external systems and culture, the diversity of stakeholders, and the strategic significance and wide-ranging influence of the project make the organizational structure and organizational behavior of megaprojects extremely complex. It is therefore difficult to adopt a single governance mechanism for the governance of megaprojects. It is the intersection of governance mechanisms in different fields such as government governance, public governance, project governance, and corporate governance, and is presented the multi-layered nature of functions such as decision-making, coordination, management and implementation in the vertical, as well as the complexity and network of organizational relationships in the horizontal, thus forming the multi-layer and network governance characteristics of hybrid organizations.

Generally speaking, due to the public nature of megaprojects, whether in developed or developing countries, the governments play a critical role in top-level decision-making and governance arrangements. They could be not only sponsors, decision makers, but also organizers, coordinators, regulators and even top-level steerers. The specific governance role of the government has different specific manifestations under different institutional systems, different temporal-spatial contexts, and different project strategic backgrounds. For example, in China, before the implementation of reform and opening up and planned economy, the government was in a dominant position in megaprojects. In recent years, the government has gradually become a decision maker and top-level coordinator, and the specific organization of the project is executed by the enterprises. Even for the megaprojects led by enterprises, such as energy projects or urban development projects, also reflect the multi-level governance, which is determined by the organization's different governance functions and governance capabilities. For example, major decision-making issues and system design are usually the responsibility of the top management teams, while the implementation of specific tasks is usually the responsibility of different subcontractors, project teams, or engineers. This hierarchical governance mechanism is not equivalent to the traditional bureaucratic mechanism, but is an inherent requirement for the governance of complex organizational systems. Therefore, from a vertical perspective, the multi-layer governance characteristics of megaprojects are very prominent.

Due to the complexity of organizational relationships and behavioral interactions, governance and network governance are sometimes synonymous, or governance consists of self-organized inter-organizational networks. If organizational structures, organizational relationships, and organizational behavior of megaprojects are further examined, it is found that there are complex networks with frequent interactions within megaproject organizations, with formal organizations and informal organizations interacting with each other. Different mission-oriented self-organizations continue to emerge or disappear, and further affect other project organizations. This is the fundamental motivation for the emergence of nonlinear collective behaviors in megaprojects.

In the context of the conflict of value and interest of various stakeholders, the asynchronous and temporary features of participation time, and the relationship heterogeneity, there are also diversity in megaproject governance mechanisms, similar to contract governance and relationship governance, which makes traditional simple hierarchical governance unable to cope with this kind of hybrid organization, and as a result, network governance becomes a necessity. This kind of network may occur at different levels of the organization, or even across levels, to form a polycentric and overlapping self-organizing network, thereby improving organizational resilience. In addition, it is worth noting that due to the openness of megaproject organizations, stakeholders may interact in a larger organizational field to form a relatively stable strategic alliance, and affects the organization formation or organizational performance of specific megaprojects. For instance, research demonstrated that megaproject innovation may cross the single-project level, forming an innovation network organization in multi-project cooperation, thereby affecting project performance [14].

14.4.2 Meta-governance: Self-Organization in a Controllable Environment

In public management, meta-governance is "the governance of governance" or "organization of self-organization", which can be interpreted as a large number of organizations and management processes within the public sector that have achieved a considerable degree of autonomy governance, and it thus is necessary to exercise certain control over the various components of governance. It is regarded as a solution to the failure of existing governance, and is more oriented towards a network of organizations composed of actors from multiple fields [15, 15]. Meta-governance emphasizes the guidance of society and economy, that is, it tends to control the behavior environment of the public sector, rather than the behavior itself; meta-governance not only recognizes the necessity of authorization and decentralization in governance, but also realizes the need for stronger central control and guidance [16]. Due to the complexity of megaprojects, it is necessary to introduce meta-governance on the basis of hierarchical governance, market governance and network governance. According to meta-governance theory, it would be meaningful to improve the efficiency and performance of multi-organization or cross-organization in megaprojects, at least in terms of strategic frameworks, playing rules and decision-making, participation and coordination, and monitoring and accountability [16, 17].

In terms of strategic framework, behavioral rules, and decision-making, megaprojects, especially public projects, are difficult to avoid the influence of the traditional bureaucracy, and this has brought universal decision-making problems such as optimism, deliberate misinterpretation, and non-professionalism [18]. However, even the participation of private enterprises could not substantially improve the quality of decision-making for megaprojects. At least there is no solid evidence that the public-private partnership model can solve the decision-making issues in megaproject. If more organizations can participate equally and democratically, especially experts and professional consultants and the public, as well as adoptmore professional tools and more transparent mechanisms, decision-makings in megaproject would be more scientific. However, this requires a systematic strategic framework and the formulation of rules of conduct.

Regarding participation and coordination, we still lack research on how different stakeholders or professional organizations intervene in megaprojects, how to manage and collaborate, and how they self-organize. Due to the high dependence between various tasks of megaprojects, close communication and cooperation between teams of different sub-tasks (such as subcontracting, special consulting, etc.) become inevitable. This makes the stakeholders or project organizations highly interactive, thus forming multiple task-specific teams, distributed organization centers, and dynamic organization networks. These are both the subject and the object of governance. Coordination is a key feature of high-performance megaproject organization. It not only occurs across different levels vertically, but also between different participating organizations horizontally; it not only occurs within the organization, but also across organizational boundaries; it could be formal or informal. Within self-organization, coordination may be efficient, but coordination between self-organizations may be fragmented, overlapping, or conflicting. Meta-governance must solve the problems of coordination, cooperation and negotiation between different levels and different participating organizations.

Last but not least, for monitoring and accountability, as the dynamic nature of the megaproject environment and the high risk of out-of-control, it is necessary to build a performance and benchmark monitoring system to control the status of the project system in real time. Once the project deviates from its original goal, the correction mechanism becomes crucial. In governance mechanisms, accountability is usually conceptualized as a mechanism to exercise control over public organizations and projects, but it is also an effective way to guide project improvement [16]. Lack of accountability is considered to be the key reason for the current out-of-control performance of megaprojects [18], and it is often considered to be caused by traditional bureaucracy. Therefore, how to build an efficient correction mechanism through a monitoring and accountability system is the key to always being "controllable" for megaprojects.

In order to ensure that the project is carried out in a controllable environment, meta-governance is a concept/idea that can be used for reference. Further, the adoption of some governance tools is also helpful, involving authority, economic and informational instruments, such as performance management, strategic management, budget control, personnel, trust and values [15, 16]. Among them, dynamic performance management not only contributes to accountability management, but also contributes to goal realization and dynamic control of megaproject systems. Governance at the strategic level helps indirect control, which provides sufficient flexibility for efficient self-organizing behavior. Effective budget control is still a classic way to prevent investment out of control, and proper personnel arrangements can also assist in balance government governance and market governance. Especially for the top-level governance organization level, personnel arrangements are often very important, which provides an essential guarantee for improving the leadership of megaprojects. Of course, in efficient governance, trust and common values are the most frequently discussed topics. They are flexible governance methods that guide, shape, and stabilize organizational behavior, and are regarded as the strongest sustainable guarantee for the high performance of megaprojects.

14.4.3 Dynamic Governance: More Resilient Self-Organizing Ability

Megaprojects have long project cycles. The internal and external environment, governance structure, and governance elements have all undergone drastic changes and co-evolution. There is a greater "tense point" phenomenon between the traditional static project governance theory within the scope of a single organization and the practice of this type of project, and it thus is necessary to learn from the theory and practical experience of other fields. According to the Evolutionary Governance Theory (EGT), governance intervention should start with a deep understanding of the contexts. It should be recognized that the community and the governance situation are highly dynamic. All governance elements (such as context, goals, actors, institutions, knowledge, power, structure, composition, governance technology, etc.) are state-dependent, and the relationship between these elements and them are constantly evolving [19]. Therefore, in a dynamic environment, the governance mechanism of megaprojects and the governance environment are coupled with each other. Van Assche, Raoul, and Martijn [20] proposed the EGT model, which consists of three parts: (1) actors, institutions, and power/knowledge configuration; (2) dependence and path creation, including path dependence, interdependence, and goal dependence; (3) Governance path, goal and theme.

Our long-term in-depth investigation of the mega event project of the Shanghai World Expo found that in order to achieve the project goals and vision, the governance structure and governance elements need to evolve, even through partial fundamental adjustments to improve adaptation, and the interaction between governance configurations and co-evolution. Under certain contexts, the evolution of mega events is path-dependent, and governance elements at different levels have continuity as they evolve at different stages. However, due to the pressure of phased projects, new paths may be developed locally [10]. In this dynamic context, megaprojects need more complex governance capabilities (governability), that is, when faced with the diversity, complexity and dynamics of environmental elements. The ability to lead to governance activities and to organize and implement governance activities is an overall governance capability [21]. In the megaproject context, this overall governance capability is complex, including traditional rigid capabilities, as well as robustness, flexibility, and resilience, so as to have the ability to passively stabilize and respond to changes.

Generally speaking, as an open and complex giant system, the organizational behavior of megaprojects is multi-level and diversified, with both traditional organizational control (other organization) and strong self-organization behavior, which reflects the complexity of system behavior. Therefore, a single governance mechanism cannot cope with such a complex system, reflecting a mixture of multilayer governance, network governance, meta governance, and dynamic governance. Since these governance mechanisms or governance strategies have different functions, how properly use them is not only a technical issue, but also a management art issue, and is deeply influenced by institutional culture. The self-organization behavior in megaprojects is not only affected by these governance integrations, but also affects the strategy of the governance portfolio and the realization of governance effects. Figure 14.3 illustrates the integrations of Hong Kong–Zhuhai–Macau Bridge governance strategies.

Disorderly self-organization can also cause self-organization failure. The purpose of governance is to achieve good order through governance mechanisms [22]. The integrated application of self-organization and dynamic meta-governance of megaprojects is to ensure the dynamic balance of the organization and operation of the project in the "controllable" environment to achieve the "controllable" strategic goals. On the one hand, self-organization is used to deal with the challenges and uncertainties in megaprojects; and on the other hand, the realization of megaproject strategic goals is achieved through governance, so that self-organization is not disorderly and chaotic. However, although this approach ensures that megaprojects do not have catastrophic consequences and are completely out of control, it still does not fundamentally solve megaprojects' performance issues. As megaprojects encounter more and more challenges, the resolution to such performance issues calls for more innovative concepts and mechanisms.

14.5 Towards More Sustainable Value Creation and Value Emergence

14.5.1 The Grand Challenge of Megaprojects Under the Next Normal

The openness of the megaproject systems makes it extremely sensitive to the external environment, so the corresponding challenge is constantly changing. When facing

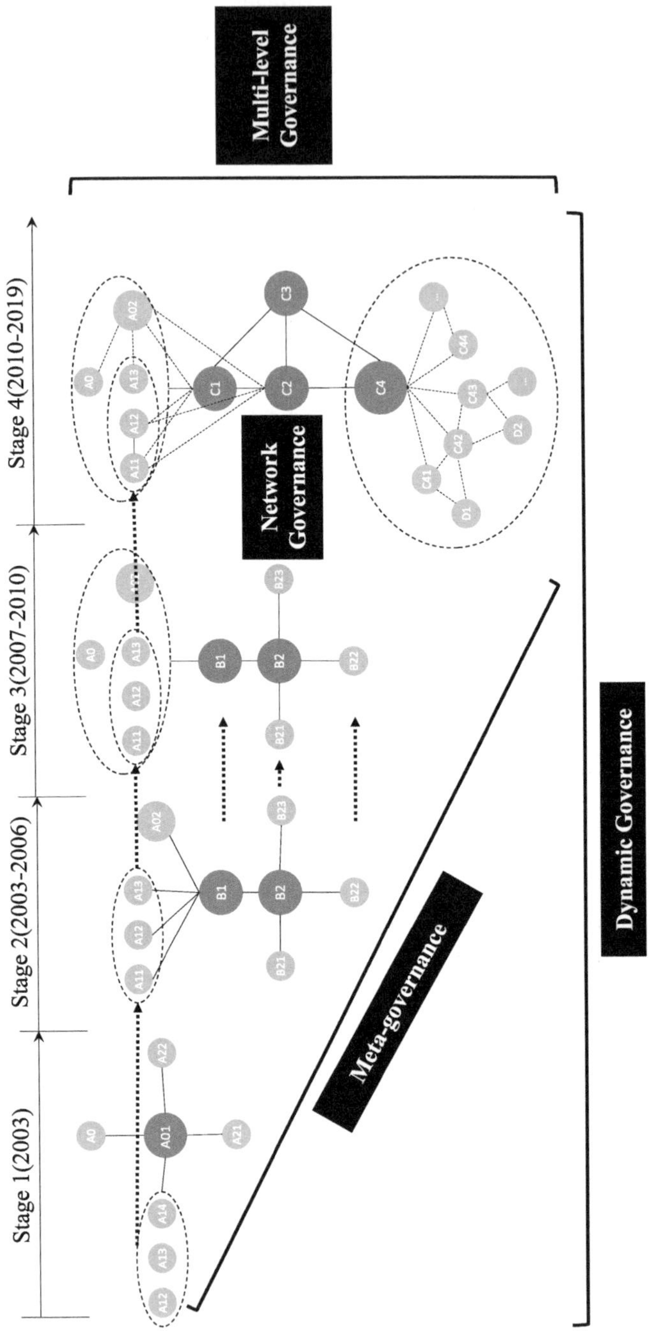

Fig. 14.3. Governance integrated strategies of Hong Kong–Zhuhai–Macau Bridge

current and future megaprojects, past experience is often unsuitable inadaptable and requires continuous innovation. From the foreseeable external environment, new demands such as new globalization in the (post-)epidemic era, application of emerging technologies, more dimensional goals, global climate change and sustainable development in the are evolving. Megaprojects will face new grand challenges in the next normal.

New globalization in the (post-)epidemic era: the continued outbreak of Covid-19 has a profound impact on geopolitics, industrial chains, supply chains, and value chains. Megaprojects, especially global megaprojects, would face more uncertain institutional complexity. With new changes in workforce, cooperation and resource supply on a global scale, the implementation of long-term megaprojects would face more uncertainties such as project termination, plan adjustments, and replacements of key equipment and suppliers. The shaping and stabilizing of the new globalization still requires a long-term process.

Impact of emerging technologies such as 5G, Internet of Things, and Artificial Intelligence: the digital era has come, and the Covid-19 epidemic further accelerated this process. There are more and more new types of megaprojects based on digital technology, such as smart city, smart transportation, data center, large-scale technology infrastructure, etc. However, we still lack experience, knowledge and control capabilities for large-scale implementation of these emerging technologies. On the other hand, the application of emerging technologies has also profoundly affected the organizational behavior of megaprojects, such as remote work, virtual teams, data and artificial intelligence-driven decision-making, building information model (BIM)-based technology applications, etc. How to integrate emerging technologies for megaproject management is a "the-future-is-already-here" challenge.

More dimensional goals: traditional project management performance evaluation standards centered on schedule, quality, and cost are increasingly being criticized. Compared with general projects, megaprojects undertake more missions and strategic goals, such as economic development, regional development, social equity, environmental protection, technological innovation, industrial cultivation, social responsibility, etc. The goals of megaprojects are more diversified, and the performance evaluation of megaprojects is more multi-level and multi-dimensional. In many cases, megaprojects exhibit greater ambitions and also contain greater risks.

Global climate change and sustainable development: climate change has made the global village a community of destiny, and the whole world must act in unison to meet the challenges of climate change and achieve ambitious carbon emission plans. Megaprojects must be more harmonious, more resilient and more sustainable with the entire society and natural system. The sustainability agenda and social responsibility must be incorporated into the management of megaprojects, and there would be new changes in the connotation and requirements of sustainability, especially in underdeveloped and developing regions. Megaprojects must consider to bring greater value to human society, not just to achieve performance goals of schedule, quality and cost.

14.5.2 Megaprojects as Value Creation Platforms

Why do megaprojects exist? This question is closely related to how to evaluate the success of a megaproject. From the perspective of the emergence of modern project management, megaprojects often exist as a strategic tool. The project itself is often not an end, but a way to achieve a larger strategy goal. Therefore, the two essential issues of whether megaprojects themselves could be successfully delivered (e.g., the Manhattan Project) and whether successful project delivery could help achieve broader strategic goals (e.g., end the World War II), determine that we must re-examine the function of megaprojects and the united mission of participating organizations.

To successfully deliver megaprojects or further achieve strategic goals by delivering megaprojects, a megaproject needs to be regarded as a value creation platform [23] as well as value creation and distribution process [24]. All participating organizations involved in the megaproject need to work together to deliver established value or jointly create higher value to ensure the success of the project as well as the success of the strategy. In this process, the flexibility, high efficiency and creative advantages of self-organization will be further presented, and the disadvantages of traditional top-down command, control or governance methods will be further amplified. To achieve value co-creation and sharing, it is necessary to unify the conflicts of interest of all stakeholders and achieve win-win results through co-governance. For example, in the construction of the Hong Kong–Zhuhai–Macau Bridge, the island tunnel project is a critical and controlling project with extremely high technical difficulty and poor construction conditions. China's domestic contractors lacked experience and technical skills, but there were legal and policy obstacles to international contractors, so it is necessary to find a new path for resource integration. Through thorough market research, an international and domestic consortium including construction, design, consulting, and multiple cooperation and subcontracting was finally formed. At the same time, cooperative partnerships based on the spirit of responsibility and contract, mutual trust, mutual understanding, and mutual respect were also developed [25]. The exploration and application of this model contributed to the successful completion of the Hong Kong–Zhuhai–Macau Bridge and improved the technical capabilities, management level and market competitiveness of participating organizations in the megaproject. Further, in this case, the large-scale steel box girder procurement and manufacturing procurement model not only satisfied the application, but also promoted industrial upgrading [25]. Participating organizations need to consider megaprojects as a platform for value co-creation and sharing rather than a distribution platform for existing fixed benefits to ensure the ultimate project strategic success.

Due to the heavy investment, large in scale and long construction period of megaprojects, stakeholders would pursue more sustainable value creation rather than one-time value distribution. The current contract mechanism of all parties involved in the project is actually a fixed resource benefit distribution mechanism, which causes the "tragedy of the common-pool resources." This ultimately damages the interests of the project and all parties, or the interests of some stakeholders are guaranteed but the overall project interests are damaged. The incompleteness of contracts provides opportunities for opportunism, which is an important source of cost overrun in megaprojects.

Integrated central control and decentralized governance could better improve the behavior of megaprojects and assist in the value creation and distribution. Therefore, megaprojects must design a whole-life-cycle sustainable value co-creation, sharing and trust mechanism in order to align the goals of self-organization and other organizations, thereby reducing transaction costs and finding a more sustainable path for value emergence.

14.5.3 Searching for a New Paradigm for More Sustainable Value Emergence

For megaprojects, past experience is not useless. The experience of some megaprojects offers much inspiration, such as multiple governance portfolio strategies. Practice has proved that such strategies could reduce the risk of project out-of-control and provide an acceptable bottom line benchmark for megaproject performance. However, to fundamentally improve the project performance of megaprojects, new paths and even new paradigms must be found.

There is no doubt that new types of organization are the direction we must consider. Self-organization, task forces, agile organizations, polycentric organizations, network organizations, integrated project delivery, smart organizations, etc. increasingly demonstrate their advantages of high efficiency and high flexibility in complex and uncertain environments. Although top-down control and governance are necessary to some extent, their drawbacks are increasingly revealed. Especially when faced with new megaprojects that are increasingly lacking in experience or solutions to learn from, such as emerging technologies and innovative megaprojects, the traditional organization model is not only difficult to achieve the success of megaprojects, but also cannot avoid failure. Therefore, subverting the traditional one-off cooperation mode of conflict of interest, forming a community of interests, exploring more sustainable value emergence paths or paradigms through value co-creation and sharing, is an important direction that needs to be explored for megaproject organizational behavior and organizational governance.

14.6 Conclusion

The megaprojects are huge "animals" that we have been fighting against for many years, but obviously we have not "tamed" this new "species". However, it is gratifying that the "practical wisdom" still provide us with meaningful "taming skills". With the deepening of research and the introduction of theories in other fields, we have formed more and more "theoretical schemes" and continuously discovered the road to high performance in megaprojects. Complexity, adaptation, self-organization, meta-governance, and value co-creation are valuable guiding marks on this path. In this respect, this chapter has formed several meaningful conclusions, as below.

First, the macro-, meso-, and micro-organizational contexts of megaprojects determine the multiple, network, three-dimensional, and evolutionary characteristics of the organizational behavior system. The complexity of megaprojects is the driving force for the emergence and evolution of self-organization, but the efficient collaboration of this efficient process requires strong leadership.

Second, organizational adaptation leads to the complexity of megaprojects, and high adaptation is also the characteristic and advantage of self-organization of megaprojects. It is this kind of flexible and highly adaptable self-organization that continuously forms collective wisdom, and solves the unanticipated challenges in a timely manner through collective behavior, and finally creates one miracle after another.

Third, megaprojects are a mixture of organizational design, organizational control, and self-organization. It may be hybrid meta-organizations of hierarchical organizations, functional organizations, project organizations, and network organizations. Both traditional organizations and new organizational forms have the basis for the existence of "legitimacy", but the new organizational forms represented by self-organization obviously have stronger vitality.

Fourth, the hybrid meta-organizational characteristics of megaprojects determine that the combination strategy of multi-layer governance, network governance, meta governance and dynamic governance is necessary. Single governance strategies could not cope with the complexity of megaproject organizational behaviors, and self-organization may also fail.

Fifth, the big challenges faced by megaprojects are constantly changing. It is necessary to regard megaprojects as value creation platforms, realize value co-creation, and value sharing through meta governance, in order to ensure the success of projects and strategies, and strive to find a new paradigm for the emergence of sustainable value.

Certainly, this research only contributes to develop a basic framework, and some findings are derived from the investigation of the megaprojects in China. We also need to further observe the self-organization phenomenon in megaprojects, and discover the "code" of self-organization, adaptation, meta-governance, and value creation in megaprojects through more cases, interviews, analysis, and even social experiments, in a way to find the key to high performance in megaprojects. Nevertheless, self-organization, in general, provides us with a new and important route to deal with the megaproject complexity and to solve the universal low-performance issues of megaprojects.

Acknowledgement This material is based in part upon work supported by National Social Science Foundation of China under Grant No. 19VDL001. We are also grateful to Dr. Xinglin Gao, Assistant

Director of the Hong Kong–Zhuhai–Macau Bridge Authority and Director of the Planning and Contract Department, for his support and assistance during this work.

References

1. Groves L (2009) Now it can be told: The story of the Manhattan Project. Da Capo Press, Boston
2. Griffin M (2007) Specifying organizational contexts: systematic links between contexts and processes in organizational behavior. J Organ Behav 28(7):859–863
3. Li Y, Lu Y, Cui Q, Han Y (2019) Organizational behavior in megaprojects: integrative review and directions for future research. J Civ Eng Manag 35(4):04019009
4. Lenfle S, Loch C (2010) Lost roots: How project management came to emphasize control over flexibility and novelty. Calif Manag Rev 53(1):32–55
5. Flyvbjerg B (2014) What you should know about megaprojects and why: an overview. Proj Manag J 45(2):6–19
6. Mitchell M (2009) Complexity: a guided tour. Oxford University Press, Oxford
7. PMI (2014) Navigating complexity: a practice guide. Project Management Institute (PMI), Pennsylvania
8. Holland J (1995) Hidden order: how adaptation builds complexity. Addison-Wesley, Boston
9. Sheng Z (2018) Fundamental theories of mega infrastructure construction management. Springer, New York City
10. Li Y, Lu Y, Ma L, Kwak Y (2018) Evolutionary governance for mega-event projects (MEPs): a case study of the world expo 2010 in China. Proj Manag 49(1):57–78
11. Turner R (2016) Gower handbook of project management. Routledge, Abingdon
12. Li Y, Lu Y, Kwak Y, Le Y, He Q (2011) Social network analysis and organizational control in complex projects: construction of EXPO 2010 in China. Eng Proj Organ J 1(4):223–237
13. Gil N, Ludrigan C, Pinto J, Puranam P (2017) Megaproject organization and performance: the myth and political reality. Project Management Institute, Pennsylvania
14. Han Y, Li Y, Taylor J, Zhong J (2018) Characteristics and evolution of innovative collaboration networks in architecture, engineering, and construction: study of national prize-winning projects in China. J Constr Eng Manag 144(6):04018038
15. Gjaltema J, Biesbroek R, Termeer K (2020) From government to governance to meta-governance: a systematic literature review. Public Manag Rev 22(12):1760–1780
16. Peters B (2010) Meta-governance and public management. The new public governance? Routledge, Abingdon
17. José N, Victor B, William V (2016) Self-Organization and the role of government: how and why does self-organization evolve in the shadow of hierarchy? Public Manag Rev 18(7):1063–1084
18. Flyvbjerg B, Bruzelius N, Rothengatter W (2003) Megaprojects and risk: an anatomy of ambition. Cambridge University Press, Cambridge
19. Beunen R, Assche KV, Duineveld M (2015) Evolutionary governance theory. Springer, Berlin
20. Assche KV, Beunen R, Duineveld M (2014) Evolutionary governance theory: an introduction. Springer, Berlin
21. Kooiman J, Bavinck M, Chuenpagdee R, Mahon R, Pullin R (2008) Interactive governance and governability: an introduction. J Trans Envir Stud 7(1):1–11
22. Williamson O (1996) The mechanisms of governance. Oxford University Press, Oxford
23. Lehtinen J, Peltokorpi A, Artto K (2019) Megaprojects as organizational platforms and technology platforms for value creation. Int J Inf Syst Proj Manag 37(1):43–58
24. Gil N, Fu Y (2020) Megaproject performance, value creation and value distribution: an organizational governance perspective. Acad Manag Discov
25. Gao X, Dai J, Ruan M (2020) Bidding planning and case analysis of Hong Kong-Zhuhai-Macao Bridge. Chinese Plan Publishing house, Beijing

Chapter 15
Evaluation of Managerial Flexibilities in Critical Path Method-Based Construction Schedules

Önder Ökmen, Marian Bosch-Rekveldt, and Hans Bakker

Abstract Scheduling of a construction project can be done by using the Critical Path Method (CPM) in case the project is composed of interrelated activities that can be combined through a network. Given uncertainties nowadays and the related need for project schedule adaptations, the question is raised whether and how "traditional" CPM-based schedules allow for flexibility in project planning and management. In order to give an answer to this question, first, the managerial flexibilities provided by CPM were evaluated at three levels, i.e., activity, path, and project. Afterwards, the CPM schedules of two different projects were examined. Finally, the first conclusion arrived was that, in spite of its criticized deterministic features, CPM contains various flexible aspects from a managerial viewpoint. Second, potential flexibilities in CPM are mainly associated with resource leveling, noncritical activities, noncritical paths, activity float times, and activity float types. Third, CPM contains complete flexibility through independent floats and resource leveling capability. Investigating the flexible features of CPM in its traditional form, this study aims to open the way to develop a more flexible schedule management approach based on CPM and its extensions, which future self-organizing teams can adjust or apply.

Keywords Critical path method · Activity criticality · Activity float times · Activity float types · Flexible schedule management

Ö. Ökmen (✉) · M. Bosch-Rekveldt · H. Bakker
Delft University of Technology, Postbus 5, 2600 AA Delft, The Netherlands
e-mail: O.Okmen@tudelft.nl

M. Bosch-Rekveldt
e-mail: M.G.C.Bosch-Rekveldt@tudelft.nl

H. Bakker
e-mail: H.L.M.Bakker@tudelft.nl

R. Ding et al. (eds.), *Research on Project, Programme and Portfolio Management*, Lecture Notes in Management and Industrial Engineering,
https://doi.org/10.1007/978-3-030-86248-0_15

15.1 Introduction

Construction projects are required to be completed in planned time, envisaged budget and predetermined scope along with ensuring the expected quality, safety, and stakeholder satisfaction. These are the basic criteria of success in almost all kinds of projects [2, 18], although different importance is given to such criteria based on the stakeholders involved [3]. The "time" portion of this multilateral trade-off problem is kept under control through well-prepared schedules. A schedule is generally prepared at the beginning of a project and updated as the project progresses.

Critical Path Method (CPM) has been the method of time scheduling in construction projects since it was first developed in the 1950's [5]. This method has found application in all sectors. In case a project can be defined through a series of interrelated activities or tasks from top to bottom and these activities can be combined through a network, CPM could be used.

In construction projects, various resources are used to realize the tasks constituting the CPM schedules. Resources can be expressed in terms of time, cost, labor, equipment, or materials. As a result, CPM turns into a tool not only for managing time but also for managing the other resources.

The success and popularity of this traditional activity network scheduling method is probably related to the information it provides to its users like project managers, owners, contractors, engineers, foremen, workers, or even lawyers. The information that CPM provides includes, but is not limited to [12, 13];

- Shortest possible project completion time.
- Critical path(s), critical activities.
- Noncritical path(s), noncritical activities.
- Activity float times.
- Early/late start/finish times of activities.

However, CPM also has limitations [1, 8, 11, 16, 19]:

- CPM is deterministic, therefore it is unable to reflect the uncertainty effect on schedules and it is not capable of modeling the dynamic character of projects.
- CPM ignores the correlations that might exist between activities and between risk factors, which causes greater uncertainty on activity durations and in turn on project duration.
- CPM assumes unlimited availability of resources.
- CPM has limited flexible features, therefore it is not adaptable enough to changing conditions and it is unable to reflect the ever-changing dynamic nature of projects.

CPM has limited capability in scheduling linear or repetitive type of projects such as multi-story building, highway, railway, and canal construction projects. Besides ensuring the network logic within each unit, also the resource continuity is required to be provided along the repetitive parts in this type of projects. Literature also contains studies proposing ways of overcoming CPM's shortcomings. These studies mostly focus on the application of probabilistic methods, risk analysis procedures, and optimization methods on traditional CPM [10, 16, 17, 19]. Besides, the literature includes studies that propose the integration of CPM into linear scheduling methods [1, 15]. None of these studies disregard CPM, but it is tried to expand and improve CPM in compliance with the uncertain, complex and dynamic character of real-life construction projects.

Given uncertainties nowadays and the related need for project schedule adaptations, the question is raised whether and how "traditional" CPM-based schedules allow for flexibility in project planning and management. In order to give an answer to this question, the managerial flexibilities of CPM were investigated in this study.

The literature on project management brings a number of different definitions for "flexibility" [9]:

- the situation of being ready for potential change in an uncommitted manner [4],
- the situation of being capable of adjusting the project to possible consequences of uncertainty without going beyond the context of the project [7, 14],
- the way of converting the irreversible decisions into more reversible decisions or postponing the irreversible decisions till more related information is gathered [14], and
- the situation of being ready and able to deal with the dynamics of a project [9].

Taking into account these definitions, "flexibility" in projects can be considered as the adaptability of a project to complex, uncertain, and dynamic conditions. A paradigm change in traditional project management seems needed for achieving such flexibility. One of the ways to provide flexibility in project management is to explore the managerial flexibilities of existing methods and subsequently investigate new approaches. Therefore, the exploration of inherent flexibilities in traditional methods of project management, such as the CPM, gains importance in order to open the way into developing new flexible schedule management approaches. In this regard, this study carries the goal of being an initial attempt to develop a more flexible schedule management approach based on CPM and its extensions, which future self-organizing teams then could apply.

First, the managerial flexibilities provided by CPM were evaluated at three levels, i.e., activity, path, and project levels. Next, the CPM schedules of two different example projects were examined. While the first project was used to reveal the managerial flexibilities at the activity and path levels, the second one was used to disclose the project-level flexibility based on resource leveling. In line with this setup, the next section presents how flexibility can be linked to different attributes of the CPM at the three different levels. Next, in order not to get stuck into theoretical expressions, the aforementioned example project applications are introduced. Finally, the discussion and conclusions are given along with recommendations for future research.

15.2 Managerial Flexibilities Provided by CPM

In this section, managerial flexibilities of CPM are evaluated with respect to the activity, path, and project levels. The links that can be set between flexibility and different characteristics of the CPM are discussed.

15.2.1 Flexibility at Activity Level

Flexibility at activity level in a CPM schedule can be evaluated based on float times belonging to noncritical activities. Activity float times provide various potential flexibilities during the management of a CPM schedule. Basic definitions of activity float types are given below [12, 13]. Furthermore, the flexibilities associated with these floats are discussed.

Total float time is the difference between activities' late start time and early start time or late finish time and late start time. It represents the amount of total flexibility of an activity that can be consumed in case of any possible delay or disruption in that activity. As long as the amount of delay stays within the limits of total float time, it does not cause delay on the envisaged project completion time. However, total float should not be considered as an absolute flexibility for an activity because some portion of it may be shared by some of the successor activities [12]. Besides, total float times can be consumed as time buffers to take limited resources into account as in the case of Critical Chain Method, which is proposed as an extension of CPM [6]. As a result, total float times do not provide full flexibility and therefore they should be consumed consciously. Furthermore, it should be emphasized that, unlike the noncritical activities, critical activities have no flexibility with respect to total float time in CPM applications because they have no float time by definition.

Free float time is the float amount up to which a noncritical activity can be delayed without causing any delays on the early start times of any of its successor activities. Furthermore, the consumption of the free float does not cause a delay on the project completion time as far as it does not contain any shared float. In case the free float contains shared float, it should be consumed with caution for not causing a delay on the project completion time and also for not affecting the activities with which the float is shared. Therefore, in a way, free float represents a kind of partial flexibility in CPM schedules.

Independent float time is the most comfortable flexibility that CPM provides at the activity level as it only belongs to the activity, which possesses it. The usage of it will neither steal from the float times of the successor or predecessor activities nor affect the early start times of the successor activities and the project completion time as well. In that sense, independent float can be perceived as the absolute and core flexibility agent of the CPM. The basic difference between the independent float and free float is that while the independent float is free of shared float and therefore it can be consumed safely, the free float, on the other hand, might contain shared float and in such a case, its consumption would affect the activities with which the float is shared and possibly the project completion time as well.

Shared float time is the float amount up to which a noncritical activity can be delayed after consuming the entire independent float that it possesses. In other words, shared float constitutes the portion of float within the total float of an activity after its independent float is consumed. The usage of the shared float causes the consumption of some of the float that also belongs to the successor or predecessor activities, as its name implies "shared". One may argue that shared float therefore does not provide full flexibility like the independent float or partial flexibility like the free float.

In a CPM schedule, the noncritical activities might not possess free, shared, and independent floats at the same time, but by definition, each activity has some float. The strategy that should be followed when managing a CPM schedule starts with the notion of float. A manager should not fall into mistakes by using the total floats carelessly as if they are only under pure possession of noncritical activities. In short, the usage of flexibility unconsciously may lead to further inflexibility in CPM schedules. This issue will be elaborated in the example applications in the next sections.

And what about the critical activities? Don't they have any flexibility from a management point of view? In the end, they have no total float as well as no other types of float. The awareness CPM brings to a project manager through the information "those activities are critical, so be careful while executing them, take the necessary precautions in advance" can also be considered as a kind of managerial flexibility. Furthermore, the critical activities, along with the noncritical activities that may turn into critical during execution of the project, can be utilized for shortening the project duration through the project crashing capability of CPM and this gives flexibility of changing the predetermined project duration if required, but in expense of an increase in project costs.

15.2.2 Flexibility at Path Level

CPM, as its name implies, is a scheduling method that operates through activity paths. It discloses the critical path(s) and noncritical path(s) to show the required workflow through related activities.

The flexibility of CPM at the path level comes from the path floats occurring due to the float times of noncritical activities lying on the paths. However, once more, conscious usage of the path floats as in the case of activity floats is a requirement because path floats are not independent from the activity floats. In other words, the type of the floats of the noncritical activities on a path determines the degree of comfort in the usage of flexibility for that path. Therefore, flexibility at path level is not a luxury of a manager to be used arbitrarily at any time.

15.2.3 Flexibility at Project Level

CPM provides the shortest possible project completion time of a project, which is especially important from the management point of view. Is there any flexibility on the calculated completion time? Traditional application of CPM can calculate only a single project duration at a time because of its deterministic nature [10, 17]. However, CPM's updating capability brings some flexibility. As far as the flow of accurate data is ensured for the completed parts of the projects and the estimations for the remaining parts are updated realistically taking into account the actual data, completion times of the projects would be calculated more precisely each time through a more flexible approach. Depending on the outcomes, project managers could take precautions to overcome deviations from the estimated project duration.

However, the flexibility of CPM at the project level is not only associated with the updateability of project completion time. It also provides other flexibilities related to its capabilities such as project compression (schedule crashing and fast-tracking), delay analysis, float allocation, dispute resolution, schedule risk analysis, progress monitoring and control, integration with linear scheduling method, multi-calendar analysis, preparation of reports at various detail levels by means of scheduling software, etc. Furthermore, CPM can also be used as a cost schedule, labor schedule, or equipment schedule. In other words, a CPM schedule provides further flexibilities when it is used as a resource schedule in terms of cost, labor, or equipment and in that case CPM can also be used for:

- leveling resources through consuming activity float times in order to stay below the maximum available resource limit,
- shifting resources from noncritical activities to critical activities in order to avoid schedule overruns,
- optimizing time against costs (i.e., time–cost trade-off analysis).

Such features increase the managerial flexibilities provided by the CPM. However, proper and conscious usage is required throughout the project in order to fully benefit from these flexibilities at the project level as it was also in the cases of activity and path levels.

15.2.4 From Ideas to Practical Application

So far, the notion of flexibility is described from a theoretical perspective on three levels: activity level, path level, and project level. The following sections illustrate such managerial flexibilities of CPM at these three levels using two example projects. First, a sewer line project is used to evaluate the flexibility at the activity and path levels. Subsequently, the flexibility of CPM at the project level is discussed in terms of its resource leveling capability through a second example project.

15.3 Example 1: A Sewer Pipeline Construction Project

Example application 1 has been implemented on a hypothetical sewer pipeline construction project. The purpose of the application is to show the flexibilities provided by CPM at the activity and path levels from a managerial point of view. Table 15.1 shows the data used in the application, i.e., activities, activity numbers, activity durations, predecessor activities, lag times, and network relationships between the activities. Taking the basic requirements of constructing a simple sewer pipeline into consideration has constituted this data.

The time schedule of the project was prepared through CPM's forward/backward pass algorithm (the reader is referred to Newitt [12] and Oberlender [13] for detailed explanations on the algorithm and computation procedures). The resulting activity-on-node network diagram is shown in Fig. 15.1. The meaning of the data given inside the activity nodes are indicated by the "activity notation" in Fig. 15.1. The CPM application revealed that:

Table 15.1 Activities and network information of example application 1—Sewer pipeline project

Activity No	Activity Name	Activity Duration (day)	Predecessor activity & network relationship
1	Workplace delivery and mobilization	3	–
2	Sewage line route application	15	1 (Finish-to-Start)
3	Manhole excavation	6	2 (Finish-to-Start)
4	Inserting manhole formworks	10	3 (Start-to-Start, + 3 days lag time)
5	Pouring manhole concrete	10	4 (Finish-to-Start, -3 days lag time)
6	Sewage line trench excavation	20	2 (Finish-to-Start)
7	Removing manhole formworks	5	5 (Finish-to-Start)
8	Installation of sewer pipes	25	6 (Start-to-Start, + 2 days lag time)
9	Control of manholes	3	7 (Finish-to-Start), 8 (Finish-to-Start)
10	Handling of trench excavation equipment	4	8 (Finish-to-Start)
11	Sewer line trench filling	12	8 (Finish-to-Start), 9 (Finish-to-Start)
12	Testing the work, handing it over to the employer (completion of work)	2	10 (Finish-to-Start), 11 (Finish-to-Start)

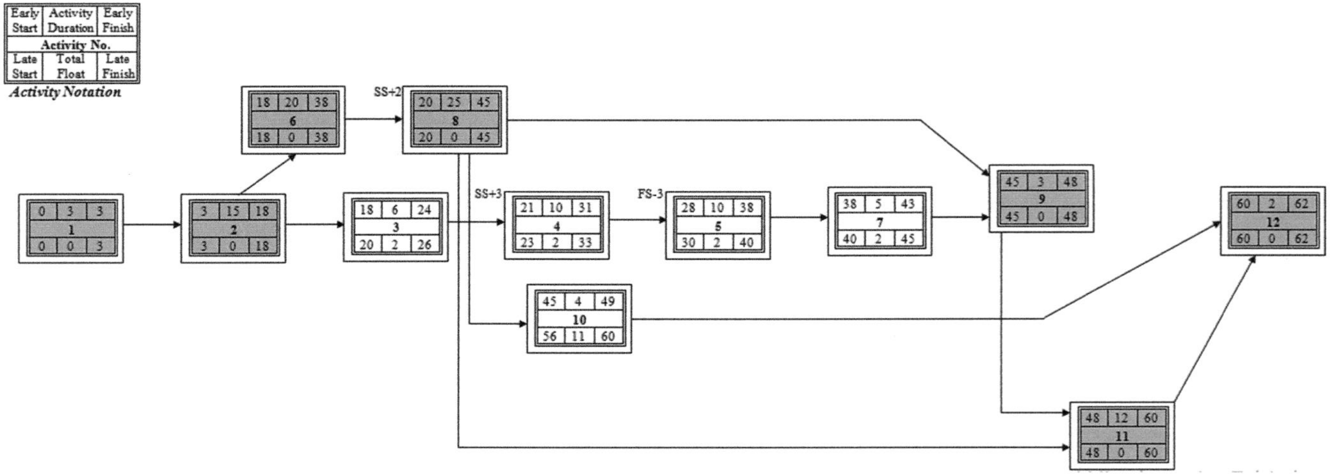

Fig. 15.1 Activity-on-node network diagram of example application 1—Sewer pipeline project

Table 15.2 Float times and criticalities of the activities and paths of example application 1—Sewer pipeline project

Activity/Path No	Free Float Time*	Shared Float Time	Independent Float Time	Total Float Time	Float Sharing Activity	Path Float Time	Criticality
Activity 1	0	0	0	0	–	N/A	Critical
Activity 2	0	0	0	0	–	N/A	Critical
Activity 3	0	2	0	2	4, 5 & 7	N/A	Noncritical
Activity 4	0	2	0	2	3, 5 & 7	N/A	Noncritical
Activity 5	0	2	0	2	3, 4 & 7	N/A	Noncritical
Activity 6	0	0	0	0	–	N/A	Critical
Activity 7	2	2	0	2	3, 4 & 5	N/A	Noncritical
Activity 8	0	0	0	0	–	N/A	Critical
Activity 9	0	0	0	0	–	N/A	Critical
Activity 10	11	0	11	11	–	N/A	Noncritical
Activity 11	0	0	0	0	–	N/A	Critical
Activity 12	0	0	0	0	–	N/A	Critical
Path 1	N/A	N/A	N/A	N/A	N/A	0	Critical
Path 2	N/A	N/A	N/A	N/A	N/A	0	Critical
Path 3	N/A	N/A	N/A	N/A	N/A	11	Noncritical
Path 4	N/A	N/A	N/A	N/A	N/A	2	Noncritical

*All the time values are in "days"

- The Activities 1, 2, 6, 8, 9, 11, and 12 are the critical activities having no float times,
- The Activities 3, 4, 5, 7, and 10 are the noncritical activities having float times,
- The Paths 1 (1–2–6–8–9–11–12) and 2 (1–2–6–8–11–12) are the critical paths,
- The Paths 3 (1–2–6–8–10–12) and 4 (1–2–3–4–5–7–9–11–2) are the noncritical paths.

The total, free, shared, and independent float times of the activities, float sharing activities, float times of the paths, and the criticality of the activities and the paths are given in Table 15.2. As shown in Fig. 15.1, the late finish time of the last activity represents the calculated project completion time, which is 62 days. Next, the flexibility is evaluated on activity level and on path level.

15.3.1 Evaluation of Flexibility at Activity Level

When the float times given in Table 15.2 are examined, it is observed that each noncritical activity does not have to possess free, shared, and independent float necessarily at the same time. However, by definition, each activity has a total float time. This is the starting point for determining the management strategy. As previously mentioned, a manager who is not fully aware of the flexibilities that CPM provides in terms of float times at activity level might mistakenly use the total floats carelessly as if they were only under the pure possession of noncritical activities. This is clarified below.

For instance, the "activity 10" has a total float of 11 days, which is relatively high when compared to the project completion time of 62 days. More important is that this total float is also the free and independent float for this activity. In other words, the "activity 10" does not share any portion of its float with any of its predecessor or successor activities. Therefore, the "activity 10" carries a large amount of flexibility for itself and for the path on which it stands. Depending on this, the flexibility that the "activity 10" possesses creates not only activity-level flexibility but also path-level flexibility.

Next to the flexibility created by the "activity 10", the flexibilities associated with the other noncritical activities of the project, i.e., the activities 3, 4, 5, and 7, should be discussed together because these activities are located subsequently on the same path. As given in Table 15.2, each of these activities has 2 days of total float. However, the float they each possess is the shared float at the same time. In other words, they share the 2 days of float. Therefore, the cumulative flexibility of these four activities sequentially lined up as shown in Fig. 15.1 is as much as their shared float amount, i.e., 2 days, rather than the cumulative of their total floats, i.e., 8 days.

15.3.2 Evaluation of Flexibility at Path Level

The flexibility of CPM at path level depends on the path floats, which is composed of the float times of the noncritical activities on the paths. However, once more, conscious usage of the path floats as in the case of activity floats is required because path floats relate to the activity floats. In other words, the type of the floats of the noncritical activities on a path determines the level of comfortable usage of flexibility for that path. For instance, the path 1-2-3-4-5-7-9-11-12 (Path 4) may seem to have 8 days of float in total (due to the cumulative floats belonging to the activities 3, 4, 5, and 7) although it actually has only 2 days of float (due to shared float among the activities 3, 4, 5, and 7).

Let's assume that the project has progressed up to the activity 5. If the activity 5 is completed with 2 days of delay by consuming its total float, no float will remain for the activity 7 which is the successor of the activity 5 because the 2 days of float was shared among the activities 3, 4, 5, and 7. In other words, consuming the 2 days of float for each activity on the Path 4 will cause "8 (*cumulative of the floats belonging to the activities 3, 4, 5, and 7*)—2 (*float shared among the activities 3, 4, 5, and 7, i.e., the real float value of the Path 4*)" = 6 days of delay on the project completion. This example shows the importance of utilizing the flexibility in the CPM schedules in the right way.

When we examine the Path 3, it is a different situation because it has a float of 11 days depending on the independent flow of the single noncritical activity on that path, i.e., the "activity 10". In other words, the flexibility associated with the "activity 10" is directly creating flexibility on the path on which it stands. Therefore, each path on a CPM schedule should be analyzed separately based on the flexibilities that the activities standing on these paths provide by taking the float types into account.

15.4 Example Application 2

In this section, the flexibility of CPM at project level is discussed in terms of its resource leveling capability through another hypothetical project. The project data used in this application is given in Table 15.3. The activities of the project are represented with letters and all the network relationships between the activities are assumed to be finish-to-start (FS) without any lag times. Table 15.3 also includes the information about the activities, activity durations, predecessor activities, network relationships between the activities and the number of workers required for each activity. This data was used in the implementation of CPM, in the preparation of the time and labor schedules, and in leveling the labor requirement. For the sake of simplicity, only one type of resource was taken into account.

First, in accordance with the information given in Table 15.3, a time schedule was created with forward/backward CPM calculations. The schedule is given in Table

Table 15.3 Activity network information and labor requirement of example Application 2

Activity	Activity duration (day)	Predecessor activity	Network	Labor requirement (worker)
A	2	–	–	4
B	4	A	FS	6
C	5	A	FS	2
D	4	A	FS	3
E	2	A	FS	7
F	6	B C	FS	8
G	3	A	FS	5
H	4	F	FS	2
I	9	F	FS	2
J	2	DE GHI	FS FS FS FS FS	3

Table 15.4 CPM solution of example Application 2

Activity	Early Start Time*	Early Finish Time	Late Start Time	Late Finish Time	Total Float Time	Criticality
A	1	2	1	2	0	Critical
B	3	6	4	7	1	Noncritical
C	3	7	3	7	0	Critical
D	3	6	19	22	16	Noncritical
E	3	4	21	22	18	Noncritical
F	8	13	8	13	0	Critical
G	3	5	20	22	17	Noncritical
H	14	17	19	22	5	Noncritical
I	14	22	14	22	0	Critical
J	23	24	23	24	0	Critical

*All the time values are in "days"

15.4 and shown as an activity-on-node network diagram in Fig. 15.2. According to the schedule:

- The Activities A, C, F, I, and J are the critical activities.
- The Activities B, D, E, G, and H are the noncritical activities.
- The Path A-C-F-I-J is the critical path.
- The Paths A-B-F-H-J, A-C-F-H-J, A-D-J, A-E-J, and A-G-J are the noncritical paths.
- The completion time of the project is 24 days.

Subsequently, labor charts showing the labor requirements along the project were prepared by the data given in Tables 15.3 and 15.4. It is assumed that the maximum number of workers that can be employed throughout the project is 9. In other words, there is a workforce of nine workers available for this job. Therefore, a labor shortage is assumed to occur when more than nine workers are needed in any working day, and some of the workers will remain idle when less than nine workers are required.

Labor requirement schedules can be prepared by a stepwise procedure consisting of three stages. In the first stage, a labor schedule is prepared according to the early start and early finish times of the activities. In this schedule, it is determined on which dates labor shortage and on which dates excess labor problems will occur. Then, a second labor schedule is prepared according to the late start and late finish times of the activities. In this schedule, again it is determined on which dates labor shortage and on which dates excess labor problems will occur. Then, by using these two labor schedules and shifting the required activities over the total float times of the noncritical activities, a leveled or optimized labor schedule is constituted. This process is called "resource levelling" [12, 13]. In this final labor schedule, the need for labor generally increases at the beginning of the project, follows a regular course in the middle and begins to decrease towards the end of the project. The

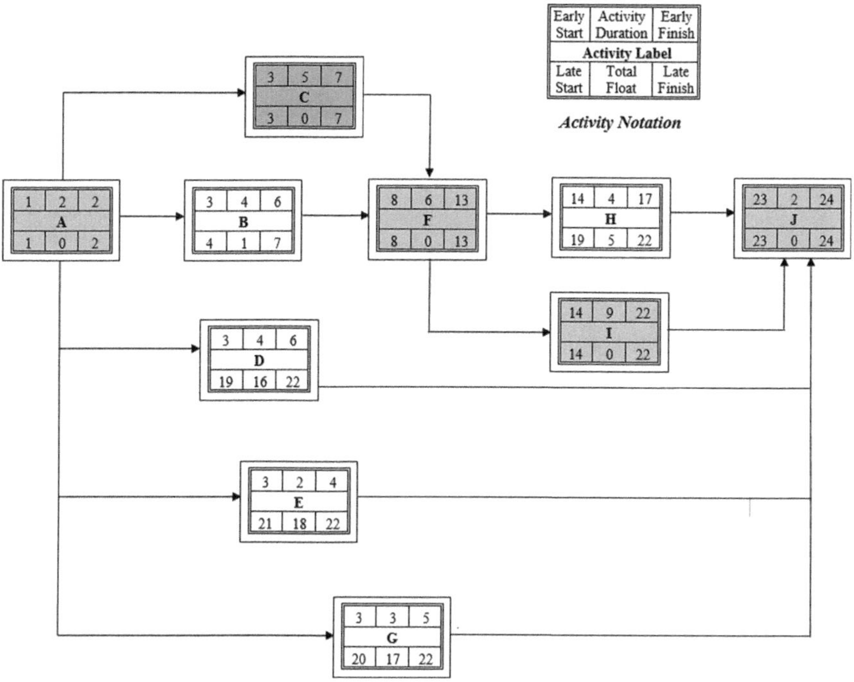

Fig. 15.2 Activity-on-node network diagram of example Application 2

number of workdays with labor shortage or excessive labor problems is minimized or eliminated.

Using the values in Tables 15.3 and 15.4, taking into account the maximum allowable labor assumption, i.e., maximum nine workers daily, and following the three-stage resource leveling procedure described above, the labor requirement charts (i.e., labor schedules) of the example project were prepared. The charts are given in Tables 15.5, 15.6 and 15.7. Tables 15.5 and 15.6 show that we are facing an over-requirement problem in terms of labor force. This problem can be resolved through benefitting from the flexibilities provided by CPM at project level, as described below.

Table 15.5 illustrates the labor requirement chart prepared according to the early start and early finish time. It shows that 23, 23, 16, and 11 workers are needed on the 3rd, 4th, 5th, and 6th days, respectively. Hence, a labor shortage due to the need for more workers than the maximum worker capacity of 9 is observed. The need for workers increases rapidly at the beginning of the 24-day total work period, and the need drops suddenly before the project reaches the end. In other words, the need for labor does not follow a regular course. Accordingly, we need some flexibility to overcome this labor shortage.

Table 15.6 illustrates the labor requirement chart prepared according to the late start and late finish time values. Now only 12 workers are needed on the 21st and 22nd days. In other words, labor shortage will still occur due to the need for more

Table 15.5 Labor requirement chart prepared based on early start and early finish times

Activity	Day																							
	1	2	3	4	5	6	7	8	9	10	11	12	13	14	15	16	17	18	19	20	21	22	23	24
A	4	4																						
B			6	6	6	6																		
C			2	2	2	2	2																	
D			3	3	3	3																		
E			7	7																				
F								8	8	8	8	8	8											
G			5	5	5																			
H														2	2	2	2							
I														2	2	2	2	2	2	2	2	2		
J																							3	3
Total worker requirement	4	4	**23**	**23**	**16**	**11**	2	8	8	8	8	8	8	4	4	4	4	2	2	2	2	2	3	3

Table 15.6 Labor requirement chart prepared based on late start and late finish times

Activity	Day																							
	1	2	3	4	5	6	7	8	9	10	11	12	13	14	15	16	17	18	19	20	21	22	23	24
A	4	4																						
B				6	6	6	6																	
C			2	2	2	2	2																	
D																			3	3	3	3		
E																					7	7		
F								8	8	8	8	8	8											
G														5	5	5								
H																	2	2	2	2				
I														2	2	2	2	2	2	2	2	2		
J																							3	3
Total labor requirement	4	4	2	8	8	8	8	8	8	8	8	8	8	7	7	7	4	4	7	7	**12**	**12**	3	3

Table 15.7 Leveled labor requirement chart

Activity	Day																							
	1	2	3	4	5	6	7	8	9	10	11	12	13	14	15	16	17	18	19	20	21	22	23	24
A	4	4																						
B				6	6	6	6																	
C			2	2	2	2	2																	
D														3	3	3	3							
E																		7	7					
XF								8	8	8	8	8	8											
G																				5	5	5		
H														2	2	2	2							
I														2	2	2	2	2	2	2	2	2		
J																							3	3
Total labor requirement	4	4	2	8	8	8	8	8	8	8	8	8	8	7	7	7	7	9	9	7	7	7	3	3

workers than the maximum worker capacity of 9, but the number of days with labor shortage has decreased from 4 to 2 compared to Table 15.5. Also the number of extra workers needed has decreased significantly. Apart from this, although the need for workers increases slowly at the beginning of the 24-day total work period, the need drops rapidly with a sudden leap towards the end, after following a regular course. In other words, the labor schedule prepared according to the late start and late finish times still needs to be leveled due to the sudden leap after the 21st day though it is superior with respect to the labor schedule prepared by the early start and early finish times. Thus, we still need additional flexibility.

In order to overcome this problem, we can benefit from the resource leveling capability of CPM. The labor requirement chart obtained by resource leveling is given in Table 15.7. According to the leveled chart, there is no day left with labor shortage. The need for workers follows a regular and balanced course throughout the 24-day period. In order to create this leveled schedule, noncritical activities were shifted over their total float times for eliminating the days of labor shortage, hence creating more balanced labor requirements. In other words, we actually did benefit from the flexibility of CPM to overcome the aforementioned over-requirement problem in terms of labor.

15.5 Discussion and Conclusions

This paper investigated the flexibilities that traditional CPM provides during the management of construction projects. The managerial flexibilities provided by CPM were evaluated at three levels, i.e., the activity, path, and project levels, first conceptually, then through two different example CPM applications. The results of the study have shown three important findings. First, CPM, in spite of its criticized deterministic features, contains various flexible aspects from a managerial viewpoint. Second, potential flexibilities in CPM are mainly associated with resource leveling, noncritical activities, noncritical paths, activity float times, and activity float types. Third, CPM contains complete flexibility through independent floats and resource leveling capability.

This research can be seen as a starting point towards the development of advanced and flexible schedule management approaches to be used in complex, uncertain, and dynamic conditions of today's construction projects. In this regard, the subject is open to development through investigating the ways of incorporating managerial flexibilities into the extensions of CPM such as the Critical Chain Method, CPM-based schedule risk analysis models, and CPM-based linear scheduling methods. Such extensions could be considered in future research.

Obviously, trying to manage the CPM schedules without being aware of the managerial flexibilities will not contribute to the aim of successfully completing complex construction projects. One purpose of this research is to raise such awareness. The findings of the research have revealed that CPM is not as rigid as it is assumed as far as the manager or the practitioner using it is aware of the means of benefitting from its flexible features. For instance, it is important to make the distinction between the activity float types in a CPM network schedule, i.e., total, free, shared, and independent. While total float times belonging to noncritical activities are probably the most well-known aspect of CPM to the practitioners, the free, shared, and independent floats that the total floats contain should be taken into account during the schedule applications in order to avoid any wrong usage of the activity float times. Otherwise, the number of critical activities would increase and the schedules would face the risk of overrunning the target durations.

The authors believe that being aware of the managerial flexibilities already existing in the traditional approaches of project management will make contributions to the managing capabilities of future self-organizing teams who will be involved in today's projects that are getting more complex and uncertain. Actually, the desired paradigm change in project management, i.e., from traditional, rigid and plan-driven towards modern, flexible, and change-driven to cope with the increasing complexity and uncertainty, does not necessitate giving up the usage of traditional methods or approaches but rather amending their implementation. In this regard, the use of traditional methods of project management, including their inherently flexible features, will empower the managerial skills of the practitioners operating in self-organizing teams. Obviously, the establishment of the self-organizing ideal through teams requires not only highly communicative and professional skills in team members but also tools and implementation guides which are fit for purpose and sufficiently flexible.

Acknowledgements We would like to introduce our thanks to the Jean Monnet Scholarship Programme which has been carried out through an agreement between the Republic of Turkey and the European Commission, and funded by the European Union within the scope of the Instrument for Pre-Accession for Turkey.

"This document has been produced with the financial assistance of the European Union. The contents of this document are the sole responsibility of Önder Ökmen, Marian Bosch-Rekveldt, and Hans Bakker and can under no circumstances be regarded as reflecting the position of the European Union."

References

1. Ammar MA (2013) LOB and CPM integrated method for scheduling repetitive projects. J Constr Eng Manag 139(1):44–50
2. Atkinson R (1999) Project management: cost, time and quality, two best guesses and a phenomenon, its time to accept other success criteria. Int J Proj Manag 17(6):337–342
3. Bakker H, Arkesteijn R, Bosch-Rekveldt M, Mooi H (2010) Project success from the perspective of owners and contractors in the process industry. In: 24th IPMA World Congress, İstanbul
4. Bateson G (1972) Ecology and flexibility in urban civilization. Steps to an ecology of mind. 494–505
5. Galloway PD (2006) Survey of the construction industry relative to the use of CPM scheduling for construction projects. J Constr Eng Manag 132(7):697–711
6. Goldratt E (1997) Critical Chain. North River Press, USA

7. Husby O, Kilde HS, Klakegg OJ, Torp O, Berntsen SR, Samset K (1999) Usikkerhet som gevinst. Styring av usikkerhet i prosjekter: mulighet, risiko, beslutning, handling. Report No. NTNU 99006, The Norwegian Centre for Project Management at the Norwegian University of Science and Technology, Trondheim, Norway. (Title in English: Uncertainty as benefit. Managing project uncertainty: possibility, risk, decision, action)
8. Jaafari A (1984) Criticism of CPM for project planning analysis. J Constr Eng Manag 110(2):222–233
9. Jalali-Sohi A, Bosch-Rekveldt M, Hertogh M (2019) Four stages of making project management flexible: insight, importance, implementation and improvement. IPMA 7th Research Conference, Zagreb
10. Khamooshi H, Cioffi D (2013) Uncertainty in task duration and cost estimates: fusion of probabilistic forecasts and deterministic scheduling. J Constr Eng Manag 139(5)
11. Koskela L, Howell G, Pikas E, Dave B (2014) If CPM is so bad, why have we been using it so long? In: The 22th International Group for Lean Construction Conference, pp 23–27
12. Newitt JS (2009) Construction scheduling: principles and practices. Prentice Hall, London
13. Oberlender G (2014) Project management for engineering and construction. McGraw-Hill Education, New York
14. Olsson NOE (2006) Management of flexibility in projects. Int J Proj Manag 24(1):66–74
15. Ökmen Ö (2013) A procedure for critical path method-based scheduling in linear construction projects. J S Afr Inst Civ Eng 55(2):12–20
16. Ökmen Ö, Öztaş A (2008) Construction project network evaluation with correlated schedule risk analysis model. J Constr Eng Manag 134(1):49–63
17. Pohl J, Chapman A (1987) Probabilistic project management. Build Environ 22(3):209–214
18. Toor S, Ogunlana O (2010) Beyond the "iron triangle": Stakeholder perception of key performance indicators (KPIs) for large-scale public sector development projects. Int J Proj Manag 28(3):228–236
19. Zhou J, Love PED, Wang X, Teo KL, Irani Z (2013) A review of methods and algorithms for optimizing construction scheduling. J Oper Res Soc 64(8):1091–1105

Chapter 16
How Construction Projects Can Be Agile

Irina Nechaeva

Abstract There is currently the growing interest of business in the problem of shorten the total length of the construction project life cycle. At the same time, the number of dispersed teams increased. The level of technology development, including BIM, allows AEC teams to collaborate effectively. Due to environment turbulence and uncertainty, developers (customers) need to have an opportunity to react on market tendencies during the project execution and implement changes in the project. The group of adaptive PM approaches (agile, hybrid) may help to solve this problem. The paper presents the comparison of typical construction project design phase and new product development. The system hybrid project management approach with mix of agile and waterfall project management is proposed.

Keywords Agile · Hybrid project management · Construction projects · Construction management · Product development · Adaptive approach

16.1 Introduction

Construction field is the most conservative industry. The resistance to new technologies, new management approaches are still high throughout the years. But the business needs dominate due to the necessity to increase the margin and financial efficiency. Companies attempt to shorten the duration of projects to achieve better results. Complexity of construction projects increases when the schedule crashing techniques are used. Different approaches, such as fast-tracking, concurrent, and parallel engineering and agile project management, are used to reduce the total length of the project life cycle from the beginning of the project to the completion during the design and construction phases.

At the same time, the changes during the project are inevitable. Project environment is turbulent and, due to the long-lasting duration of construction projects which

I. Nechaeva (✉)
National Research University Higher School of Economics, 20 Myasnitskaya ulitsa, 101000 Moscow, Russian Federation
e-mail: irina-nechaeva@list.ru

R. Ding et al. (eds.), *Research on Project, Programme and Portfolio Management*,
Lecture Notes in Management and Industrial Engineering,
https://doi.org/10.1007/978-3-030-86248-0_16

can take 3–6 years, companies need to be ready to react on changes. Some researchers such as Highmish [4], Sanchez and Nagi [9], Howell et al. [5] consider company's agility as a response to turbulent project environment.

The key point to successfully react on changes and decrease the duration of construction project considering complexity is to create a controlled hybrid or agile project management system.

16.2 Hybrid Project Management Approaches

Usually, a type of project determines which approach to use—agile or traditional (waterfall/predictive).

When a project has specific non-changeable goals at the start and all stakeholders understand the final deliverables before it is created, the best way is to use waterfall approach.

When project requirements and scope can't be defined in full, the speed and frequency of changes are unpredictable, it is better to use adaptive or agile techniques.

Agile practice guide [7] assumes that there is no necessity to use only one approach for an entire project. Projects can combine elements of different life cycles in order to achieve the certain goal. A hybrid approach appears as a combination of predictive, iterative, incremental and/or agile approaches. Predictive approach is a traditional project management approach where are fixed requirements to the scope and product and the client gets the result only once—at the end of the project. In contrast, agile approach characterized by dynamic nature of requirements and include incremental and iterative delivery to provide results frequently.

In some projects, it is possible to use waterfall and agile approaches simultaneously for synergy effect. Nowadays, there are several hybrid proposed: iterative agile, agile-fall, water-scrum, Water-scrum-fall, Scrummer-Fall, Water-Agile-Fall [3, 11, 12].

PMI also demonstrates several combinations of agile and predictable project management used together during the life cycle [7]: agile development followed by a predictive rollout; combined agile and predictive approaches used simultaneously; a largely predictive approach with agile components; a largely agile approach with predictive component.

Assumption to mix and combine different approaches seems logical, because in some projects separate phase can be equal to a small project. Hybrid approach is beneficial for both a customer and a contractor. It allows the client to get the result early for future usage and the contractor to get a feedback before the whole work is finished and a rework is needed.

Theocharis G. et al. in their research conclude that different combinations of waterfall and agile became reality and provide a «win–win» situation between the requirements of project managers to stability at estimation, planning, and control, on the one side, and the freedom to select best practices by developers, on the other side [11].

The literature review demonstrates the trend of hybrid approaches applications for more effective project management. Therefore, the adoption to specific industry the combination of traditional and agile methods is required.

16.3 Agile in Construction

The application of agile approaches in construction area raises intensive debates in business and scientific society.

16.3.1 Practice of Agile Application in Construction

Nowadays, there are heated debates in scientific and business communities about the possibility of agile application in construction field due to industry conservatism. However real cases with agile approach confirm its' viability and effectiveness.

Daneshgari demonstrates agile implementation by electrical constructor and company processes transformation from traditional project management to proactive and agile [2].

Naim and Barlow note that agile approach is about effectiveness and customer satisfaction by product and result [6].

Streule Th., Miserini et al. represent on the real case the possibility of Scrum implementation in design, which results in the documentation timeframe [10]. The case study shows improvements stated in better communication and collaboration, better information flow, faster project development.

It sounds surprisingly, but there is an example when adaptive project management was used for optimization of project design solutions on the project of nuclear power plant [12]. The project consisted of speedy creation of numerous hypothesis to decrease the volume of building, then to do quick check of each hypothesis and select the most effective solution. The application of agile was supported by top-management due to schedule tightness and continues changes. The aim of the subproject was achieved in time and without reducing safety and reliability of the nuclear power plant.

One more case demonstrates the application of Scrum in design [1, 10]. The total duration of design decreased. As a result of scrum usage, the team noticed some significant changes in team collaboration: increase of transparency, improvement of communications and interactions, improvement of information flow, more rapidly development. The most important advantage is a possibility to observe the job of other team members and understand why they do their job this way. Also team members have an opportunity to expand their knowledge in adjacent fields.

Based on the described cases [10, 12], we can see that conscious application of agile approaches can provide excellent results in construction projects. The development of complex system hybrid approach has become topical. It will allow to fill the gap how agility might be incorporated with traditional approach in construction projects and to provide flexibility to companies.

16.3.2 Building Design as a New Product Development. Case study

New product development projects constitute an iteration process with several phases and stage gates with a continuous coherent design of product features.

The result of design phase in construction project is a design documentation with a set of detailed characteristics of the future building in accordance with customer requirements. The process of building design is like product development. Both represent a development of new product, but construction design is more predictable from the view of norms and regulations. If the design phase is divided into sub-stages, for example, in accordance to RIBA concept design, developed design, technical design what is a part of the whole life cycle: (0) strategic definition, (1) preparation and brief, (2) concept design, (3) developed design, (4) technical design, (5) construction, (6) handover and closeout, (7) in use [8], we see that the level of detail increases. In this way, the design process might be agile concerning the opportunity to implement changes throughout the design phase and provide intermediate benefits to client via incremental deliverables. BIM technologies simplify the documentation change process and allow design companies to provide additional value. The speed of design development increases due to early-received feedback. Such possibility to implement changes during the design phase can significantly increase the efficiency of collaboration between companies as well as the client satisfaction level of the final delivery. Consequently, the adaptation of agile project management approaches to the application at the design phase in construction might be really effective and interesting for all parties.

Current processes of new building development were studied during the design stage of multi-story residential building in Moscow, Russia. The project team included the customer, the architectural bureau (Architect), design company, responsible for structure and MEP design. The sequence of product development at different design stages was observed.

Different stages of design phase present a sequence of continious development of building design documentation: from concept (Fig. 16.1a) till working, technical documentaion (Fig. 16.1c). The evolution of product design development in construction is presented in Fig. 16.1. The stage-to-stage detail of elements shows the relevance between building design and new product development.

Described sequence design process represents the traditional project management approach. In this case, the customer receives the design documentation at the end of stage and only after that provides a feedback. All intermediate results are provided by Architect on request as well as the customer feedback or comments.

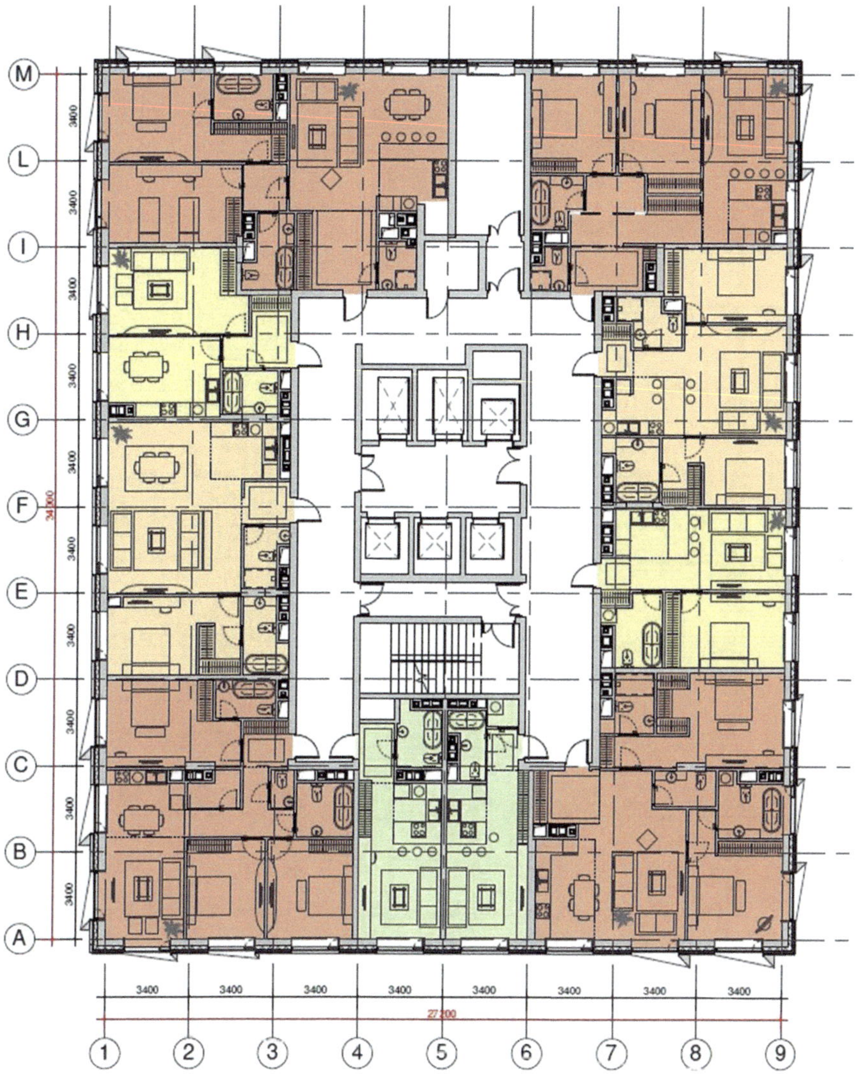

Fig. 16.1 Evolution of product design development

In the way to improve collaboration between parties and to work on product development together, some iterations within the concept design stage were implemented.

At the concept design stage, Architect forms and proposes different options of floor plans for future building—determines massing of the buildings, external limits, core elements, and MEP shafts in preliminary view (Fig. 16.2a). Based on this proposal after the first iteration, customer can analyze how the requirements are performed, calculate the loss factor of the building, check apartment mix, and evaluate financial model feasibility. To get feedback at this stage is beneficial for both. Parties don't lose

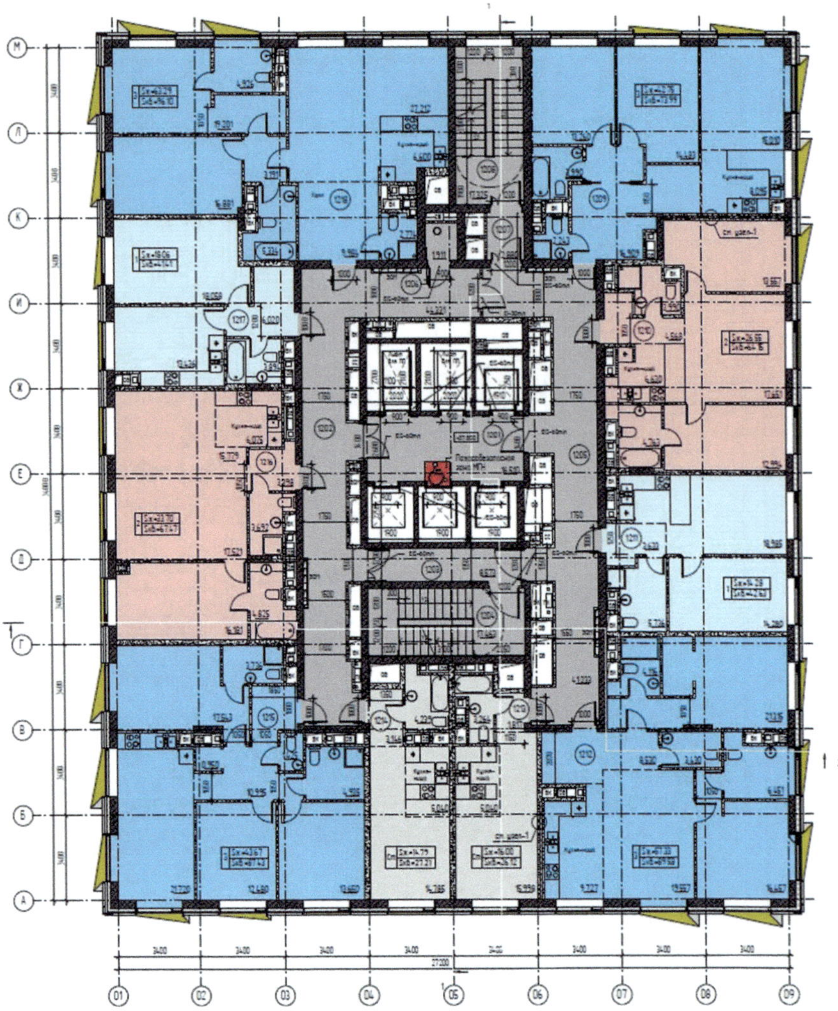

Fig. 16.1 (continued)

time waiting for a month to get unsatisfactory result and redo a lot of job. Through the options, customer can choose the most appropriate, which is taken by Architect to more detailed design (Fig. 16.2b).

Such iterations of interchanges between customer and Architect are repeated till the desired result is produced. If during the analysis of concept drawing a customer wants to make some changes, it will be easy to implement. As a delivery of the concept design stage a customer will get a design documentation which in full correlates with requirements. Intermediate incremental delivery in the form of concept design might be used by the customer in different ways—to consult with brokers, start to develop

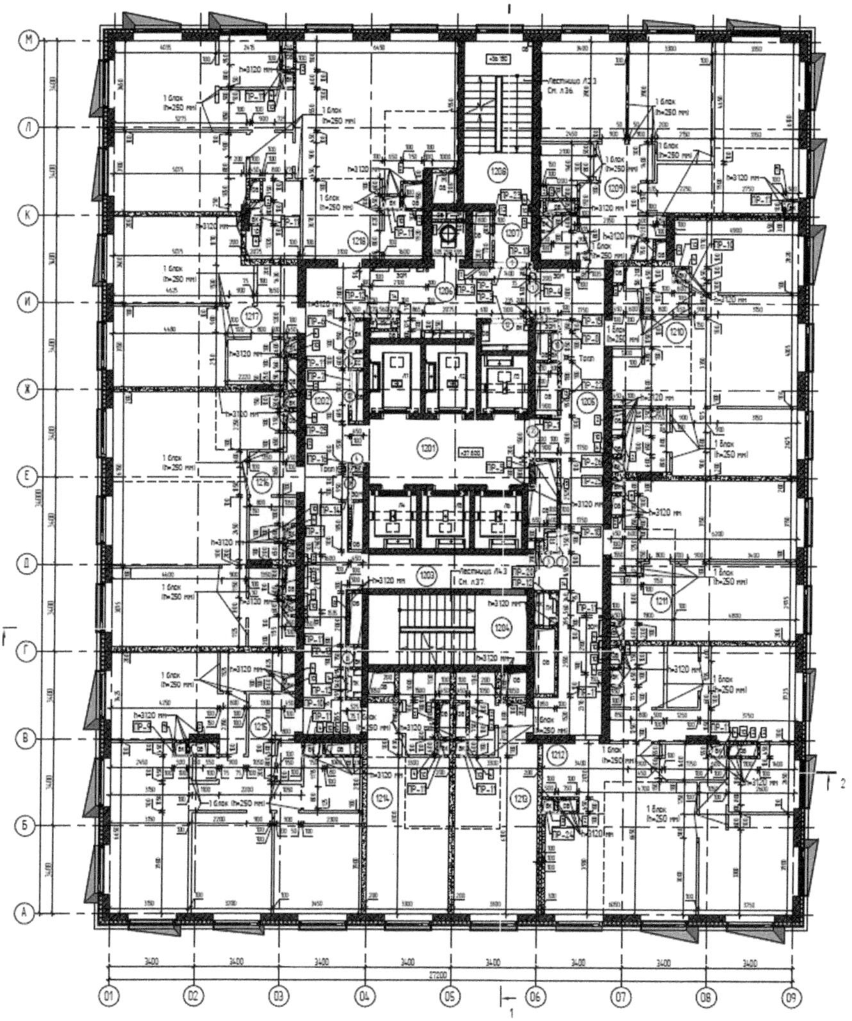

Fig. 16.1 (continued)

marketing materials, etc. Also the increased level of BIM application allows different parties work with model simultaneously.

Described process of creating the product of construction project during the design phase demonstrates characteristics of agility:

- Incremental progressive development and detailed elaboration of design documentation and project features;
- Employment of interim results;
- Getting customer feedback for improvement during the next iteration or stage.

Fig. 16.2 Example of iterations during the design phase

Based on the current design stages with the noticed peculiarities the combination of traditional and agile project management approaches is a logical evolution in construction project management.

16.4 System Hybrid Project Management Approach

As mentioned above, the traditional waterfall approach to management construction projects is widespread. At the same time, new product development is equal to the design stage. To sum up all of the above, there is a great potential to combine agile and waterfall to create controlled product development and project management in construction field.

Described process of creation product in construction and peculiarities in this field need to be connected to the development of a controlled hybrid project management system which allows to react on changes and continuously improve the product. To get more benefits of predictive, iterative and incremental methods, we propose to mix all of them.

The project management hybrid system for construction project presented in Fig. 16.3 consists of several elements:

1. Predictive approach for high-level planning. It is crucial for complex engineering projects to have a schedule with determined milestones. Stage-gate process as a key point from phase-to-phase transfer is recommended to keep as waterfall with milestones for proper control (Fig. 16.4).

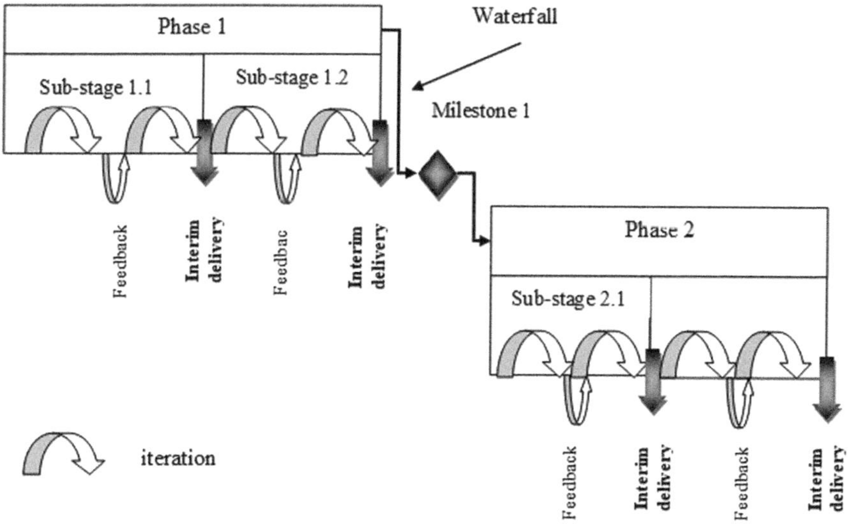

Fig 16.3 System of hybrid project management. Source: own elaborations

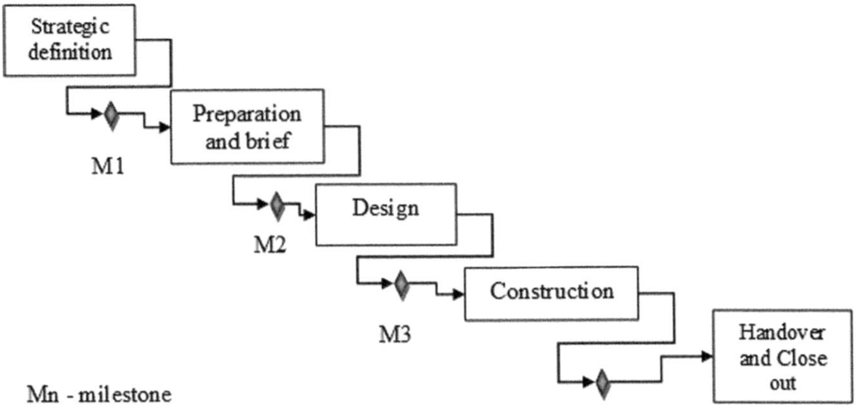

Fig 16.4 Waterfall from phase to phase. Source: own elaborations

2. Each phase divided on stages or sub-stages (correlates with WBS). For example, different parts of design phase can be sub-stages such as concept design, detail design, technical design. Incremental delivery transfers to the customer at the end of sub-stage. It isn't necessary to wait for years to get and/or accept intermediate results.
3. Iterative sprints or short-term plans during the sub-stages. Continuous process of demonstration results in getting client's feedback. Within each phase—design or construction—project team can determine the necessary number of iterations depending on design stage, customer needs, high-level schedule, volume of works.

The proposed hybrid project management approach allows to get intermediate valuable results. Possible size of iterations within the design and construction phases and types of interim delivery provided in Table 16.1. The size of iterations should be adopted to specific project. It is recommended to implement the iterations in contract terms between the architectural bureau and customer to avoid misunderstanding and misapplication.

16.5 Conclusion

The need of business to react on changes and uncertainty of environment and the complexity of construction projects require more flexibility of processes. The current design phase in construction project is equal to new product development design phase. The existing processes in construction field are iterative by its nature. Customer and designer know the requirements to the product but don't know how the building will look like till the end of phase. The leading product development is crucial for

Table 16.1 Characteristics of hybrid elements during design and construction phases

Phase	Stage/sub-stage	Size of iterations	Types of interim delivery	Value of interim delivery
Design	Concept design	weeks	– preliminary floor plans; – preliminary apartment plans and sizes, room location; apartment mix	– to consult with marketing or broker agency to get feedback about plans or the whole project; – check financial effectiveness; – check the apartment mix and area
	Developed Design	1–1,5 months	– final floor plans – final facades – final MEP solutions	– start sales; – start tender procedures
	Technical Design	2–3 months	– working documentation divided by scope of works	– start of construction work. For example, for concrete works you don't need to wait for electrical documentation
Construction	Underground works/Concrete works/Façade works/ MEP systems/Landscape	month	– Physical works finished by contractor, confirmed by quality audit	– show a progress to buyers/ banks/state control regulator

its success. Few cases of agile application in construction shows that projects can be agile even in the most conservative industries, such as construction.

This research demonstrates the system view on the combination of traditional and agile project management methods with the application in construction projects. The system hybrid approach to project management in construction systemises different existing approaches of project management in the way to add some agility to construction projects and construction firms and provide them with more evident opportunity to react on changes. Hybrid approach allows companies to keep a stage-gate process between construction project life-cycle stages and, at the same time, to develop a product with the early feedback from the stakeholders. The agility could be a company's advantage. The expected benefits of system hybrid approach implementation in construction projects are reduction of total project duration including the design stage, collaborative product design and development, reduction of rework due to early feedback and reaction on changes, better project control.

The proposed hybrid methodology was applied by the author in practice at the concept design stage on the residential project in development company in Russia. Implementation of agile approach and working through iterations on architectural and landscape concept development proved its applicability and effectiveness. The duration of design stage were 1.5–2 months shorter in comparison with traditional waterfall approach.

Further development of system hybrid project management approach in construction needs approbation on pilot project with investigation and measurement of effectiveness and getting feedback from different project stakeholders. It would be needed to look at the level of agility through the design companies and study their ability to change processes. The adoption of the hybrid approach to the company or project specific should be conducted. Also, the peculiarities of iterations should be included it tender documentation and contracts.

The implementation of hybrid approach in construction field provides companies with the opportunities to decrease the project duration, improve collaboration and lessen rework. At the same time, it may face restrictions from the team and contractors. It is necessary to take into account the risks that may arise from application of agile approach during the design phase: increased number of requirements changes, increased cost of design as a result of possible variations and more labor involvement, risk of growth of design duration instead of reduction. To avoid and mitigate risks there is a requirement to clearly provide the value of the approach to avoid misuse in case of continuous changes in product requirements. As any transformation process, the transfer to system hybrid project management approach should be controlled.

References

1. Cervone HF (2011) Understanding Agile project management methods using Scrum. OCLC Systems&Services: International digital library perspectives 27(1):18–22
2. Daneshgari P (2010) Agile construction for the electrical contractor. Jones and Bartlett Learning
3. Funtov NV, Paramonov DV, Malozemov SN (2017) Adaptive management in nonadaptive industry scientific research and development. Russian J Project Manag. 6(1):25–36
4. Highsmith JA (2002) Agile software development ecosystems, vol 13. Addison-Wesley Professional
5. Howell GA, Laufer A, Balland G, Uncertainaty and project objectives. Project Appraisal 8(1):37–43
6. Naim M, Barrow J, An Innovative supply chain strategy for customized housing. Construction management and economics, pp 593–602
7. Project Management Institute (2017) Agile practice guide, Project Management Institute, Pennsylvania
8. RIBA:(2013) Plan of work 2013. RIBA, London
9. Sanchez LM, Nagi R (2001) A review of Agile manufacturing systems. Int J Prod Res 39(16):3561–3600
10. Streule T, Miserni N, Bartlome O, Klippel M, Garcia de soto B (2016) Implementation of scrum in the construction industry. Proc Eng 164:269–276
11. Theocharis G, Kuhrmann M, Münch J, Diebold P (2015) Is *Water-Scrum-Fall* reality? On the use of agile and traditional development practices. In: Abrahamsson P, Corral L, Oivo M,

Russo B (eds) Product-focused software process improvement. PROFES 2015. Lecture notes in computer science, vol 9459. Springer, Cham
12. West D, Gilpin M, Grant T, Anderson A (2011) Water-scrum-fall is the reality of agile for most organizations Today/Forrester Research, Inc.

Chapter 17
Open Innovation in Practice—Challenges and Results in Telecommunications

Jovana Mihailovic, Marija Todorovic, and Vladimir Obradovic

Abstract The aim of this paper is to analyze the performance of open innovation in practice. The focus is on utilizing open business models in telecommunication sector. Mobile operators are challenged both in technical field (due to constant data increase) and customer experience and are struggling to assure their position as content providers. In order to achieve this they need to have strong strategic focus to increase productivity, innovate, and make changes in their business model. Paper gives general overview of management practises on open innovation. It identifies challenges that occur in practice during open business model implementation. It highlights the relationship between strategy and open innovation in telecommunication sector. Case study was conducted in a telecommunication company to analyze the bond between open innovation activities, self-organization, strategic orientation, and innovation performances in practice.

Keywords Open innovation · Mobile communication · Project · Self-organization · Strategy

17.1 Introduction

Mobile communication industry is highly turbulent and competitive environment. Digitalization is popular and frequently used term nowadays and the terms internet of things, virtual reality, artificial intelligence, big data, and machine learning are all

J. Mihailovic (✉)
A1 Srbija, Bulevar Milutina Milankovica 1Z, 11000 Belgrade, Serbia
e-mail: j.mihailovic@a1.rs

M. Todorovic
Faculty of Organizational Sciences, University of Belgrade, Jove Ilica 154, 11000 Belgrade, Serbia
e-mail: marija.todorovic@fon.bg.ac.rs

V. Obradovic
Faculty of Organizational Sciences, University of Belgrade, Jove Ilica 154, 11000 Belgrade, Serbia
e-mail: vladimir.obradovic@fon.bg.ac.rs

R. Ding et al. (eds.), *Research on Project, Programme and Portfolio Management*, Lecture Notes in Management and Industrial Engineering,
https://doi.org/10.1007/978-3-030-86248-0_17

often used in this context. It is phenomenon which affects both the private, professional, and business spheres. All these solutions are based on connectivity and efficient data exchange via high-performance networks, which is the core business of mobile operators and therefor great opportunities and potential for telecommunications providers. This means operators are confronted with a dynamically increasing demand for bandwidth. Accordingly, companies in the telecommunications sector are undergoing a fundamental shift to adapt to a digital world and have been developing from telephony service providers into data companies for some time now.

By being the "backbone" of digitalization mobile operators become just infrastructure providers, which impact scientifically their income since they went from major service provider (offering voice and messaging services) to data package distributor. Global telecommunication revenues declined by 4% between 2014 and 2015 [22]. Mobile broadband prices, as a percentage of gross national income per capita, dropped by half between 2013 and 2016 [22]. The traditional sources of income for telecom operators are showing signs of becoming obsolete [39]. Messaging platforms such as WhatsApp, Viber, Skype, Facebook messenger, and many others, have drastically decreased the voice traffic and messaging services, the core service of mobile operators. As an example, one study published by Juniper research predicted that consumer migration to OTT (Over The Top) messaging services might cost network operators nearly $104 billion this year which is equivalent to 12% of their service revenues [23].

On the other hand, at this moment, the industry could become more than the provider of infrastructure which enables connectivity and data exchange. Due to its position and influence, mobile operators could create additional value across a range of industries. Based on that fact, modern telecommunications companies are intensively working on expanding their business fields and skills and are finding ways to becoming end-to-end service providers in "digital world". Fact that operators move from selling data bundles (gigabits) to selling services- and applications-centric benefits shows that they are connecting more and more to the digital lifestyle of their customers [2]. It is not rare nowadays that mobile operators offer OTT services like music and video streaming services, cloud storage, location-based services, etc. Mobile operators are working on expanding their portfolio or more precisely, they are in certain way undergoing through portfolio transformation. These new products and services are offered by provider itself or through different kinds of partnerships with OTT service providers [3]. OTT service providers are usually startups that have idea they commercialize through application. Today in emerging markets, there are many opportunities for mobile operators and startups to collaborate [1]. It is good for companies and startups to collaborate since corporation has resources, scale, power, and the routines, while startup has promising ideas, organizational agility, the willingness to take risk, and aspirations of rapid growth [42].

Open innovation is a trend nowadays that enables companies to be more adoptive, agile and flexible according to market and customer demands.

Self-organization consists of many interacting components that have partial or no global system knowledge [38]. The advantage of self-organization is seen in various points: the possibility of the network itself to combine and rearrange its skills without the necessary intervention of a central control and strengthening commitment, which intensifies the exchange of tangible and intangible assets and generates learning, which is a requisite for innovation [36]. A key feature of self-organization is high decentralization and it's also a key feature of open innovation [26]. There are many definitions of open innovations in literature. Open innovation (OI) refers on a cooperation with external organizations or individuals in order to provide access to new technologies and specific technical, scientific, and/or research competencies to improve the innovation performance of an organization [9]. The OI implementation present the usage knowledge flows to accelerate internal innovations and create market opportunities for external use of innovation. Open innovation paradigm starts from the premise that companies has a capacity and should use external ideas and knowledge in addition to internal ideas, to improve company's technology [5]. The efficiency and effectiveness of open innovations depend on a company's absorptive capacity to gather external ideas, competencies and knowledge and apply it in a commercial way [41]. By pooling intellectual capital in a system, open invention can provide superior services and/or product and create a competitive advantage for a company comparing to those provided with only internal capacity [7].

First hypothesis of this paper is that for mobile operators open innovation (implemented via project) is a successful method to increase company's productivity.

Most of mobile operators have a similar goal and vision for the future, to be a successful service and content provider to customers. Strategy defines the means, policies, and activities which the organization will pursue to achieve its goals [31]. Driven by a market trend, on a path towards the goal, operators need to innovate both in terms of business model and in terms of products. Implementing open innovation involves changes in different organizational areas: organizational strategy and organization, social networks, evaluation system and knowledge management [10]. Development and implementation of each product is a project itself. Kenny defined projects as means of implementing strategy and stated that project goals should link directly to the strategic goals [25]. Main idea of strategic project management is that project management teams should focus on projects business aspects (time, budget, and performance goals), and to support company's business strategy and sustainability [33]. Sustainable project management is significant for all phases of the project life cycle, from initiation to closing the project [40]. Applying open innovation is more a matter of business strategy and for open innovation adaptation internal environment is more important than external [20]. The organization's strategy is responsible for the choice and funding of a specific project and its goals, but it also has consequences in terms of what is valued and how outcomes are achieved [11]. As Shenhar stated, project management is a complex activity and a risky organizational adventure, which could rarely fail due to lack of expertise in the project application program and needs to be focused on cultural and strategic elements of projects [35]. Only a project management system that "fits" with its strategic drivers is able to maximize the value contribution of projects [11]. Innovation management culture can be described by the degree to which management encourages risky innovation projects and self-organization for solving complex problems, in other words to which extent it is allowed to exchange ideas without direct involvement of management [27]

Ghezzi, Balocco & Rangone explained strong relationship between open innovation and strategy by characterizing OI as [17]:

- having long-term impacts (OI initiatives permanently change companies' offer and resources)
- involving different organizational units (for example, the strategic, financial, marketing, and sales units)
- requiring significant resource allocation (a significant budgets is required to establish processes that combine both internal and external sources of innovation)
- influencing performance because the commercialization of OI reflects on the companies' margin.

The second hypothesis of the paper is that open innovation gives best results when it is connected with strategy.

The overall objective of this paper is to investigate open innovation practice and application in mobile communication sector. Through a case study conducted in mobile operator company in Serbia, it investigates a link between open innovation activities, strategic orientation, innovation implementation through projects, agile and self-organized teams and innovation performance.

Paper is structured in the following way: chapter two reviews literature on open innovation topic, its performance in practice, and major challenges; chapter three describes research methodology. Case study and results are presented in chapter four. Chapter five concludes the paper.

17.2 Theoretical Background

17.2.1 Open Innovation in Telecommunication Sector

Telecoms are facing a turbulence due to increasing traffic, number and variety of devices, and complexity of services and architecture. Operators are challenged both in terms of customer experience and investments needed to secure services and network quality. Primary goal most of mobile operators have nowadays is to create value for the customer. Their long-term vision is to ensure their position as content provider and to offer services to their customers other than just data and voice. This strategic orientation is a result of market and customer demand for new products and services, it is driven by greater return on investment but also on existential problems since margins for voice and data services have dropped significantly in the last years [29].

Monitor Deloitte predicted four possible scenarios that could happen to mobile operators in the future depending on infrastructure and services ownership. In the first model, operators possess the whole system from technological assets to end to end customer services. In the second mode, mobile operators still hold the network infrastructure, but have lost end user control. Third scenario defines mobile operators as end service providers, who don't control network any more, and in the fourth scenario, the most pessimistic one, operators have lost the full control over all the elements [12]. No matter which case happen, every scenario triggers changes in business model.

Hagel & Singer [19] proposed a model of Unbundling Corporation. They isolated three kinds of businesses in a company—a customer relationship business, a product innovation business, and an infrastructure business. These entities employ different types of people, have different economic, competitive and cultural imperatives and as such can function much better if separated. Osterwilder & Pigneur [32] proposed this model for mobile operators—to outsource operation and maintenance to network equipment providers so that they could focus on core competencies, customer relationship. This change could make them more flexible, innovative, since they will become significantly smaller companies.

Ghezzi, Cortimiglia and Franc [18] assumed that external and internal changes influence the company's strategy and affect business model elements, as a result strategic re-planning process should be triggered. Operators rarely have enough knowledge to develop new product on their own. Open innovation is a model that can give operator access to external knowledge they lack and introduce agility in terms of creating new products. Xu & Chen [43] concluded in their research that for both mobile operators and OTT service providers the biggest profit on both side can be achieved through cooperation strategy. Creating and developing networks with external organizations offer flexibility, speed, innovation, and the ability to adjust smoothly to varying market and new strategic opportunities [13].

Telecommunications platforms don't only require investments in network, they need investment in programming and marketing skills to develop a user-friendly environment, which can also be acquired from external companies. Mobile companies should be in contact with external information coming from different partners (and different industries), recognize what could bring them value and integrate it within an improved offer.

17.2.2 Open Innovation and Performance Relationship

Telecommunications is fast-changing industry, and this is one of the drivers that lead mobile operators to use open innovation. Hung and Chou [21] concluded that technological turbulence considerably and positively moderates the effect of inbound open innovation on firm performance.

The first enabler that explains success in open innovation is the ability of the firm to develop and maintain external connections [37]. Once established, self-organized innovation networks may have sustainable competitive advantage because their rapid learning are hard to replicate by competitors [34]. The result of research that included 223 companies in Asia confirmed that open innovation implementation is significantly and positively associated to organizational learning, R&D

activities and production of an innovative products and/or services, customer-related processes, acceptance of new product and services on the market, and financial performances [6]. Additionally, companies which utilize open innovation more widely in their projects report higher satisfaction with open innovation [8]. Caputo, Lamberti, Cammarano & Michelino [4] discussed that open innovation can have beneficial effect on company's performance only up to some point—when the recourse to external resources increases, the management complexity of such resources is higher than the benefits, thus resulting in an increase of costs. The failure of open innovation system in telecommunication industry was recorded in the mobile operator Wind. It was due to managerial and interoperability issues between third parties and operator and unclear strategic vision and mission [17]. Frankenberger, Weiblen & Gassmann [15] argued that level of tie strength, centrality, and shared vision between company and its partners are the key elements that define the success of the alliance. The conclusion was that for open business models with high-solution customer centricity, strong ties to partners lead to superior firm performance and in business models with low-solution customer centricity, weak ties lead to superior firm performance.

17.2.3 Challenges of Implementing Open Innovation

Projects are mechanisms for implementing new processes, products, and changes. Every innovation can be seen as a project [30]. Innovations should be designed and managed strategically in the sense of firm scope, technology design, and intellectual property strategy, relation with partners and internal organization [16]. One of the main characteristics of open innovation is the participation of external parties in the innovation process, that's why effective and successful management of OI model is very important concept and great challenge. In the very competitive and fast-changing industry like mobile communications, managers have to deal with challenging market as well as to combine customer needs on one side and developers on the other and to maintain pleasant atmosphere that stimulates innovation. Creating partnerships is an essential and time-consuming issue in open innovation [20]. Another important issue for open-oriented companies is how to attract the contributors, and then how to sustain their participation over time [7]. Internal organizational challenges are perceived as most difficult to manage, because there is no dominant model that suggests how firms should organize open innovation internally [8]. Kelly, Schaan, & Joncas [24] conducted a survey in which they investigated the challenges and issues that appear during the first year of the life of alliance between companies. Communication barriers, cultural differences, and uncertainty over responsibilities were identified as biggest problems. Caputo, Lamberti, Cammarano & Michelino [4] also concluded in their research that time spent in managing relationships with partners hinders the efficiency of companies. Similarly, the research wrote by Ghezzi, Balocco & Rangone [17] identified problems that appear in OI environment—lack of a cooperation plan at the highest management level, lack of managerial commitment, unappropriated partner selection, lack of a communication with partners, organizational culture towards change and lack of control over company's resources (information and knowledge). Research performed by Du, Leten and Vanhaverbeke, [14] showed that the relationship between open innovation co-operators and project's performances are directly influenced by the process of project management in the company. Research results showed that in companies where there is no formal way of managing projects, a cooperation with external partners are associated negatively with project's performance [28]. The exception in this research is a cooperation with science-based partners.

Designing the open business model is seen as big challenge. First to define the extent to which each business model parameter should be "opened" to third parties, then its integration with the pre-existing "closed" models [17].

17.3 Research Methodology

The goal of this paper is to investigate open innovation in telecommunication industry and for that purpose a case study in a telecommunication company that operates in Serbia was conducted. Case study method can illustrate a topic within an evaluation in a descriptive mode and it allows researchers to present a real-life business situation and retain a specific characteristics of a management processes [44]. The main method of data collection was in-depth interviews with employees involved with open innovation process. The list of questions is presented in Appendix.

To reflect the wide image of innovation practices, employees from different sector and with different roles were selected. Selected interviewees also had different roles in the project and belonged to different firm levels. Interviews lasted between one and one and a half hours. The interviews were recorded and transcribed. Data was complemented through documentation provided by the interviewees and with public material.

17.4 Case Study Open Innovation in a telecommunication Company

The company is a private mobile operator, owner of the mobile telephony license for GSM and UMTS networks in Serbia. The company is a member of a Group which has over 24 million customers across seven countries in the region of Serbia and is the one of the biggest telecommunication providers in the world. The company has more than 2.1 million customers in Serbia, 24% market share. The company's product portfolio has shown the most divers offer including different digital solutions for residential and business customers. In public, the company is perceived as youthful and innovative operator.

Employees from different sector and with different roles were selected. Case study was conducted in digital transformation and innovation unit and in technical department. Innovations in technical department are focused more on improving internal processes and improving customer experience and require strong technical background from employees, while innovations in digital transformation and innovation department are focused on developing new offers for customers and expanding product portfolio with new, innovative solutions. Selected interviewees belonged to different firm level: CTO, team manager and experts, but they were all highly involved in open innovation projects as project manager (PM) or project owner (PO).

17.5 Results

There are several reasons why mobile operator use open innovation, one is to increase the public awareness and perception of mobile operator being an innovative solution provider, to improve customer experience, to facilitate internal processes and to increase profit with new solutions. The problem was recognized by company's CTO:"I would like to emphasize two things which are important for telco industry – first is a mindset: workforce that telco seeks is flexible, agile, self-organized, they want to learn constantly and are not afraid of changes, they like to change things and innovate. IT and telco skills are converging in order to deal with complex problems that will arise in the future. My team and myself are at the moment focused on operational efficiency ".

Results of implementing open business model are in general very positive. Project owner: "When you work with someone who is from different industry, who has different views and domain knowledge, it opens up your mind for new ideas and approaches in your own business". PM: "Percentage of profit gained by innovative projects has increased several times in the past few years". CTO: "We have managed to save investments but at the same time have ensured higher sales volumes new customers and have improved quality through faster and predictive corrective actions in the network, additionally we have created sellable product which can bring additional revenues to the company because it can be sold to any mobile network operator in the world".

They all agreed on one major problem with using open innovation, and that is the speed of realisation, internal processes are too slow, it takes a lot of time for company to respond to innovation and make it happen. CTO: "It takes a lot of time for company to understand the working model with innovation. PO:"Internal resources are problem"; PM: "So far we are not organization that is ready to accept innovation at the right way (based on internal processes). We are still focused on core competences, voice and internet since they generate the biggest revenue. It is hard to introduce, develop and sell new product. Hard to embed it in existing system".

There was not a direct link between open innovation and corporate strategy. Innovative solutions are part of strategy in the context of digitalization. On the other hand, revenue growth is essential part of strategy, and innovative solutions are not generating as much revenue as regular services with voice and data and because of that priority is given to projects that attract more customers and have higher margins. It was strategic vision and strong management support that made realisation of the projects possible (PM). Regarding the innovations in technical department CTO explained: "It was not written clearly in strategy but technologically it was a need. After two years of work and research in the analytics field strategy is talking about analytics". She also pointed out that it is easier to justify your resources when projects are in relation with strategy, but on the other hand because of corporate goals to be efficient, lean, optimal with labor it is very hard to stimulate innovation (because people are full with daily work).

Innovation projects, are managed in a different way than other projects in the context of pressure and deadline, was the conclusion of all interviewees. Project teams are consisted of self-organized teams that gather individuals with different knowledge and skills. Interviewees also agreed that they don't have a specific working model that defines way of establishing, developing and maintaining relationship with partner, and that the relationship and way of working depends on partner and is quite flexible. PM: "Everything I have done so far I have done from the scratch. For most of the cases there are no predefined steps". CTO: "Making a model means you know what you will get, and now we don't know what will happen and what will appear. You need to be enough open, otherwise you might miss opportunity". Team manager: "When communicating and cooperating with startups, we don't have a strict collaboration model because at this moment we want to stay as flexible as possible. With little budget we try to take maximum and to stay opened for all possible suggestions, ideas and solutions". PO: "Partially we have our playground, development in which we cannot predict result and in such cases, we work on goodwill, but if we know what is the target than we have precisely defined project".

Previous experience is of big help in most of the cases. PM: "If there are certain similarities, procedural activities stay the same or modify as little as possible which speeds up the development process".

Interestingly even though all interviewees were applying open innovation in practice, most of them were not familiar with the term "open innovation".

17.6 Discussion and Conclusion

Mobile operators have clear strategy for technical improvements. Statistical data showed growth of data traffic in the past years, and predictions are that this trend will continue. Mobile operators are challenged to innovate and to make changes in their traditional business model. They have different goals: to create new product, improve their image as content provider and innovator, ease internal processes, and ameliorate customer experience. Open innovation is one of possible solutions mobile operators can apply to rise innovation activities. Additionally, self-organized innovation networks may create competitive advantage because their rapid learning are hard to replicate by competitors. Literature review has shown mostly positive results in implementing open innovation, however as rather new concept for mobile operators there are not so many research about use of open innovation in telecommunication industry.

Open innovation was shown to be successful way to stimulate internal innovation. Case study conducted in mobile operator showed that finding external partners and developing product with them had positive results in both technical and innovation department. With not so high expectations, in collaboration with one startup, engineers managed to create a product, a data analytics tool used for prediction of traffic and customers in radio access network. Product was helpful in their daily work, it improved customer experience but is also was a first tool, owned by company, that could be offered and sold to other operators as technical solution. Product and marketing department reported also positive experience with open innovation, they stated that digital solutions created with a partner are not as profitable as traditional voice and data services but are showing signs of growth in the last years. Results obtained in case study have confirmed the first hypothesis of the paper that for mobile operators open innovation is a successful method to increase company's productivity.

Strategy plays an important role in open innovation projects was a conclusion of case study. Digital solutions are not generating as much profit as standard services, but as strategically important they have a support for realization and implementation. For the innovation in the technical department, there was not a direct link between strategy and the product obtained. In general, strategic focus was to excel in your own job and in favor of that new product was created. But it was the success of this solution that gained strong company's support and it influenced strategic orientation. As CTO reported, when project is in line with strategy, it is easier to justify costs and provide funds. Case study therefore confirmed the second hypothesis that open innovation gives best results when it is connected with strategy.

Open innovation projects are realized in a more relaxed atmosphere, with not so strict deadlines and procedures and as such were led more by strategic project management principles than traditional project management. Teams working on innovation were usually self-organized teams who had more flexibility in product creation, product improvement, and learning process. On the other hand, the main problem, reported by all interviewees was the inertness and speed of internal procedures and processes. Not all employees are flexible and open enough for new, more dynamic ways of working. As CTO stated that the challenge of management was to propagate the innovation spirit and idea of openness to lower hierarchical levels so that everyone could understand it. Even though they had proven success in implementing open innovation, majority of interviewees didn't know what the concept of open innovation was. The author believes that internal employee education about open innovation and its principles could only improve its efficiency and results.

The research had several limitations. Digital solutions and products of open innovation model used in company are rather new and it is very hard to analyze results. First, because there are not so many projects to make a statistical analysis, and second, those are projects with newer date and it might be too early to judge results. The issue of having a small sample could be exceeded by expanding this research to not just more operators (because including all three operators in Serbia in research would still result in a small sample) but also to all other players in mobile industry such as equipment vendors, consultant companies, and OTT service providers and this could be a topic for future research.

The case study conducted in the research analyzed results and challenges of implementing open innovation in telecommunication company, a relationship between open innovation activities, strategic orientation, and innovation performance and as such findings on this study can help manages and directors refine internal innovation processes as well as the collaboration strategies of their company. The case study is conducted in telco industry, but findings could be of use in any industry that is challenged by new technologies. Wherever there is a possibility to use external sources of knowledge and expertise to expand, innovate and excel in business (today usually data analytics and IT-based companies offer the expertise that are useful) companies should consider introducing open innovation in their practice.

17.7 Ethical Approval:

All procedures performed in this study involving human participants were in accordance with the ethical standards of IPMA Serbia and with the 1964 Helsinki declaration and its later amendments or comparable ethical standards, which was confirmed in the Permission No. 3–12/2020 issued by IPMA Serbia Ethical Committee.

Informed consent: An informed consent was obtained from all individual participants included in the research and the data used in this study are completely anonymized.

Acknowledgements This paper is a result of: the Project No. 179081 funded by the Ministry of Education and Science of the Republic of Serbia: Researching Contemporary Tendencies of Strategic Management Using Specialized Management Disciplines in Function of Competitiveness of Serbian Economy and the International Research Programme 2020–2030 on the subject of Capabilities for delivering projects in the context of societal development (CaProSoc)—Alma Mater Europaea (AME) and Serbian Project Management Association—IPMA Serbia.

Ethical Statement

Conflict of Interest: The authors declare that they have no conflict of interest.

Appendix—Semi-Structured Interview Guide

Basic information.

What is your role in the company?

What is the business strategy of your firm?

What is the firm's culture and how it empowers innovation?

Describe the project?

What is the purpose of project?

What is your role in the project

Early steps, the development of idea

Where did the idea come from?

What is the driver of idea/product (ex. market, users)?

Why did your company decide to invest in project?

The development of the project

How did you select the partner? Did you have a predefined selection model?

Do you have a predefined communication model?

Are the roles and responsibilities clearly defined, please explain in detail?

Are project development steps clearly defined?

Describe phases of the project and participants?

Who manages project and takes major decisions?

Does project management differs compared to other project in the company?

Are project goals clearly defined?

Is deadline clearly defined?

Innovation and strategy

What is the relationship between company's strategy and project?

Is project supported by top management? Explain.

Does management support help and speeds up the realisation of project?

Open innovation and results.

Do you have predefined success parameters?

What are positive sides of cooperation?

What are results of project? Did it fulfil expectations?

Did the productivity of company increase thanks to cooperation strategy?

Are you happy with the way project was managed and how it is supported by the company?

Which kind of problems you faced during project and what was their cause?

If there was something you could change what would it be?

Are you planning to continue cooperation with the same partner?

Are results motivating you to continue with new innovation?

Do you know what open innovation is?

References

1. Ajadi S, Bayen M, Accelerator E (2017) Building synergies: how mobile operators and startups can partner for impact in emerging markets. https://www.gsma.com/mobilefordevelopment/programme/ecosystem-accelerator/building-synergies-how-mobile-operators-an

2. Allot (2014) App-centric operators on the rise—allot mobile trends 2014. http://investors.allot.com/releasedetail.cfm?releaseid=859652
3. Bercec (2016) BEREC Report on OTT services. http://berec.europa.eu/eng/document_register/subject_matter/berec/reports/5751-berec-report-on-ott-services
4. Caputo M, Lambert E, Cammarano A, Michelino F (2016) Exploring the impact of open innovation on firm performances. Manag Decis 54(7):1788–1812
5. Chesbrough H (2006) Open business models: how to thrive in the new innovation landscape. Harvard Business School Press, Boston
6. Cheng C, Huizingh E (2014) When is open innovation beneficial? The role of strategic orientation. J Prod Innov 31(6):1235–1253
7. Chesbrough H, Appleyard M (2007) Open innovation and strategy. Calif Manage Rev 50(1):57–76
8. Chesbrough H, Brunswicker S (2013) Managing open innovation in large firms. Fraunhofer Institute for Industrial Engineering, Stuttgart
9. Chiaroni D, Chiesa V, Frattini F (2010) Unravelling the process from closed to open innovation: evidence from mature, asset-intensive industries. R&D Manage. 40(3):222–245
10. Chiaroni D, Chiesa V, Frattini F (2011) The open innovation journey: how firms dynamically implement the emerging innovation management paradigm. Technovation 31(1):34–43
11. Cooke-Davies T, Crawford L, Lechler T (2009) Project management systems: moving project management from an operational to a strategic discipline. Proj Manag J 40(1):110–123
12. Deloitte M (2017) To be or not to be |The future of telco business model, https://www2.deloitte.com/content/dam/Deloitte/de/Documents/technology-media-telecommunications/Deloitte-TMT-Telco-2030.pd
13. Dittrich K, Duysters G (2007) Networking as a means to strategy change: the case of open innovation in mobile telephony. J Prod Innov 24(6):510–521
14. Du J, Leten B, Vanhaverbeke W (2014) Managing open innovation projects with science-based and market-based partners. Res Policy 43(5):828–840
15. Frankenberger K, Weiblen T, Gassmann O (2013) Network configuration, customer centricity, and performance of open business models: a solution provider perspective. Ind Mark Manag 42(5):671–682
16. Gawer A, Cusumano MA (2013) Industry platforms and ecosystem innovation. J Prod Innov 31(3):417–433
17. Ghezzi A, Balocco R, Rangone A (2016) The open innovation—strategy nexus: findings from the mobile telecommunications industry. Manag Res Rev 39(5):569–598
18. Ghezzi A, Cortimiglia M, Frank AG (2015) Strategy and business model design in dynamic telecommunications industries: a study on Italian mobile network operators. Technol Forecast Soc Change 90:346–354
19. Hagel J, Singer M (1999) Unbundling the corporation, Harvard Business Review
20. Huizingh EK (2011) Open innovation: state of the art and future perspectives. Technovation 31(1):2–9
21. Hung K, Chou C (2013) The impact of open innovation on firm performance: the moderating effects of internal R&D and environmental turbulence. Technovation 33(10):368–380
22. ITU (2017) ICT Facts and Figures 2017. http://www.itu.int/en/ITU-D/Statistics/Pages/facts/default.aspx
23. Juniper R (2017) PR Newswire. http://www.prnewswire.com/news-releases/consumer-migration-to-ott-services-to-cost-mobile-operators-more-than-100bn-in-lost-revenues-this-year-610109315.html
24. Kelly M, Schaan J, Joncas H (2002) Managing alliance relationships: key challenges in the early stages of collaboration. R&D Manag. 32(1):11–22
25. Kenny J (2003) Effective project management for strategic innovation and change in an organizational context. Proj Manag J 34(1):43–53
26. Kiemen M (2011) Self-organization in Open Source to support collaboration for innovation. In: Proceedings of the XXII ISPIM conference

27. Kratzer J, Meissner D, Roud V (2017) Open innovation and company culture: internal openness makes the difference. Technol Forecast Soc Change 119:128–138
28. Lichtarski MJ (2018) Project-driven heterarchy: an empirical study on project temas' learning abilities and creativity. European Proj Manag J 8(1):10–16
29. Meffert J, Mohr N (2017) Overwhelming OTT: Telcos' growth strategy in a digital world. https://www.mckinsey.com/industries/telecommunications/our-insights/overwhelming-ott-telcos-growth-strategy-in-a-digital-world
30. Milutinovic R, Stosic B (2013) Key elements of innovation project management in services management. J Theory Pract Manag 18(69):65–74
31. Mintzberg H (1994) Rise and fall of strategic planning. Free Press, New York, New York
32. Osterwalder A, Pigneur Y (2010) Business model generation. Wiley, New Jersey
33. Patanakul P, Shenhar A (2011) What project strategy really is: the fundamental building block in strategic project management. Proj Manag J 43(1):4–20
34. Rycroft R, Kash D (2004) Self-organizing innovation networks: Implications for globalization. Technovation 24(3):187–197
35. Shenhar A (1999) Strategic project management: the new framework. In: Management of engineering and technology. Technology and innovation management PICMET'99. IEEE, Portland, pp 382–386
36. Silva L, Guerrini M (2018) Self-organized innovation networks from the perspective of complex systems. J Organ Chang Manag 31(5):962–983
37. Sisodiya S, Johnson J, Grégoire Y (2013) Inbound open innovation for enhanced performance: enablers and opportunities. Ind Mark Manag 42(5):836–849
38. Steiner A, Morel L, Camargo M (2014) Well-suited organization to open innovation: empirical evidence from an industrial deployment. J Innov Econ 13(1):93
39. Sujata J, Sohag S, Tanu D, Chintan D, Shubham P, Sumit G (2015) Impact of Over the Top (OTT) services on telecom service providers. Indian J Sci Technol 8(s4):145–160
40. Todorović M, Obradović V (2018) Sustainability in project management: a project manager perspective, in Sustainable growth and development in small open economies. In: Ljumović I, Éltető A (eds) Institute of world economics, centre for economic and regional studies of the hungarian academy of sciences. ISBN 978-963-301-664-0 (e-book)
41. Vanhaverbeke W, Vrande V, Cloodt M (2008) Connecting absorptive capacity and open innovation. SSRN Electron J
42. Weiblen T, Chesbrough H (2015) Engaging with startups to enhance corporate innovation. Calif Manag Rev 57(2):66–90
43. Xu X, Chen R (2015) Competition, cooperation, and pricing: how mobile operators respond to the challenge of over-the-top. Int J Mark Stud 7(6)
44. Yin RK (2003) Case study research: design and methods. Thousand Oaks, Sage CA

Index

R. Ding et al. (eds.), *Research on Project, Programme and Portfolio Management*, Lecture Notes in Management and Industrial Engineering,
https://doi.org/10.1007/978-3-030-86248-0

Printed by Printforce, the Netherlands